NATURAL DISASTER RESEARCH, PREDICTION AND MITIGATION

EARTHQUAKES

TRIGGERS, ENVIRONMENTAL IMPACT AND POTENTIAL HAZARDS

NATURAL DISASTER RESEARCH, PREDICTION AND MITIGATION

Additional books in this series can be found on Nova's website under the Series tab.

Additional e-books in this series can be found on Nova's website under the e-book tab.

NATURAL DISASTER RESEARCH, PREDICTION AND MITIGATION

EARTHQUAKES

TRIGGERS, ENVIRONMENTAL IMPACT AND POTENTIAL HAZARDS

KOSTAS KONSTANTINOU
EDITOR

nova publishers
New York

Library of Congress Cataloging-in-Publication Data

Earthquakes : triggers, environmental impact, and potential hazards / editor, Kostas Konstantinou.
p. cm.
Includes index.
ISBN 978-1-62081-883-1 (hardcover)
1. Earthquake hazard analysis. I. Konstantinou, Kostas.
QE539.2.S34E29 2012
551.22--dc23
2012017843

Published by Nova Science Publishers, Inc. † New York

CONTENTS

PREFACE

Earthquakes represent a natural phenomenon that throughout human history has caused destruction of man-made structures and dramatic loss of life. The study of earthquake phenomena has seen a rapid growth over the past few decades and has contributed significantly towards mitigating the effects of large earthquakes. This edited volume consists of a collection of ten chapters that address three topical areas related to earthquake research. The contributed chapters in this volume will be of use to a range of professionals such as Earth scientists, engineers and civil protection officers.

Chapter 1 - Four different geologic records are compared for evidence of convergent tectonic strain in the South Central Cascadia Margin (500 km in length), Oregon, USA. The strain records represent widely different time scales, including late Neogene, late Pleistocene, late Holocene, and modern time scales. Modern strain records are taken from global position system (GPS) data complied from three representative localities, positioned north, central and south in the study area at 1-50 km distance landward of the coast. These modern GPS data show convergent shortening rates of 5-9 mm yr^{-1} over 40 km west–east baselines. The southernmost GPS stations also demonstrate an along coast shortening trend that is consistent with oblique convergence in the southernmost part of the study area. Late Neogene records of convergent strain are represented by west and east dip angles of middle Miocene to Eocene sedimentary rocks. Background dip angles (10-20°) increase to 30-60° at distances of less than 90 km from the deformation front in the central part of the study area. The elevations of late Pleistocene high sea level stand deposits (80–120 ka in age) are measured at 76 sites along the study area coastline. Background variability of the sea level high stand deposits (±4 m 1–σ) increases to ±11 m 1–σ at distances of less than 75 km from the deformation front. The late Neogene and late Pleistocene deformation is assumed to reflect inelastic strain associated with local upper plate folds and faults near the interplate coupled zone. Late-Holocene records of episodic tidal marsh subsidence (0.5–1.5 m) in 33 representative core sites from 7 estuaries establish the regional trough of elastic coseismic subsidence. The distance between the trough of maximum coseismic subsidence (1.5 m subsidence) and the deformation front narrows from 150 km in the north to 100 km in the south of the study area. These across margin distances demonstrate a relatively wide zone of episodic interplate coupling throughout the length of the South Central Cascadia Margin. All four coastal records, including modern GPS strain, late Neogene deformation, late Pleistocene deformation, and late Holocene episodic subsidence reflect convergent strain associated with subduction of the Juan De Fuca oceanic plate under the North American continental plate.

Chapter 2 - Geological aspects, as well as some historical antecedents related to seismic hazard of the Villavicencio City [Colombia, South America] are shown. This city, with a population of 430,000 inhabitants, is located along the Eastern Frontal Fault System (EFFS), in a faulted zone, with the Andes Cordillera region at the west, and the South American Craton at the east. The analysis of historical and instrumental earthquake records, associated to strong motion effects in cities like Villavicencio, Bogotá, Neiva, and Bucaramanga indicates that instrumental networks should be improved in order to evaluate the seismic regime at local and regional scales. The systematic monitoring of seismic activity along Algeciras, Guaicáramo, and Servitá faults would represent an important chance for understanding the deformation front related to the continental collision and the basinal geometry of the Colombian Llanos Foothills. On the other hand, field data suggest the presence of large areas with active tectonic evidence that are interrupted by quiet segments, and that could represent a reorientation of the local stress field or its accumulation for promoting scenarios of strong earthquakes. Large tectonic structures passing near Villavicencio, as the Algeciras Fault, have a long history of strong earthquakes on this city (e.g. August 31, 1917). Similarly, the earthquake data corresponding to the Servitá Fault present several events that affected Villavicencio, Bogotá [1743 to Present – day], as well as other small towns. The Guaicáramo Fault also has presented strong activity in several segments near to Villavicencio [e.g. 1923, 1924, and 1995 earthquakes].

The vulnerability of Villavicencio city regarding the seismic hazard is analyzed, as well as the role carried out by local institutions. A description of a possible post-seismic scenario and certain recommendations to mitigate the effects of an earthquake are presented.

Chapter 3 - On 9 July 1956 a large earthquake occurred near the island of Amorgos, southern Aegean, causing extensive damage while it was followed by a tsunami that inundated the coasts of nearby islands also reaching the Turkish coast. This chapter aims first at providing a thorough review of previous studies about this earthquake and then developing a seismotectonic model for its occurrence by combining all available information. Analysis of historical seismograms of the Amorgos event shows that its focal mechanism is a purely normal fault and that the fault plane exhibits a low-angle (< 30 degrees) dip. This analysis also revealed a relatively deep centroid (45 km) placing the high slip area inside the upper mantle, unlike previous estimates which suggested a shallow focus (~10 km). This relatively large depth seems to be supported by numerical modeling of the tsunami generation and propagation that points to submarine landslides rather than co-seismic rupture of the seafloor as its cause. In addition, the crustal rheology of the area implies that the Amorgos earthquake could have nucleated near the depth of peak strength (33 km) rupturing up to a depth of 7 km without breaking the seafloor. However, these considerations are at odds with geological observations of high-angle faulting near the epicentral area of the earthquake. The geological history of the southern Aegean can offer clues so as to resolve this apparent controversy. Amorgos lies at the eastern edge of a metamorphic core complex that started being exhumed 23-19 Ma due to the rollback of the subducting African lithosphere. A direct consequence of this exhumation is the development of large scale detachment faults consisting of upper crust high-angle and lower crust shallow-angle normal faults. The existence of these faults is based on geological observations of exhumed metamorphic rocks and thermo-mechanical numerical models of the exhumation process. Such a detachment fault can explain the geometry and nucleation depth of the Amorgos event thus resolving all the aforementioned discrepancies.

Chapter 4 - The authors review previous numerical simulations of interaction between fault segments on a plate interface based on a friction law, comparing coupled events with single events on the amount of coseismic and preseismic slip. The authors find out the characteristics of coupled event: (i)In case of multiple fault segments nearby enough to have short delay time of rupturing, coseismic slip is increased nearly in proportion to the size of coupled fault segment while preseismic slip is almost the same as single event; (ii) In case of those distant enough to have long delay time of rupturing, seismic moment magnitude is almost the same as single event while preseismic slip is significantly increased by several tens percents.

The authors also investigate the changes in slip rate and recurrence interval of repeating earthquakes using 3-D numerical simulations of a subduction zone involving large and small asperities. Their results reveal that the postseismic slip of a large earthquake can trigger an event at a nearby small asperity, although the event may differ to other repeating earthquakes at the asperity because the slip velocity of the event tends to be slower than that of other repeaters; the slip distribution also differs. The authors discuss several coupled earthquakes in Sumatra with short-term and long-term delay times, and point out that the slip amount estimated from repeating earthquake analyses tends to be smaller than that estimated from GPS analyses for large amounts of postseismic slip, as observed for postseismic activity following the 2003 Tokachi-oki earthquake. These results may be applied to the postseismic slip off Kanto and off Sanriku after the 2011 off the Pacific Coast of Tohoku Earthquake, which may trigger megathrust aftershocks in the near future. In southwest Japan, Dense Oceanfloor Network System for Earthquakes and Tsunamis (DONET) may be useful to detect the pre-seismic change of a possible M9 class coupled megathrust earthquake composed of Tokai, Tonankai, Nankai and Hyuga-nada asperities.

Chapter 5 - In this chapter, the authors investigate the statistical seismicity in view of complex systems, namely, interoccurrence time statistics and intermittency of earthquakes. To show statistical properties of earthquakes, the authors analyze three different kinds of data; i) natural earthquake data, ii) synthetic data, and iii) time sequence analysis based on a one-dimensional map. For i), Japan Meteorological Agency (JMA), Southern California Earthquake Data Center (SCEDC), and Taiwan Central Weather Bureau (TCWB) are analyzed. For ii), the authors investigate synthetic data produced by the two-dimensional (2D) spring-block model. For iii), intermittency and correlation of earthquakes are studied using JMA data and the constructed map. First, the authors focus on the interoccurrence time statistics, statistical properties of time intervals between successive earthquakes by analyzing both natural data and the synthetic data. The authors show that the probability distribution of interoccurrence time depends on the threshold of magnitude, mc and is described by the superposition of the Weibull distribution and log-Weibull distribution for all earthquake data. As mc increases, the ratio of the Weibull distribution increases. In other words, the distribution changes from the Weibull to the log-Weibull distribution according to an increase of mc, i.e., the Weibull - log-Weibull transition. This result reinforces the view that the interoccurrence time statistics contain the Weibull statistics and the log-Weibull statistics. In synthetic data, the distribution changes from the pure log-Weibull regime to the Weibull regime via the superposition regime, i.e., a sharp Weibull - log-Weibull transition. For natural data, the distribution presents a smooth Weibull - log-Weibull transition which means that the distribution varies from the superposition regime to the pure Weibull regime. For large earthquakes, the Weibull distribution with the exponent $a1 < 1$ is preferred, which indicates that the sequence of earthquakes is not a Possion process. The authors give a geophysical interpretation of the Weibull - log-Weibull transition. Second, to characterize an intermittency and the correlation of interoccurrence times in earthquakes, the authors analyze the JMA catalog. The authors demonstrate that interoccurrence times are not independent and identically distributed random variables. However, the authors find that the tails of all conditional probability distribution functions of interoccurrence times are universal. The tail characterizes intermittency of earthquakes. It is shown that forecasting of the occurrence of large earthquakes is more difficult than that of small earthquakes.

Chapter 6 - Triggering and synchronization (i.e. adjustment of rhythms of oscillating objects due to their weak interaction) are encountered in various fields, from mechanics to biological and social processes. Thus, it is only natural that triggering and synchronization phenomena are observed in many geophysical fields, as the Earth is embedded in the oscillating field of different origin, with extremely wide range of frequencies, from seconds to months and years. There are a lot of observations on the significant impact of small external forcing on the seismic regime, namely on the seismicity induced by: tides, reservoir exploitation, big explosions, magnetic storms, strong electrical pulses, wave trains of remote strong earthquakes, oceanic microseisms, etc, which means that these phenomena are connected with nonlinear interactions of objects, namely with initiation of instability in systems that are close to the critical state.

Early observations of triggering/synchronization were not supported by strict validation of procedures and should be considered as mainly qualitative or even unreliable. The situation in the field changed dramatically in the last decade after implementation of dense seismic networks equipped by high sensitivity broadband devices, which allow reliable identification of dynamic triggering/synchronization of earthquakes. Besides, new analytical tools for quantitative assessment of the strength of triggering/synchronization developed in the last decades made the procedure of revealing these effects much more robust procedure.

In the chapter the results of laboratory experiments on the electromagnetic (EM) and mechanical triggering and synchronization of mechanical instability (slip) in the spring-slider system are presented. Slip events were recorded as acoustic emission bursts.

Besides, field data on the forcing of local seismicity by strong electromagnetic pulses, water regime variations and remote earthquakes are analyzed as well as the potential of synchronization technique for earthquake prediction.

Chapter 7 - The seismo-ionosphere coupling phenomena, namely the ionospheric GPS TEC (Total Electron Content) disturbances observed before strong (M>5) earthquakes have been discussed. These precursors have the form of stable local long-living TEC disturbances at the near-epicenter and often at the magnetically conjugated areas. They do not propagate along meridians in contrast with the ionospheric disturbances related with magnetic activity. Basing on a few case studies of relative TEC deviations before the recent strong seismic events (Jan. 12, 2010 Haiti, Jan. 1 and Jan. 2 2011 Argentina and Chile, March 11 2011 Japan earthquake) the authors discovered 'ban-time' and terminator effects which were not reported in other authors' researches. These new effects are: the sunrise terminator income triggers the TEC anomalies weakness up to their almost full destruction with subsolar point arrival, i.e. at the local time near-noon hours. The sunset terminator passage is often followed by the anomalies renewal.

The authors propose the possible physical mechanism of the TEC anomalies forming, namely the vertical transport of the F2-layer ionospheric plasma due to the electromagnetic

[E×B] drift under influence of the electric field of seismic origin. This drift transfers plasma to heights where the ion loss rates are lower (if plasma drifts upwards) or higher (if plasma drifts downwards). As result, the F2-region electron density and TEC increases (or decreases). The consistency of this hypothesis was checked using the first-principle 3D time-dependent Upper Atmosphere Model. Seismogenic electric fields were modeled in two ways: (1) by setting additional electric potentials; (2) by setting vertical electric current flowing between the Earth and the ionosphere at the near-epicenter area as input for the electric potential equation. The model TEC calculation results are close to the GPS observed TEC disturbances if the model electric currents needed to create them are of about 10^{-8} A/m^2.

Chapter 8 - Deep borehole observation can eliminate effectively various Earth's surface interferences, providing considerable advantage over other geophysical measurements made at the ground, and improve greatly our capability to detect very small deformations and vibrations in the surrounding strata. This new technology will promote consequently our understanding of presently active geological processes and its relation to earthquakes and tectonics.

A new RZB multi-component borehole instrument is an integrated packages consisting of six channels for the strainmeters (4 horizontal components, 1 vertical component, and 1 for system reference), two channels for the tiltmeters, three channels for broadband strain seismometers, and one channel for the ground temperature. Furthermore, another four channels provide the assistant observation including well temperature, well water level, atmospheric pressure and electronic compass.

The objective of setting four strainmeters in the horizontal plain is to measure changes in four diameters apart from each other at 45° of the cylinder, which allows us to record shear strain and areal strain changes. Each measurement unit for strain, tilt, and broadband strain seismometer as well as ground temperature consists of transducers and a circuit board distributing data highway and power to individual gauges assembled in a stainless-steel pressure cylinder with length of 1m and outside diameter of 102mm. A cable, connecting the sensors up hole, provides simultaneously a digital data connection via an RS-485 link and an analogue signal transmission line for seismometers. All data are stored in digital format in the acquisition system at the ground with 1-Hz sampling rate for strain, tilt and temperature, 200-Hz sampling rate for broadband strain seismometer. Flexible combination of diversified sensors and their duplicate parts enables us to make good project that can adapt to differing circumstances.

Because of its principle of operation, multi-component borehole instrument is permanently cemented in the bottom of an open hole to assure good coupling with the surrounding rock in order to get meaningful information about changes in the Crustal stress state. These new borehole instruments have been installed at stations in Beijing and Fujian, Sichuan, Yunnan provinces. The instruments can run at high sensitivity to provide strain resolution of 0.1 nanostrains with basic dynamic range of ±500 microstrain and tilt resolution of 1 nanorad with basic dynamic range of ±500 microrad. Up to now these instruments have observed good tidal signals, strain-steps associated with large earthquakes, and strain changes of geophysical interests.

Chapter 9 - In this chapter a new well established model is presented and commented, that saves human lives because it helps avoiding collapses in the reinforced concrete constructions during strong earthquakes. Using this model the initial formation of plastic hinges in the beams and the final concentration of the damages only in the beams is accurately assured. The model also assures that the columns and the beam-column joints can remain intact. This model can be applied for the design of modern R/C constructions, as well as for the design of strengthening schemes of old R/C constructions by the use of reinforced concrete jackets. The model can also predict the type of earthquake damages of old structures but also the earthquake damages of the modern structures. Further by the use of this model a new technique can be applied which helps avoiding the serious disadvantages related to the currently used FRPs for the strengthening of old R/C structures. In this direction a mixed type strengthening system is proposed which includes strengthening of the beam – column joints with a conventional reinforced concrete jacket and the other parts of the structures with FRPs. This mixed type system is calculated by the use of the new proposed model.

Chapter 10 - Recognition of the growing risk from earthquakes faced by communities in many countries, and particularly those situated on the Ring of Fire, has stimulated interest in promoting people's capacity to co-exist with the hazardous aspects of their environment. From a community perspective, one important contribution to this goal involves encouraging the adoption of preparedness measures that increase the ability of people and communities to cope with and adapt to the environmental impact of earthquakes. However, despite the attention and financial resources devoted to encourage people to do so, the fact that levels of preparedness in communities at risk from earthquakes are generally low makes it important to determine why this is the case and to identify ways of encouraging sustained community preparedness. This chapter contributes to understanding the factors that influence household and community preparedness for earthquakes by identifying how people interpret their risk and make decisions about how to manage the risk pose by living in seismically active areas.

This chapter discusses how people's decisions about preparing for earthquakes derives less from the objective use of information and more from how community members interpret earthquake risk and information about it. It describes how interpretive processes at the person (outcome expectancy), community (community participation, collective efficacy), societal (empowerment, trust) and cultural levels of analysis interact to predict levels of earthquake preparedness. The chapter accommodates the fact that earthquakes occur in many countries and acknowledges that the practical utility of any theory is increased if it can be demonstrated that it applies across cultural and national boundaries.

The cross-cultural applicability of the theory is discussed by comparing findings from research into earthquake preparedness in countries that differ in cultural orientation; New Zealand, Taiwan, and Japan. The theoretical (e.g., identifying the degree to which earthquake preparedness is culturally equivalent) and practical (e.g., providing a common basis for collaborative learning, research and risk management intervention between countries) implications of the cross-cultural equivalence of the model are discussed. The chapter also discusses how cultural specific processes such as the Hakka Spirit in Taiwan and Jishubo in Japan influence how community processes are enacted to manage earthquake risk. Finally, these perspectives are combined to discuss how the sustained effectiveness of strategies designed to develop community capacity to deal with the environmental impact of earthquakes can be increased by integrating risk management activities with community development strategies.

In: Earthquakes
Editor: Kostas Konstantinou
ISBN: 978-1-62081-883-1

Chapter 1

Coastal Tectonic Strain and Paleoseismicity in the South Central Cascadia Margin, Oregon, US

Curt D. Peterson, Kenneth M. Cruikshank and Mark E. Darienzo*

Department of Geology, Portland State University, Portland, OR, US

Abstract

Four different geologic records are compared for evidence of convergent tectonic strain in the South Central Cascadia Margin (500 km in length), Oregon, USA. The strain records represent widely different time scales, including late Neogene, late Pleistocene, late Holocene, and modern time scales. Modern strain records are taken from global position system (GPS) data complied from three representative localities, positioned north, central and south in the study area at 1-50 km distance landward of the coast. These modern GPS data show convergent shortening rates of 5-9 mm yr^{-1} over 40 km west east baselines. The southernmost GPS stations also demonstrate an along coast shortening trend that is consistent with oblique convergence in the southernmost part of the study area. Late Neogene records of convergent strain are represented by west and east dip angles of middle Miocene to Eocene sedimentary rocks. Background dip angles (10-20°) increase to 30-60° at distances of less than 90 km from the deformation front in the central part of the study area. The elevations of late Pleistocene high sea level stand deposits (80–120 ka in age) are measured at 76 sites along the study area coastline. Background variability of the sea level high stand deposits (±4 m 1–σ) increases to ±11 m 1–σ at distances of less than 75 km from the deformation front. The late Neogene and late Pleistocene deformation is assumed to reflect inelastic strain associated with local upper plate folds and faults near the interplate coupled zone. Late-Holocene records of episodic tidal marsh subsidence (0.5–1.5 m) in 33 representative core sites from 7 estuaries establish the regional trough of elastic coseismic subsidence. The distance between the trough of maximum coseismic subsidence (1.5 m subsidence) and the deformation front narrows from 150 km in the north to 100 km in the south of the study

* Department of Geology, Portland State University, PO Box 751, Portland, OR 97207-0751.

area. These across margin distances demonstrate a relatively wide zone of episodic interplate coupling throughout the length of the South Central Cascadia Margin. All four coastal records, including modern GPS strain, late Neogene deformation, late Pleistocene deformation, and late Holocene episodic subsidence reflect convergent strain associated with subduction of the Juan De Fuca oceanic plate under the North American continental plate.

INTRODUCTION

The South Central Cascadia Margin, defined here by the southern onehalf of the subducting Juan De Fuca plate (Figure 1) demonstrates convergent tectonic strain over different time and length scales. Late Holocene records of coastal subsidence [Darienzo and Peterson 1990; Darienzo et al., 1994] and associated nearfield paleotsunami inundations [Atwater et al., 1995; Peterson and Darienzo, 1997; Peterson et al., 2008] indicate prehistoric occurrences of megathrust ruptures in the South Central Cascadia Margin.

Unlike the well developed sequences of coseismic coastal subsidence in Willapa Bay and Grays Harbor in the North Central Cascadia Margin [Atwater et al., 2004] (Figure 1) the records of episodic coastal subsidence diminish to below detection in some small estuaries of the South Central Cascadia Margin [Briggs, 1994; Peterson et al., 2000].

These poorly developed records of coastal subsidence have led some investigators [Nelson, 1987; 1992; Shennan et al., 1998] to argue against through going megathrust ruptures in the South Central Cascadia Margin.

A likely contributing factor to the along coast variability of episodic coastal subsidence in the South Central Cascadia Margin is the orientation of the coastline, relative to the buried subduction zone trench, or deformation front (Figure 1). The across margin distances between the shoreline and the deformation front decrease from about 120 to 75 km with distance south along the 500 km coastline in the study area. Along coast changes in plate convergence angles and oceanic plate thermal gradients [Hyndman and Wang, 1995] add further difficulties to the assessment of seismic hazards in the South Central Cascadia Margin.

In this chapter four different geologic records of convergent plate tectonic strain are examined in the South Central Cascadia Margin (Figure 1). These records include 1) modern strain rates from continuously monitored GPS stations, 2) late Neogene deformation of coastal sedimentary rocks, 3) vertical deformation of late Pleistocene marine terraces, and 4) late Holocene records of episodic coastal subsidence. The coastal deformation is related to 1) inelastic strain in the upper plate deformation belt and 2) elastic strain recorded in the trough of cosesismic subsidence, which slightly overlaps the seaward extents of the upper plate deformation belts. These relationships can be used to define the interplate coupled zone, which episodically ruptures to produce megathrust earthquakes and associated nearfield tsunami in the South Central Cascadia margin.

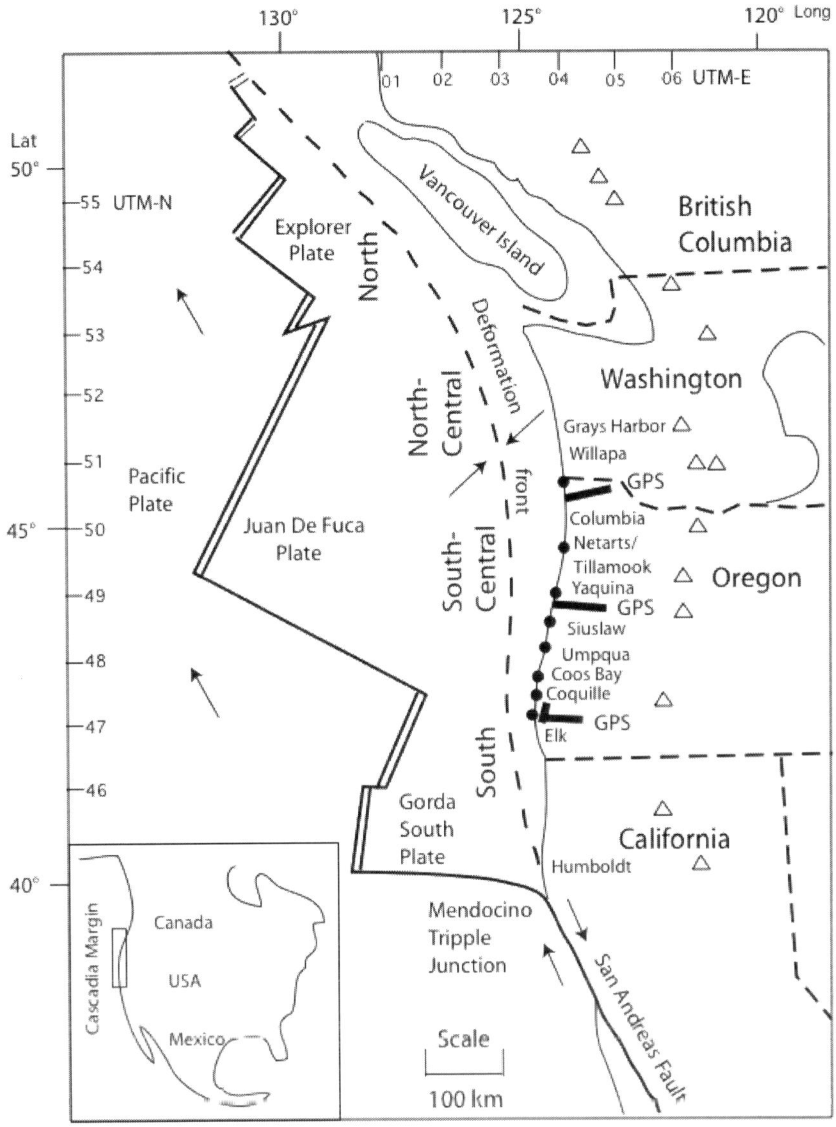

Figure 1. Cascadia margin coastline (solid line), buried trench or deformation front (dashed line), and continental volcanic arc (open triangles). Arrows represent directions of plate convergence with a reported convergence rate of about four centimeters per year in the central Cascadia margin. Convergence becomes increasingly oblique with proximity to the San Andreas transform fault located south of the Mendocino triple junction. For this paper the Central Cascadia margin is defined by the extent of the Juan De Fuca Plate segment. The South Central Cascadia margin is contained in the State of Oregon, located south of the Columbia River. Tidal basins recording late Holocene records of episodic coastal subsidence (solid circles) in the South Central Cascadia Margin are named. Modern GPS strain traverses (bold lines) are shown for the north, central, and southern parts of the study area. Map coordinates are in degrees latitude and longitude and in 100 km intervals in UTM zone 10.

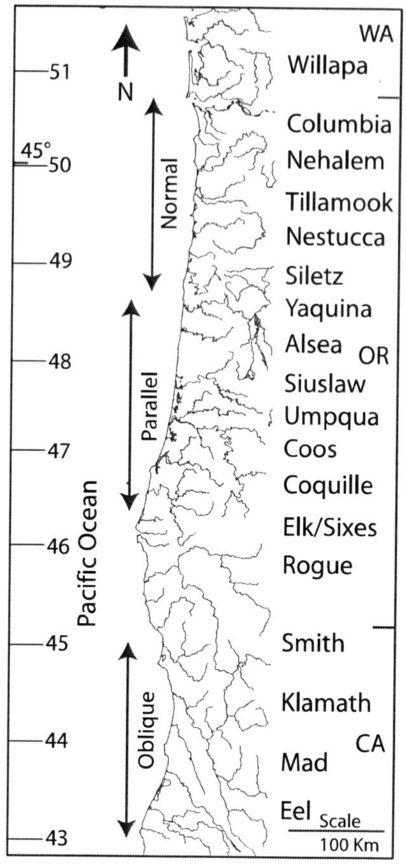

Figure 2. Coastal river valley orientations in the South Central and South Cascadia margin. Coastal drainages transition from oblique to parallel to normal or perpendicular with increasing distance north along the 800 km long coastline. See Figure 5 in the Results section for larger scale maps of the coast parallel drainages in the south central part of the study area. Map coordinates are in degrees latitude and 100 km intervals UTM zone 10.

Study Area and Previous Work

Observations of coastal stream alignments and uplifted Pleistocene marine terraces in the South Central Cascadia Margin led to early searches for compressive strain accumulation that could be recorded in uplifted estuarine deposits [Peterson and Scheidegger, 1984]. Events of episodic coastal subsidence were subsequently found in late Holocene tidal wetland deposits, which are developed in submerged river valleys of the South Central Cascadia Margin [Darienzo, 1987; Darienzo and Peterson, 1990; Darienzo et al., 1994]. Strikes and dips of coastal sedimentary rocks are exposed in river and road cuts that follow the coastal river valleys landward from the estuaries [Baldwin, 1956, 1961, 1974; Snavely et al., 1972a,b; Newton, 1980]. For these reasons the coastal strain records that are discussed in this chapter are loosely organized around estuaries and their associated tributary drainages (Figure 2).

In the South Cascadia margin, which is located landward of the Gorda Plate segment (Figure 1) the coastal river valleys trend northwest. The coastal rivers intersect the coastline at oblique angles (Figure 2). The river valleys are approximately aligned with northwest–southeast trending thrust faults. The thrust faults are thought to reflect oblique subduction, which is expected to occur north of the Mendocino triple junction. Late Holocene records of episodic coastal subsidence or uplift in Humboldt Bay in the South Cascadia Margin (Figure 1) are localized on upper plate structures, which are likely to be associated with the active thrust faults [Clarke and Carver, 1992; Li, 1992].

The southern part of the South Central Cascadia Margin contains coastal streams (Figure 2) that are characterized by trellis patterns. These streams are aligned with north–south trending folds. The coast parallel trends or meanders of the stream valleys only extend 5-20 km landward of the shoreline, before realigning to east–west dendritic patterns in the Coast Range. Late Holocene marsh deposits in some of the fold axis valleys, such as South Slough in Coos Bay, demonstrate episodic subsidence [Darienzo and Peterson, 1987; Nelson et al., 1996]. However, other salt marsh deposits in the lower reaches of Coos Bay and in nearby estuaries of the Umpqua and Siuslaw Rivers contain little or no record of episodic wetland submergence [Briggs, 1994].

Coastal streams in the northern part of the South Central Cascadia Margin form dendritic drainages that are generally aligned normal or perpendicular to the coastline (Figure 2). Salt marsh deposits in the submerged lower reaches of these river valleys show evidence of episodic subsidence, but the cyclic emergence and submergence values are modest (< 1.0 m relative to paleosea level datums) [Barnett, 1997; Peterson et al., 2000]. Some investigators have argued that such weak subsidence records are indistinguishable from aseismic submergence processes and that they do not necessarily demonstrate through-going megathrust ruptures in the South Central Cascadia Margin [Shennan et al., 1998].

West and McCrumb [1988] used low-uplift rates (0.1–0.2 mm yr^{-1}) of coastal marine terraces (80 ka) in the Central Cascadia Margin to argue against megathrust coupling. However, offshore seismic reflection mapping, as compiled by Peterson and Gray [1986] reveals a prominent fold and fault belt that approximately parallels the deformation front (Figure 3). The deformation belt extends from the deformation front or buried trench to the landward reach of the fold and fault belt, as mapped by offshore seismic profiling. The deformation belt extends 110 km east of the deformation front offshore of the Columbia River, but appears to intersect the coastline south of Coos Bay, at ~ 80 km from the deformation front in the southern part of the study area [Peterson and Gray, 1986].

Oblique, northwest trending strike slip faults cut across the offshore deformation belt (Figure 3). Those oblique faults are left lateral [Goldfinger, et al., 1992a]. Northwest trending faults are also mapped in onshore Miocene rocks in the northern part of the study area [Niem and Niem, 1985; Peterson and Gray, 1986]. However, the onshore faults are right lateral strike-slip faults [Wells, 1990] so they are not connected to deformation in the offshore fold and fault belt. The discontinuity between the offshore and onshore strike slip faults occurs within a distance of 60-90 km from the deformation front in the central study area.

The most prominent features in the offshore deformation belt are north–south trending fold axes (Figure 3). Extensive seismic profiling, side-scan sonar, and submersible diving in the early 1990s confirmed an abundance of Quaternary folds and faults in the offshore deformation belt [Goldfinger, et al., 1992b]. Late Pleistocene, shore-parallel folds are mapped to within 10-20 km of the coastline in northernmost part of the study area, but they appear to obliquely intersect the coast in the vicinity of Coos Bay in the southern part of the study area.

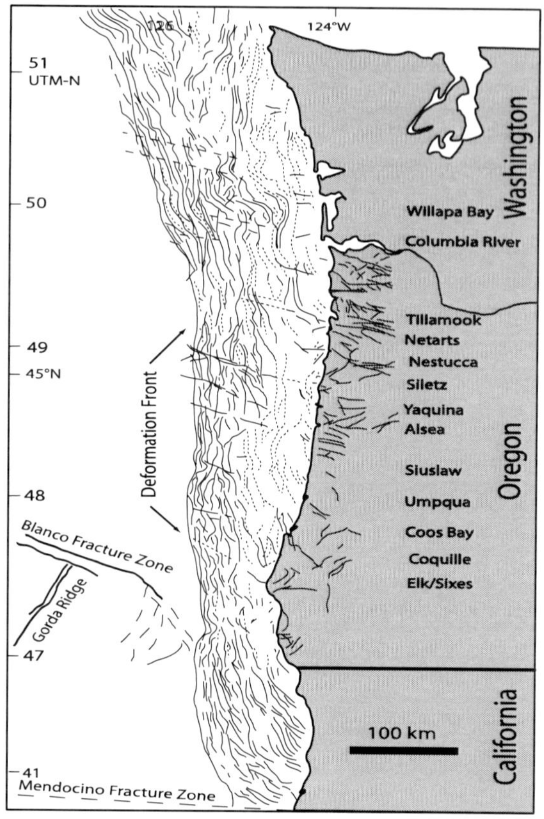

Figure 3. Map of folds (lines) and faults (bold lines) in the offshore fold and thrust belt (between shoreline and deformation front) and in Tertiary bedrock formations (located landward of the shoreline) in the Central and South Cascadia Margin. Onshore Tertiary faults are redrafted from Peterson and Gray [1986]. Pleistocene folds and faults (lines) and inferred folds and faults (dashed lines) in the offshore deformation belt are from Goldfinger et al., [1992a; 1996; 1992b] as compiled by McNeil, et al. [1998]. Map coordinates are in degrees latitude and 100 km intervals UTM zone 10.

The most current investigations of late Holocene paleoseismic indicators in the South Central Cascadia Margin have focused on continental slope turbidites [Goldfinger et al., 2008] and on overland paleotsunami inundation [Peterson et al., 2010]. These indirect records of subduction zone seismicity indicate the occurrences of megathrust ruptures with mean recurrence intervals of ~ 350–500 years. Some of the possible ruptures are correlated along the full length of the South Central Cascadia Margin, suggesting earthquake magnitudes of Mw 8.5-9.0. How can these paleoseismic indicators be reconciled with interpretations of little or no interplate coupling in parts of the South Central Cascadia Margin?

Results

In this chapter the geologic records of tectonic strain are compared to reconcile the different interpretations of megathrust coupling and coseismic rupture in the South Central Cascadia Margin. The first report on modern GPS strain records in the South Central Cascadia Margin are followed by long term records of inelastic coastal deformation associated with intermittent megathrust coupling. The chapter transitions to elastic records of strain with overviews of late Holocene records of cyclic interseismic uplift and coseismic subsidence. These records of the elastic strain cycles define the trough of coseismic subsidence, located above and landward of the interplate coupled zone. The chapter concludes with a return to the modern strain evidence as provided by GPS stations located 1-50 km landward of the coast.

Modern GPS Strain Records

Three GPS localities are selected for modern GPS strain records in the vicinities of the Columbia River (Seaside), Yaquina–Alsea Rivers (Newport), and Coquille (Bandon) in the South Central Cascadia Margin (Figures 1 and 2). The three localities represent the north, central, and southern parts of the study area. Several GPS stations are used for each locality (triangular networks) to verify results between adjacent baselines. Additional GPS stations (Cape Blanco and Coos Bay) that are located are either side of Bandon are also selected to test for north–south shortening in the southern part of the study area. The GPS station positions are analyzed for changes in station to station distance through time (3 years) depending on the duration of consistent continuous coverage (PNGA, 2011; NOAA, 2011). The GPS traverses or baselines are selected to include stations that are located at the coast and at about 40–50 km distance inland from the coast (Table 1).

The data from the continuously operating GPS stations (CORS, 2011; PANGA, 2011) were accessed and baselines between pairs of stations were processed in the period 2006 to 2011 using utilities in the GPStk software package [Tolman et al., 2004]. The Vecsol component of the GPStk package solves for the distance between stations; using this result rather than coordinates allows determination of strains without ambiguities introduced by coordinate transformations. Statistical analysis on the change in baseline lengths was performed using R statistical software [R Development Core Team, 2011] and the MASS analysis library [Venables and Ripley, 2002]. An Analysis of Variance (ANOVA) test on the regression shows that the slope of the change of baseline with time is significant ($\alpha = 0.05$, null hypothesis is that the slope of the regression is zero, alternate hypothesis is that slope is different from zero), indicating there is a significant trend. All the baselines presented here showed a significant trend in the change in distance between the two stations (Table 1, Figure 4).

These data (Table 1) show that there is significant west–east shortening throughout the length of the South Central Cascadia Margin. The change in west–east baseline length in the Seaside locality, near the Columbia River, is 7–8 mm yr^{-1} for the 50–55 km long baselines (Figure 4). Apparent cycling in the strain rates might be due to seasonal atmospheric variability. The change in west–east baseline lengths in the Newport locality in the central part of the study area is about 5 mm yr^{-1} for 38–44 km long baselines. The change in west–east baseline length in the Bandon locality in the southern part of the study area is about 9 mm for the 39 km long baseline.

Table 1. Summary of preliminary strain observations for selected parts of the Oregon coast

'From' Station	'To' Station		Baseline Length (m)	Change in length (mm a^{-1})	Strain Rate (a^{-1})
Northern Oregon					
Seaside (P407)	Whakiakum, WA (P408)		50853	-8.132	-1.6×10^{-7}
Seaside (P407)	Vernonia, OR (P409)		54878	-7.116	-1.3×10^{-7}
Central Oregon					
Ona Beach, OR (ONAB)	Alsea, OR (P374)		41238	-4.969	-1.2×10^{-7}
Newport Airport, OR (P367)	Alsea, OR (P374)		43749	-4.883	-1.1×10^{-7}
Alsea, OR (P374)	Halsey, OR (HLSY)		38349	-3.201	-8.3×10^{-8}
Southern Oregon					
Cape Blanco, OR (CABL)	Bandon Airport, OR (P364)		30913	-7.542	-2.4×10^{-7}
Cape Blanco, OR (CABL)	Coos Bay, OR (P365)		67070	-10.937	-1.6×10^{-7}
Bandon Airport, OR (P364)	Powers, OR (P363)		38653	-8.832	-2.3×10^{-7}
Bandon Airport (P364)	Coos Bay, OR (P365)		36190	-6.794	-1.9×10^{-7}
Site	Station ID	Latitude	Longitude	UTM East	UTM North
Whakiakum, WA	p408	46.2004	-123.3764	470961	5116378
Seaside, OR	p407	45.9546	-123.9306	427886	5089427
Vernonia, OR	p409	45.8511	-123.2393	481422	5077536
Newport Airport, OR	p367	44.5852	-123.9384	425505	4937306
Ona Beach, OR	onab	44.5154	-124.0732	414702	4929674
Alsea, OR	p374	44.3819	-123.5904	452967	4914457
Halsey, OR	hlsy	44.3775	-123.1089	491325	4913803
Coos Bay, OR	p365	43.3953	-124.2533	398505	4805479
Bandon Airport, OR	p364	43.0902	-124.4091	385314	4771789
Powers, OR	p363	42.8599	-124.0538	413907	4745798
Cape Blanco, OR	cabl	42.8361	-124.5633	372238	4743799

Trends for some baselines are shown in Figure 4. The baselines all became significantly shorter over the 2007–2011 period, with strain rates ranging from -8.3×10^{-8} to -2.3×10^{-7}.

a^{-1} which yields changes in lengths in the order of 5 mm over a 40 km baseline.

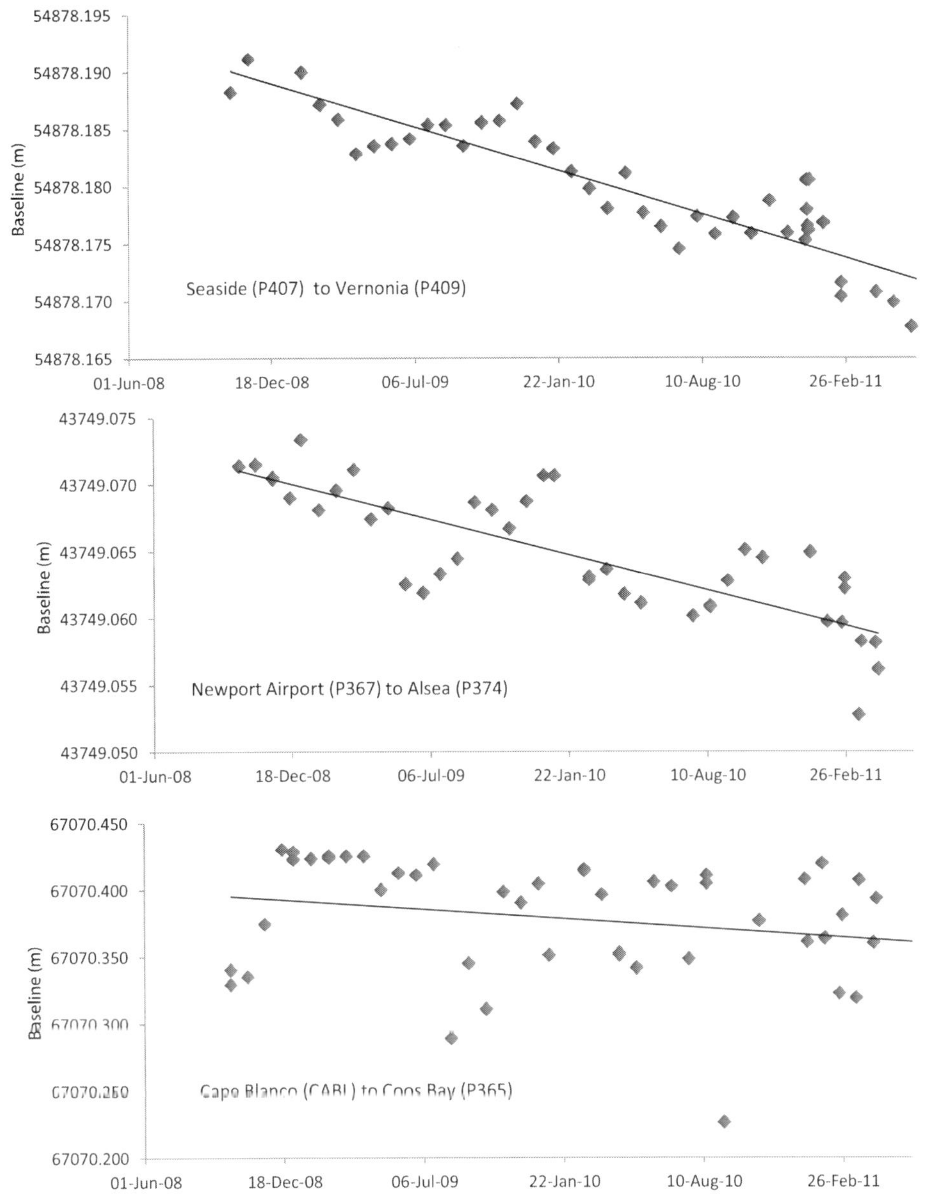

Figure 4. Representative trends in baseline lengths for the Seaside-Columbia River GPS locality (top plot), Newport-Yaquina/Alsea locality (middle plot), and Cape Blanco-Coquille-Coos Bay locality (bottom plot). See Figure 1 for approximate locations and Table 1 for calculated strain results). The west–east baselines all show significant shortening (a=0.05).

To test for south–north strain in the south coast, possibly associated with oblique convergence or south–north compression from the San Andreas transform fault (Figure 1) two south–north baselines are analyzed. Both end to end baselines (31-36 km in length), record about 7 mm yr^{-1} of shortening, suggesting that the accumulated south-north strain in the south coast is occurring regionally, rather than in a specific upper plate structure.

Late-Neogene Strain Records

Late-Neogene regional uplift of an Eocene embayment terminated marine sediment deposition and created the present Coast Range in the South Central Cascadia margin. Miocene strata in the northern Coast Range drainages give way to older Eocene strata in the southern drainages. River valley erosion has dissected the Miocene-Eocene marine strata, which have been deformed by late Neogene uplift, rotation and/or compression [Niem and Niem, 1985; Wells, 1990; Black, 1993]. The record of west–east crustal shortening from convergent strain is of greatest concern to this study. Narrow traverses in selected drainages are compiled for dip angles within 45° of west or east in Miocene–Eocene strata located between 1 and 30 km of the coastline. Such dip angles could reflect north–south trending folds, ramps, and/or thrust faults. The compiled bedrock strikes and dips were originally mapped for oil and gas surveys [Baldwin, 1956; Baldwin, 1961; Snavely et al., 1972a,b; Baldwin et al., 1973; Baldwin, 1974; Schlicker and Deacon, 1974, Newton, 1980].

A preliminary examination of the Eocene strata in the south central drainages including the Siuslaw, Tahkenitch, Umpqua, Ten Mile, and Coos Drainages (Figure 5) showed modest west–east dip angles (mean 13° ± 10° 1–σ N=246) when averaged over the ~ 25 km long traverses. The lowest dip angles (3–6° W) that are found at the eastern ends of the south–central drainage traverses are thought to reflect broad late Neogene uplift of the Coast Range. Bedrock strata in the northern drainages are dominated by Miocene sedimentary rocks [Snavely et al., 1972a,b; Niem and Niem, 1985]. Averaged west–east dip angles (mean 13° ± 5° 1–σ N=52) from the northern Siletz traverse (Figure 3) are similar to those in the south central drainages. However, there are no consistent changes in west–east dip angles with distance from the coast in the Siletz traverse [Snavely et al., 1972a,b] or in the northernmost Tillamook drainages of the study area [Niem and Niem, 1985]. Bedrock strata in the southernmost drainages, south of Coos Bay (Figure 3), contain both Tertiary and Mesozoic rocks [Baldwin et al., 1973; Baldwin, 1974]. East and west dip angles from the late Eocene sedimentary rocks within 25 km of the coast in the Coquille drainage (Figure 3) are substantially greater (mean 51° ± 15° 1–σ N=50) than those in the central drainages. There are no consistent changes in west–east dip angles within 25 km distance from the coast in the Coquille drainage [Baldwin et al., 1973]. Only the south–central drainages of the study area both 1) contain low dip angles (< 10° dip) within 25 km of the coast and 2) demonstrate increasing dip angles with increasing proximity to the coastline (Figure 5).

The five south central traverses selected for documentation of late Neogene convergent strain records (Figure 5) demonstrate patterns of increasing west–east dips from east to west.

Anomalous dip angles are clustered within and between the five localities, but the overall trend is that of increasing fold and/or fault deformation towards the coastline. For example, west–east dip angles increase from about 5° to 30° over a westward distance of 20 km in the northern Siuslaw locality (Figure 5a). Dip angles increase from about 5° to 20° over 15-20 km distances in the Tahkenitch, Umpqua, and Ten Mile localities (Figures 5b, 5c, 5d). Compiled west–east dip angles increase from 10° to 70° over a 30 km seaward distance in the southernmost, Coos Bay, traverse (Figure 5e). These seaward increasing trends of west–east dip angles demonstrate inelastic strain in the upper plate that is associated with convergent plate motions during late Neogene time.

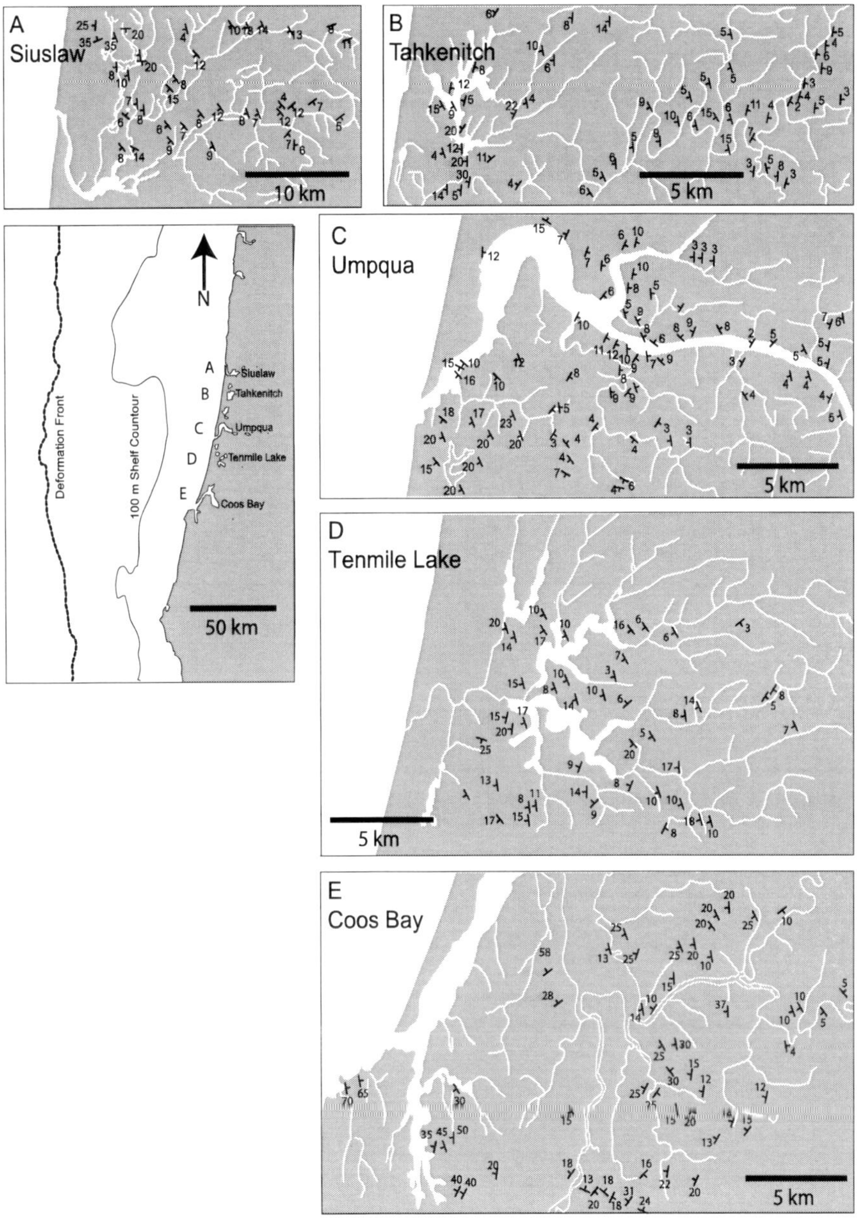

Figure 5. Five coastal drainage traverses showing west–east dip angles of Miocene–Eocene sedimentary rock strata. Siuslaw (A), Tahkenitch (B), Umpqua (C), Ten Mile (D), and Coos Bay (E). Reported dips within 45° of west or east are compiled to represent north–south trending folds or faults at about 1–25 km from the coastline.

Late Pleistocene Strain Records

Uplifted marine and fluvial terraces in the central Cascadia margin have been mapped to investigate possible regional tectonic deformation [West and McCrumb, 1988], local fault offsets [Kelsey, 1990; McInelly and Kelsey, 1990], and/or upper plate segment boundaries

[McNeil, et al., 1998; Personius, 1995]. However, few active faults have been formally established at the surface, suggesting that the topographic deformation is largely associated with subsurface structures. In this chapter the along coast length scales of deformation are used to differentiate regional uplift at hundreds of kilometers in along coast distance from localized terrace warping at tens of kilometers in along coast distance.

Table 2. Elevations of the lowest marine terrace deposits from exposed sections measured in sea cliffs and bay cliffs in the central Cascadia margin

Measured Section	UTM (North)	Elevation (m MSL)	Deformation. Front (km)	Modern Setting	Deposit Sequence	Sec. Ref	Age
Copper Point	5159000	15	148	BC	Ls/TF	1	~80 (2)
Ramsey Point	5157000	10	146	BC	Ls/TF	1	
Pickernell	5156000	7	145	BC	Ls/TF	1	
Long Beach	5134000	8	133	HSC	Ls/TF	-	
Indian Beach	5087000	13	118	SC	Col/Lg	2	
Cannon B.	5080000	11	119	SC	Col/Lg	2	
Arcadia	5077000	8	119	SC	Col/Lg	2	
Hug Point	5075000	13	119	SC	Col/Lg	2	
Arch Cape	5074000	6	119	SC	Col/Lg	2	
Cove Beach	5071000	11	118	SC	Col/Lg	2	
Short Sand	5067000	8	117	SC	Col/Lg	2	
Rockaway	5050000	9	115	HSC	Col/Lg	-	
Garibaldi	5041000	15	116	BC	Ls/Lg	-	
Tillamook	5039000	7	118	BC	Ls/Lg	-	
Cape Mears	5038000	11	114	SC	Co/Lg	2	
Happy Camp	5032000	12	112	SC	Co/BB	4	
Hanson	5030000	10	113	BC	Ls/Lg	2	
Fish	5027000	13	114	BC	Co/Lg	2	
Niflis	5027000	12	114	BC	Co/Lg	2	
Cape Lookout	5023000	9	112	SC	Co/BB	2	
Scout Camp	5017000	8	114	SC	Dn/BB	4	
ATV	5017000	10	114	SC	Dn/BB	4	
Sand Lake	5015000	3	114	BC	Dn/Lg	4	~80 (4)
N. Kiwanda	5009000	4	113	SC	Dn/BB	4	
S. Kiwanda	5008000	10	112	SC	Dn/Pf	4	
N. Daley Lk	4998000	7	110	HSC	Co/Lg	5	
Roads End	4984000	6	112	SC	Dn/BB	5	
21st	4981000	4	112	SC	Dn/BB	4	
Canyon C.	4978000	5	112	SC	Dn/BB	4	
Spahish Head	4976000	5	112	SC	Dn/BB	4	
Glen Eden	4970000	4	111	SC	Dn/BB	4	
Fogerty	4966000	10	106	SC	Dn/BB	5	
Depoe	4965000	13	107	SC	Dn/BB	4	
Boiler Bay	4964000	15	102	SC	Dn/Pf	5	
Otter Rock	4955000	19	101	SC	Dn/BB	5	
Beverly B.	4953000	28	101	SC	Dn/BB	4	
Wade	4951000	24	101	SC	Dn/BB	4	
S. Yaquina	4947000	16	101	SC	Dn/BB	5	

Measured Section	UTM (North)	Elevation (m MSL)	Deformation. Front (km)	Modern Setting	Deposit Sequence	Sec. Ref	Age
Nye Beach	4943000	10	100	SC	Dn/Pf	4	
Newport	4944000	17	100	BC	Dn/Lg	9	~80 (2)
LightHouse	4942000	11	100	SC	Dn/BB	4	
Henderson	4938000	16	99	SC	Dn/BB	4	
Grant Creek	4937000	8	99	SC	Dn/BB	4	
N.Beaver	4931000	6	98	SC	Dn/BB	4	
S. Ona	4929000	4	99	SC	Dn/Pf	4	
Seal Rocks	4927000	7	99	SC	Dn/BB	4	111 (4)
Quail St.	4926000	4	99	SC	Dn/BB	4	
Driftwood	4924000	5	99	SC	Dn/BB	4	
Brucley Cove	4924000	4	99	SC	Dn/BB	4	
Patterson	4918000	7	98	SC	Dn/BB	5	
N. Yachts	4908000	8	97	SC	Dn/BB	4	
Yachats	4907000	6	97	SC	Dn/BB	4	
S Yachats	4907000	6	96	SC	Dn/BB	4	
S. Neptune	4899000	7	95	SC	Dn/Lg	5	
Washburn	4889000	6	95	SC	Dn/BB	5	
Heceta Head	4889000	4	95	SC	Co/BB	-	
Three Mile	4844000	20	97	SC	Dn/BB	-	
Umpqua	4843000	27	97	BC	Dn/BC	-	
Umpqua	4836000	12	96	BC	Dn/BC	-	
Hauser	4813000	9	90	BC	Dn/BC	-	
Fossil Point	4801000	7	85	BC	Dn/Lg	5	
Bastendorf	4800000	9	82	SC	Dn/BB	5	
Lighthouse	4799000	12	80	SC	Dn/BB	5	~80 (6)
Cape Arago	4795000	35	76	SC	Pf	5	
N. Sacci B.	4792000	36	78	SC	Dn/BB	5	
Whisky Run	4786000	23	76	SC	Dn/BB	4	
Golf Course	4785000	22	75	SC	Dn/BB	4	
Coquille P.	4774000	24	68	SC	Dn/BB	4	
DevilsKitchen	4771000	12	67	SC	Dn/BB	5	
N. New River	4767000	5	66	SC	Dn/BB	5	
S. New River	4763000	6	66	SC	Dn/BB	5	
Floras Lake	4750000	9	60	SC	Dn/BB	5	~80 (7)
Blacklock	4748000	40	59	SC	Dn/BB	5	
Cape Blanco	4744000	57	56	SC	Dn/Pf	5	~80(7,8)
SCapeBlanco	4743000	56	54	SC	Dn/Pf	4	
Port Orford	4733000	24	56	SC	Dn/BB	4	

Settings: Sea Cliff (SC); Bay Cliff (BC). Deposit Sequence: Beach Backshore (BB), Beach Cobble (BC), Colluvium (Co), Dune (Du), Loess (Ls), Lagoon (Lg), Platform (Pl), Tidal Flat (TF). Measured Section and Terrace Age References. Clifton [1994], (2) Kennedy, et al. [1982], (3) Mulder [1992], (4) Peterson, et al. [2006], (5) Peterson, et al. [1994], (6) McInelly and Kelsey [1990], (7) Muhs, et al. [1990], (8) Kelsey [1990], and (9) Ticknor [1993]. The sea cliff and bay cliff sections shown here were measured by tape and/or EDM total station and were surveyed to timed mid-swash position (within about ± 0.5 m MSL) [Peterson, et al., 1994; Peterson, et al., 2006].

Most of the marine terraces in the South Central Cascadia Margin are overlain by Pleistocene dune and loess deposits, which are truncated by broad deflation surfaces

[Peterson et al., 2006]. This mantling of the uplifted terraces by dune sheet deposits, up to 30 m in thickness, was unknown to West and McCrumb [1988] at the time of their mapping. The eolian deposits post date the terrace ages by tens of thousands of years [Peterson et al., 2007]. The dune sheets mantle the terraces, filling in valleys and building ramps over buried terrace risers. Some of the broad dune deflation surfaces have been previously misidentified as marine terrace topographic surfaces.

In this chapter the elevations of beach and lagoon deposits that are exposed in modern sea cliffs (Table 2) are used to represent the sea level high stand datum at 80 ka (oxygen isotope stage 5a). The uplifted terraces are dated in key localities by aminostratigraphy [Kennedy et al., 1982], uranium series [Muhs et al., 1990], and thermal luminescence [Peterson et al., 2007] (Table 2).

The use of sea cliff sections to measure terrace height slightly underestimates maximum high-stand elevations taken at platform backedges [Bradley and Griggs, 1976] but this method yields a consistent terrace elevation proxy along this straight narrow coastal plain. Where long-term uplift rates are low, the 80 ka terrace might reoccupy previous high stand platforms at 105 ka (stage 5c) and/or 120 ka (stage 5e). In any case the lowest preserved high stand beach or lagoon deposit is used to establish vertical deformation of the latest Pleistocene marine terrace datum.

The high stand deposits exposed in bay cliffs and sea cliffs in the northern part of the study area consist of lagoon or bay wetland mud [Mulder, 1992; Clifton, 1994], representing elevations of 0–2 m above the 80 ka paleosea level datum. The majority of marine high stand deposits that are exposed in sea cliffs along the central and southern parts of the study area consist of beach backshore sand. Modern backshore sand deposits reach 4–5 m above mean sea level in the study area [Peterson et al., 1994], but the regressive deposits exposed in the sea cliffs likely reflect slightly lower elevations, thereby yielding elevations that are probably 1–3 m above the 80 ka paleosea level datum.

The bay cliff and sea cliff sections provide abundant exposures of the lowest marine terrace (76 sites recorded), except where covered by the thick dune sheets between the Siuslaw and Coos Bay drainages, (4820000-4880000 UTM-N; Table 2). Elevations of the upper contact of the lagoon or beach high stand deposits at 80 ka are measured by leveling to predicted tidal level (± 0.5m). The measured eleavations are corrected to mean sea level (MSL) which is about 1.0 m above the regional 0 m NAVD88 datum in the study area.

A plot of terrace elevations along the South Central Cascadia Margin shows increasing modern elevations of the 80 ka high stand deposits from Willapa Bay (~10 m MSL) to Cape Blanco (57 m MSL) at the Elk drainage (Figures 3 and 6).

The coastal record of late Pleistocene marine terrace deformation is extended slightly beyond the limits of the South Central Cascadia Margin (Figure 1) to identify any terrace warping trends (tens of kilometers length scale) at the ends of the study area. The densely sampled terrace sections show increasing variability of the 80 ka deposit elevations with increasing distance south along the study area coastline (Figure 6). In the northern bay and sea cliffs the 80 ka high stand deposits range from about 5 to 15 m elevation. In the southern sea cliff exposures the 80 ka high stand deposits range from 5 to at least 35 m elevation. Broadly even regional uplift in the northern coastline transitions to local warping in the south coast. The late Pleistocene terrace warping is likely associated with folds and/or underlying faults that obliquely intersect the coast (Figure 3) in the southern part of the study area.

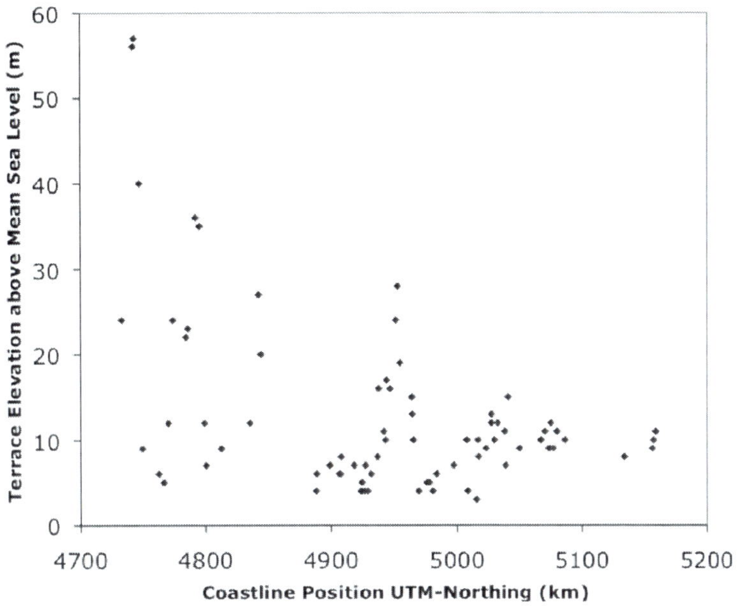

Figure 6. Plot of marine high-stand deposit elevations versus position in the central Cascadia margin from measured sections in sea cliffs and bay cliffs (see Table 2 for measured section data). Cape Blanco (~ 4740 km UTM) is located at the Elk River drainage, and Willapa Bay (~ 5150–5160 km UTM) is located just north of the Columbia River mouth (Figure 2).

Late Holocene Strain Records

Episodic coastal subsidence as recorded in late Holocene tidal wetland deposits has been documented in many estuarine sites in the South Central Cascadia Margin [Darienzo, 1991; Nelson, 1992; Darienzo et al., 1994; Witter et al., 2003, among others]. Paleotidal level datums recorded by plant macrofossils and microfossils are used to establish episodic events of land level submergence or emergence, which are superimposed over longer-term trends of net sea level rise in the South Central Cascadia Margin [Darienzo and Peterson, 1990]. In this chapter selected core site records are compiled from estuaries with the greatest landward extent of tidal influence in the study area. Significant tidal influence in the lower Columbia River estuary extends over 100 km landward from the coastline, a distance sufficient to establish the limit of recorded subsidence at site CRLV2, located at Longview, Washington (Figure 7).

The wetland subsidence records are based on downcore changes in plant macrofossil type and abundance, as detailed elsewhere [Barnett, 1997; Peterson et al., 2000]. In short, the amounts of abrupt paleosubsidence (0-2 m vertical subsidence) are based on abrupt upcore decreases in peat content and/or tree roots, which reflect vegetated wetland burial by intertidal mud. Two of these authors were the first to use microfossil paleosalinity indicators, including diatoms and foraminifera, to confirm abrupt subsidence records in Netarts Bay (Figure 8) [Darienzo, 1987; Darienzo and Peterson, 1990]. Other investigators subsequently used these salinity indicators in Cascadia Margin wetlands [Briggs, 1994; Nelson et al., 1996; Atwater and Hemphill-Haley, 1996; Barnett, 1997] to show freshwater wetlands submerged and then buried under saline intertidal mud. However, salinity wedges in the fluvially dominated estuaries in the South Central Cascadia Margin only reach about one third of the landward distance of tidal influence. The geographical limitations of salinity indicators greatly bias or reduce the spatial distribution of paleosubsidence records that are recorded in fluvially dominated estuaries in the South Central Cascadia Margin.

In this chapter the macrofossil records of vegetated wetland burial are used to establish three conditions of episodic subsidence, including 0 ± 0.5 m, 1 ± 0.5 m, and 1.5 m ± 0.5 m subsidence. At least two abrupt subsidence events that occurred during the past 2,500 years are used to establish the maximum amounts of cyclic abrupt subsidence in a core site. An interpretation of abrupt subsidence is based on sharp upper contacts (< 1cm width) between the subsided wetland and the overlying bay mud. With the exception of some new core site data from the Columbia and Coquille estuaries (Figure 2) the core site coordinates, subsidence records, and radiocarbon dates used for this study have been previously reported [Briggs, 1994; Barnett, 1997; Petersen et al., 2007]. For this study the core records are compiled from west to east traverses in estuaries with tidal reaches that extend at least 10 km inland from the coast in the South Central Cascadia Margin. These estuaries include the Columbia, Netarts and Tillamook Bays, Siuslaw, Umpqua, Coos, and Coquille (Figure 2).

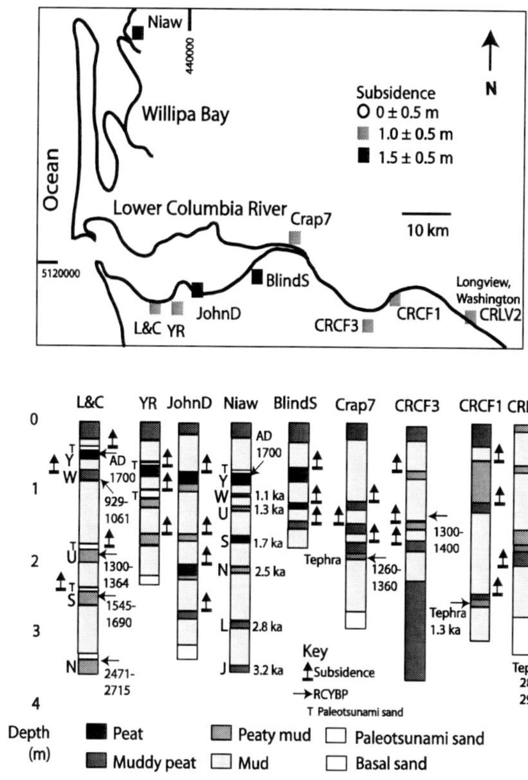

Figure 7. Map of core sites and selected core logs for the Lower Columbia River and Willapa Bay. Several abrupt submergence events in the lower reaches of the Columbia River estuary including core sites LandC and YR are associated with paleotsunami sand layers. Coseismic subsidence events in site LandC are correlated to subsidence events in Willapa Bay at site Niaw (Niawiakum) (Atwater and Hemphill-Haley, 1996; Atwater et al., 2004). The greatest amounts of subsidence (1.5 m) in the lower Columbia River occur between John Day Creek (JohnD) and Blind Slough (BlindS). This position corresponds to the trough of maximum coseismic subsidence (~ 1.5 m subsidence), located 30–40 km landward of the river mouth.

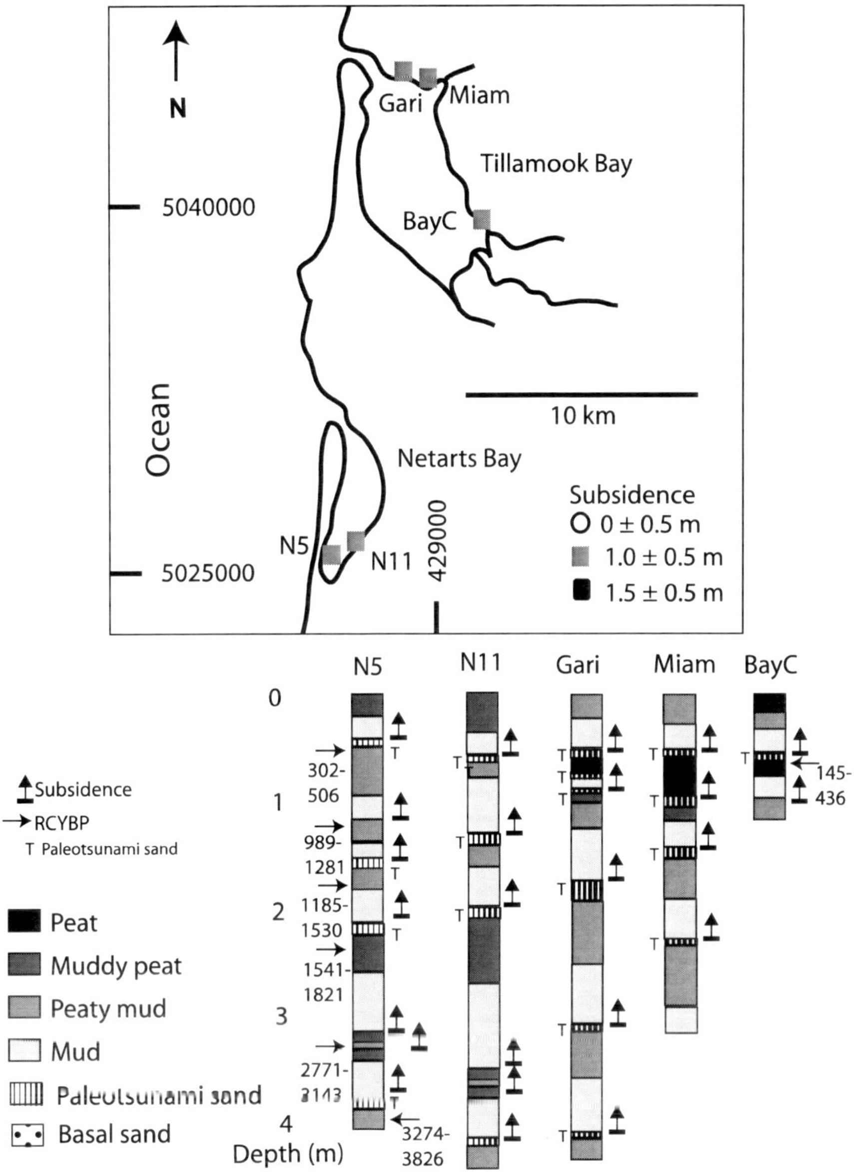

Figure 8. Map of core sites and selected core logs for the Netarts and Tillamook Bays. Abrupt wetland subsidence events are associated with paleotsunami sand layers showing correlation to regional megathrust ruptures in the Cascadia Margin. Seven subsidence events are recorded during the last ~ 3,200 years in Netarts Bay, demonstrating a mean recurrence interval of 430 years. Subsidence events in Netarts Bay are estimated to generally range from 0.5 to 1.0 m in abrupt submergence, though the last subsidence event in Tillamook Bay might reach 1.5 m.

Radiocarbon dates published by Jurney [2001] and Witter et al. [2003] in the lower Columbia and Coquille estuaries, respectively, are used to augment core logs collected in those localities by the authors of this chapter. One core log (Niaw) from the eastern margin of the tidally dominated Willapa Bay [Atwater and Hemphill-Haley, 1996] is presented for comparison to the fluvially dominated reaches of the lower Columbia River (Figure 7).

Table 3. Radiocarbon ages of selected core site intervals from South Central Cascadia Margin tidal basins

Core Site Depth (cm)	Conventional Age RCYBP	Calibrated 1s Age RCYBP	Calibrated 2s Age RCYBP	Beta Analytic, other ()Lab ID
CRCF3 (125)	1470 ± 30	1330 – 1380	1300 – 1400	294493
CRAP7 (180)	1380 ± 50	1280 – 1320	1260 – 1360	294489
LandC (60)	1080 ± 40	937 - 1052	929 - 1061	121421
LandC (180)	1438 ± 19	1309 – 1341	1300 – 1364	(UB)4499
LandC (240)	1698 ± 15	1562 – 1682	1545 – 1690	(QL)4922
LandC (340)	2488 ± 21	2494 – 2705	2471 – 2715	(UB)4497
CRLV2	2450 ± 30	2365-2691	2360– 2710	304399
BayC (55)	250 ± 50	150 – 427	141 - 464	89165
N5/11 (56)	350 ± 60	318 – 480	302 -506	24933
N5/11 (114)	1220 ± 60	1060 – 1239	989 – 1281	24934
N5/11 (190)	1450 ± 80	1287 – 1413	1185 – 1530	24935
N5/11 (220)	1760 ± 60	1569 – 1766	1541 – 1821	24521
N5/11 (360)	2820 ± 70	2846 – 3036	2771 – 3143	25030
N5/11 (420)	3290 ± 100	3405 – 3635	3274 – 3826	24523
S 216 (127)	1320 ± 60	1180 - 1298	1084 - 1338	67450
S 216 (153)	1540 ± 70	1373 - 1518	1303 - 1560	58117
S 216 (167)	1450 ± 70	1296 - 1400	1271 - 1519	67451
S 216 (210)	1910 ± 70	1739 - 1930	1634 - 2033	58118
S 222 (40)	30 ± 90	modern		58119
S 222 (322)	1910 ± 70	1739 - 1930	1634 - 2033	58120
U 301 (78)	400 ± 60	329 - 512	314 - 523	67455
U 301 (290)	2530 ± 80	2489 - 2746	2362 - 2755	67456
U 321 (205)	1420 ± 80	1277 - 1403	1179 - 1517	67460
U 328 (255)	2450 ± 70	2362 - 2697	2353 - 2716	58121
U 332 (76)	1270 ± 90	1088 - 1287	980 - 1331	58122
U 332 (130)	2820 ± 70	2846 - 3060	2771 - 3143	58123
U 332 (180)	2960 ± 60	3005 - 3239	2957 - 3330	58124
U 337 (260)	2150 ± 80	2008 - 2305	1950 - 2337	67458
U 337 (375)	3140 ± 70	3266 - 3446	3164 - 3555	67459
U 339 (89)	1040 ± 60	913 - 1054	794 - 1067	58125
U 339 (179)	1630 ± 60	1416 - 1596	1390 - 1693	58126
U 339 (225)	1850 ± 60	1715 - 1864	1619 - 1924	58127
JNEY (45)	400 ± 50	342 -532	319 -544	110469
JNEY (86)	1080 ± 60	986 – 1170	956 – 1256	110470
C 408 (80)	650 ± 70	557 - 669	525 - 696	27675
C 408 (232)	2350 ± 90	2184 - 2685	2153 - 2712	27743
C 408 (329)	2760 ± 80	2778 - 2945	2744 - 3074	34278
C 408 (550)	3620 ± 160	3707 - 4149	3561 - 4413	34279
C 448 (75)	1800 ± 60	1628 - 1819	1569 - 1869	58137
C 448 (205)	2830 ± 100	2797 - 3076	2754 - 3238	58138
C 449 (65)	1000 ± 50	800 - 963	788 - 1501	58134
C 449 (98)	1390 ± 60	1269 - 1354	1179 - 1404	58135
C 449 (139)	1780 ± 60	1619 - 1811	1553 - 1862	58136
C 465 (68)	1970 ± 70	1827 - 1996	1737 - 2113	58128
C 465 (95)	3080 ± 60	3161 - 3398	2994 - 3479	58129
C 465 (420)	4560 ± 70	5054 - 5437	4976 - 5465	58130
C 470 (72)	1020 ± 70	799 - 1051	744 - 1070	67437
C 470 (121)	1400 ± 80	1192 - 1393	1150 - 1517	67438
C 470 (190)	1660 ± 80	1419 - 1692	1373 - 1774	67439
C 470 (250)	2900 ± 80	2929 - 3202	2848 - 3321	67440
C 470 (305)	3700 ± 90	3901 - 4154	3777 - 4402	58133

Core Site Depth (cm)	Conventional Age RCYBP	Calibrated 1s Age RCYBP	Calibrated 2s Age RCYBP	Beta Analytic, other ()Lab ID
C 476 (115)	1780 ± 90	1574 - 1819	1446 - 1923	67441
C 476 (210)	1930 ± 70	1815 - 1987	1707 - 2042	58139
C 478 (140)	1740 ± 100	1539 - 1810	1415 - 1872	67443
C 478 (395)	2920 ± 60	2970 - 3161	2882 - 3257	67444
FAHY (77)	240 ± 40	150 -420	143-430	221465
Coquille521 (245)	2430±40	2360 - 2680	2350 – 2710	305787
Elk/A2 (295)	2980 ± 60	3069 – 3262	2977 – 3339	73248

Dates from Columbia River (LandC) [Journey, 2001], Columbia River (CR) (this chapter), Tillamook Bay (BayC) [Barnett, 1997], Netarts Bay (N) [Darienzo and Peterson, 1990]. Siulsaw (S), Umpqua (U) and Coos Bay (C) [Briggs, 1994; Barnett, 1997; Peterson et al., 1997], Coquille (FAHY) (this chapter), and Elk (this chapter). Radiocarbon date calibration (Siuslaw, Umpqua, Coos) (CALIB 5.02. Stuiver, et al., 2005) and all other dates (CALIB 6.02. Stuiver, et al., 2011). See Figures 7–12 for core site positions and lithologic logs.

One dated core section from the Elk River estuary, located just south of the study area (Figure 1) is presented here for the first time to compare to more extensive work in the adjacent Sixes estuary reported by Kelsey et al [2002]. Radiocarbon data for all of the late Holocene wetland sites shown here are presented in Table 3.

The core logs that are shown in this chapter are generally taken from continuous vibracores or pound cores (7.5 cm diameter) that were split, photographed, logged, and subsampled under laboratory conditions. Plant macrofossil type and relative abundance are based on standardized observation and on loss on ignition, respectively, as detailed elsewhere [Darienzo, 1991; Barnett, 1997]. Diatoms are used to confirm paleosubsidence estimates based on peat content and/or presence of tree roots in the lower saline reaches of the estuaries. Diatom data are presented elsewhere for Netarts [Darienzo, 1991], Tillamook [Barnett, 1997], Siuslaw, Umpqua, Coos [Briggs, 1994], and Coquille [Witter et al., 2003]. In this chapter the plant macrofossils are used to extend the paleosubsidence records well landward of the ocean coastline in freshwater tidal reaches of the estuaries. For example, both the numbers of events and amounts of recorded paleosubsidence increase with increasing distance from the coast in the Siuslaw estuary, with at least 3–4 events recorded near Mapleton, Oregon (Figure 9).

In this chapter the presence of tree roots or peat (>50% organic content) are used to designate paleotidal elevations of at least 1.5 m paleomean sea level (pMSL). Barren mud or slightly rooted mud occur at 0 m pMSL in the study area estuaries [Darienzo and Peterson, 1990; Barnett, 1997]. Muddy peat at 1.0 m pMSL and peaty mud at 0.5-0.75 m pMSL represent intermediate paleotidal level indicators. Upcore abrupt transitions between these four different paleotidal indicators indicate the amount of submergence (0 ± 0.5 m, 1.0 ± 0.5 m, and 1.5 ± 0.5 m) used in this chapter. Most importantly peat or tree root to mud transitions represent 1.5 m of submergence. Core sections lacking transitions between any two of the other indicators represent 0 m or no episodic submergence within the detection limits of the plant macrofossils. For example, the general absences of two or more recorded subsidence events are well established in the lowermost reaches of the Umpqua River estuary in the central part of the study area (Figure 10).

Radiocarbon ages of peat samples provide loose constraints on core depth age and recorded event recurrence intervals during the inferred subsidence events (Table 3). Event correlations based on deposit radiocarbon ages are problematic in the southern part of the study area due to the modest and spatially variable aspects of the subsidence events.

Paleotsunami inundations, as established by landward thinning sand sheets at paleosubsidence contacts, provide much better correlations of local subsidence to regional megathrust ruptures, such as in the South Slough syncline (sites JNEY and 408) in the lower reaches of Coos Bay (Figure 11).

West–east core traverses in the south central part of the study area demonstrate increasing numbers of subsidence events with distance landward from the coast in the Siuslaw and Umpqua estuaries (Figures 9 and 10). The most striking example of increasing episodic coastal subsidence with increasing distance landward of the coast occurs in the Coquille estuary. The core sites 521–523 at Coquille, Oregon, contains twice the number of subsidence events for the same depth interval as the core sites FAHY and 504/503 in the lower reaches of the Coquille estuary (Figure 12).

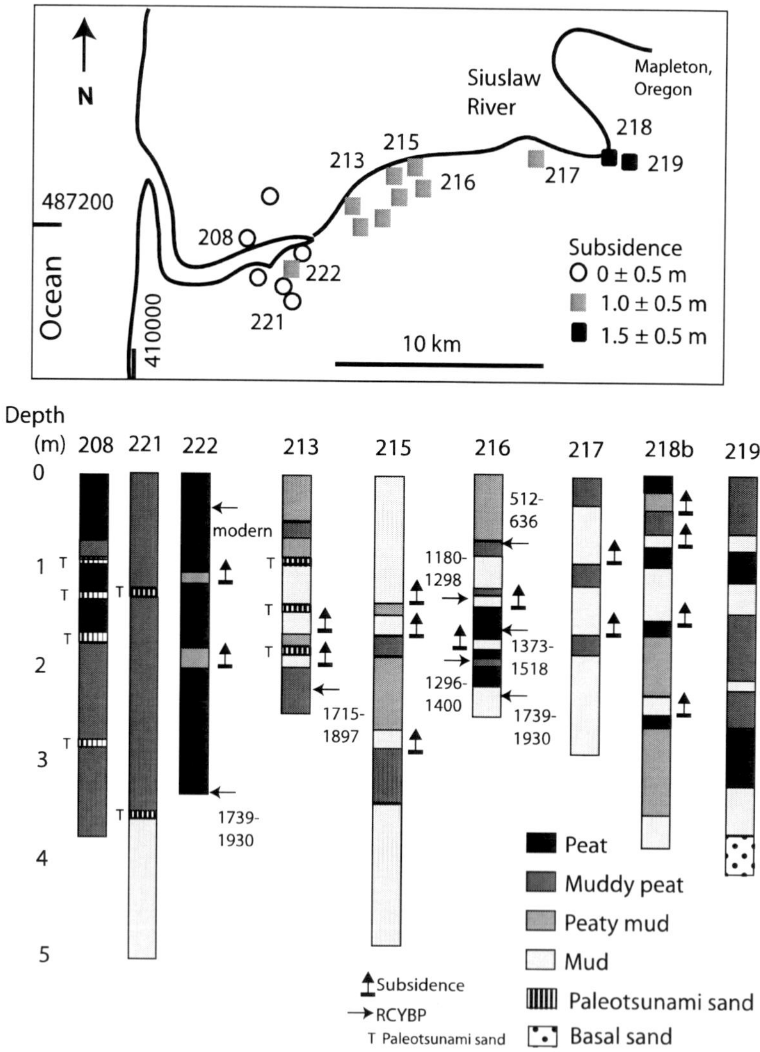

Figure 9. Map of core sites and selected core logs for the Siuslaw River estuary. Abrupt wetland subsidence events increase in number and in relative submergence with distance landward (25 km) from the coast. Three to five subsidence events are recorded during the last 1700–1900 years in the upper reaches of the Siuslaw estuary.

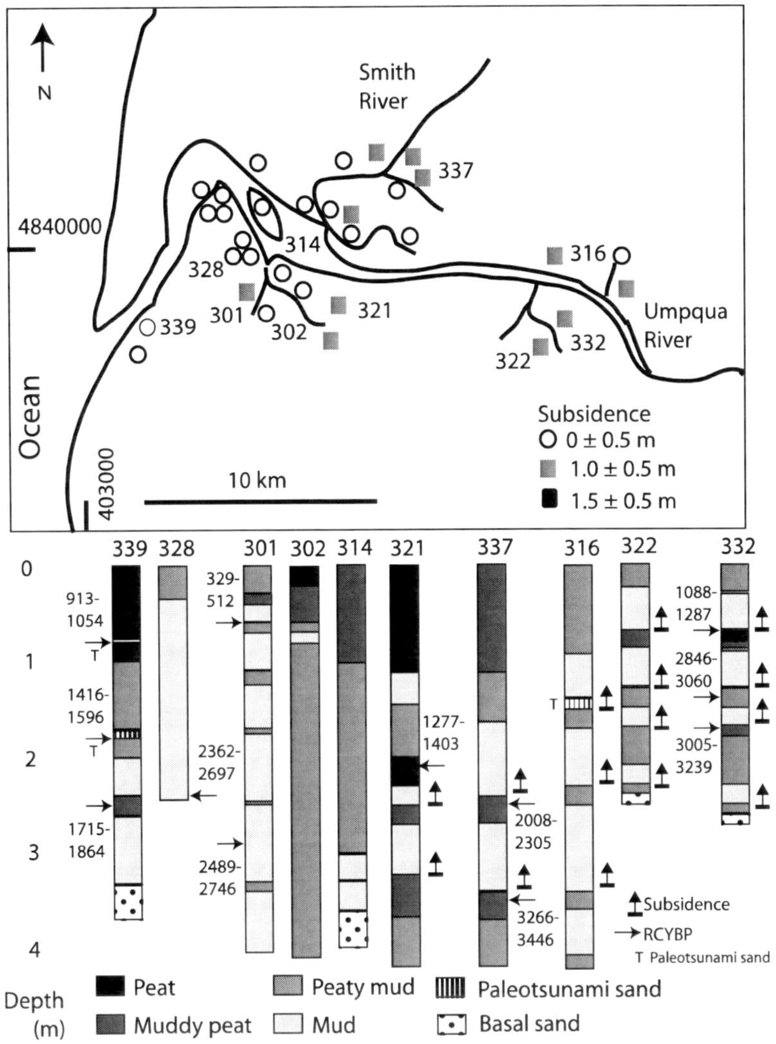

Figure 10. Map of core sites and selected core logs for the Umpqua River estuary. Abrupt wetland subsidence events increase in number with distance landward (20 km) from the coast. Four subsidence events are recorded during the last ~ 3,500 years in the upper reaches of the Umpqua estuary. A thick paleotsunami sand sheet overlies a dated peat (1416–1596 BP) at site 339 near the mouth of the estuary. An undated paleotsunami deposit is recorded at river bank site 316, located nearly 30 river kilometers from the tidal inlet.

One representative core log (Elk/2A) is shown for the Elk River tidal wetlands at Cape Blanco (Figure 12). The Elk River Drainage is located in the South Cascadia Margin, landward of the Gorda Plate segment (Figure 1). It is discussed briefly in this chapter due to its proximity to the Coquille estuary, at the south end of the South Central Cascadia Margin. The paleosubsidence records in the Elk River wetlands are very weak, with subsidence amounts at or below the threshold of detection (<0.5 m abrupt submergence).

By comparison, Kelsey et al., [2002] report substantial amounts of episodic subsidence in the nearby Sixes River wetlands. It is not presently known whether the greater subsidence estimates in the Sixes River wetlands could be influenced by paleotsunami input of marine/brackish diatoms or whether the larger subsidence values reflect anomalous stain associated with an upper plate structure, as suggested by Kelsey et al., [2002]. Locally enhanced episodic subsidence has been established in the South Slough and Pony Slough synclines of Coos Bay (Figure 11). The focus of the elastic strain investigations reported in this chapter are at the regional scale rather than the local scale, so the localized subsidence records in the lower reaches of the southern estuaries are not further discussed here.

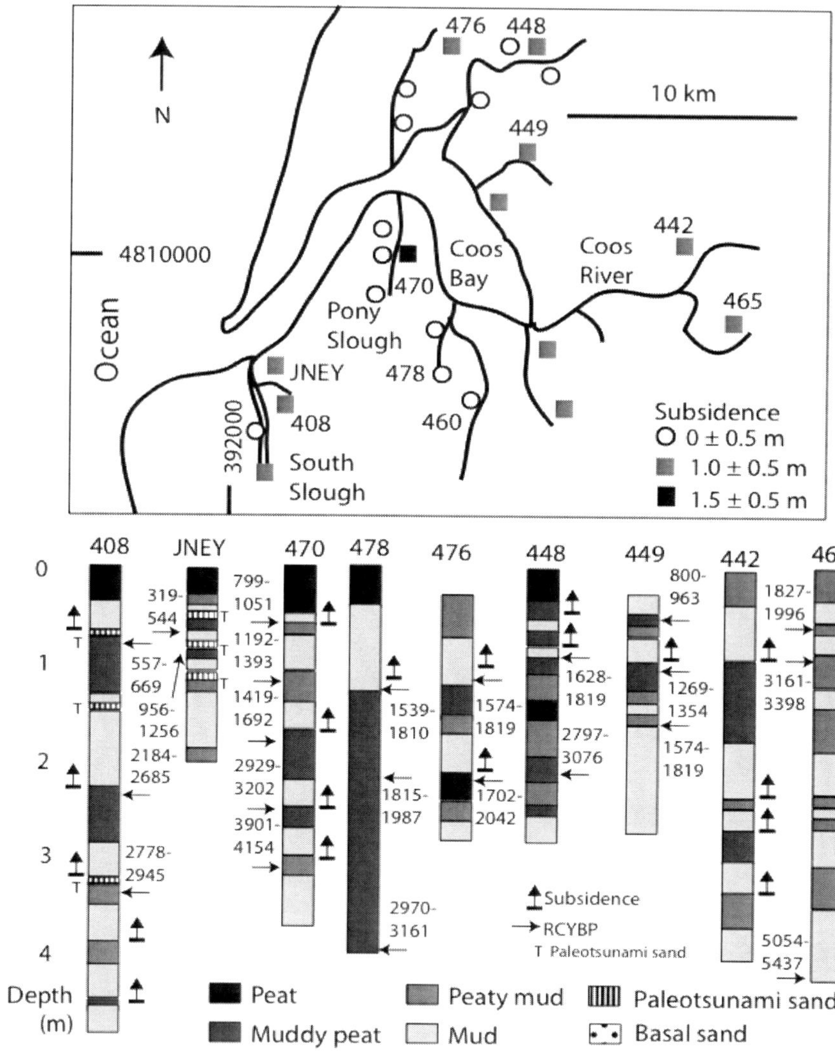

Figure 11. Map of core sites and selected core logs for the Coos Bay and South Slough estuaries. The distribution of recorded subsidence events in Coos Bay is complex, with modest subsidence events (1.0 ± 0.5 m) or no subsidence (0 ± 0.5 m) in core sites that are comingled between distances of 5 and 25 km from the coast. However, episodic subsidence is locally enhanced in mapped fold axes in South Slough and Pony Slough [Newton, 1980]. Paleotsunami inundation is recorded with at least 3 of 5 abrupt subsidence events in sites JNEY and 408 near the Coos Bay tidal inlet, showing the localized subsidence to be correlated to regional megathrust ruptures.

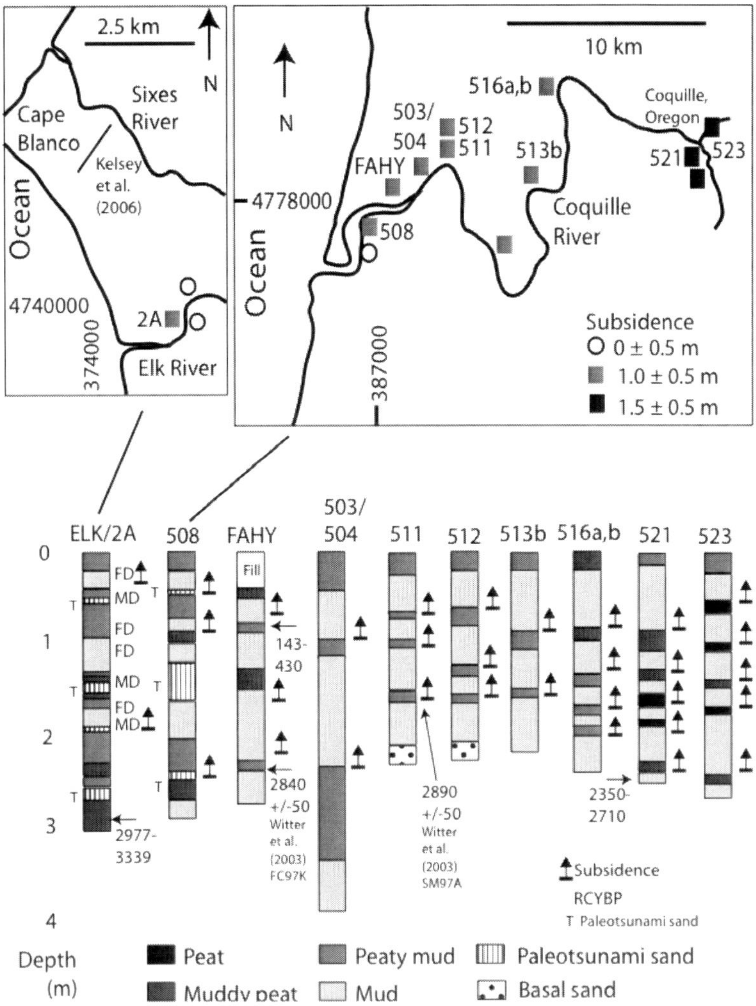

Figure 12. Map of core sites and selected core logs for the Coquille estuary and Elk River wetlands. Abrupt wetland subsidence events in the Coquille estuary increase in number and in relative submergence with distance landward (20 km) from the coast. Two events approaching or reaching 1.5 m are recorded in the most landward sites 521–523. Un-calibrated basal radiocarbon dates, which are used to estimate section ages in sites FAHY and 511, are from nearest core sites, FC97K and SM97A respectively, as reported by Witter et al., [2003]. The record of subsidence in the Elk River estuary is weak. Marine diatoms (MD) likely associated with paleotsunami inundations interrupt the otherwise freshwater diatom assemblages (FD) in the small Elk River wetlands at site Elk/2A. Kelsey et al. [2002] have reported records of episodic subsidence in core sites located in a traverse in the nearby Sixes River (solid line).

Discussion

Late Neogene Coastal Deformation

Long-term convergent deformation in the Cascadia margin is reflected, in part, by north trending folds and/or monoclines associated with west-east compression of the upper plate. Compiled west- and east-dip angles of coastal Tertiary formations in the south-central

Cascadia margin decrease with distance landward of the coastline (Figure 5). A plot of those bedding dip angles versus their landward distance from the deformation front or buried trench at equivalent latitudes is shown in Figure 13.

A regional dip angle of 3° to 6° to the west occurs at landward distances of at least 120 km from the deformation front in the central part of the study area (Figure 13). The regional dip is associated with Neogene uplift of the Coast Range. Larger dip angles of 10° to 20°, and 20° to 30° respectively, are mapped at distances of 110 and 100 km from the deformation front. The seaward increase in dip angles of the Tertiary strata indicate long–term convergent strain accumulation, associated with interplate coupling of the Cascadia megathrust, at distances of up to 110 km from the deformation front in the South Central Cascadia Margin. The inelastic deformation could either reflect overlying plate compression from an underlying strongly coupled zone and/or deformation propagated a short distance landward of the strongly coupled zone in the upper plate. The duration and/or frequency of strain accumulation during the late Neogene are not addressed here, but Quaternary records of inelastic strain are provided by late Pleistocene uplift and warping of marine terraces (see below).

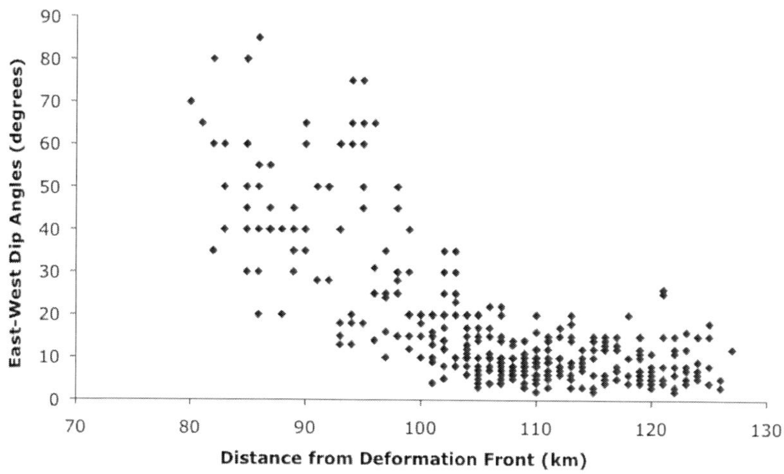

Figure 13. Tertiary strata dip angles as a function of across margin distance to the deformation front or buried trench. Graphed data are from Figure 5.

Late Pleistocene Coastal Deformation

The along coast variability in the elevations of the uplifted 80 ka marine terrace(s) serves as a proxy for inelastic upper plate deformation in the central Cascadia margin during the late Quaternary. In this chapter the regional scale of coastal terrace uplift, possibly related to broad under plating or other deeper processes, is discriminated from local terrace warping. The local vertical warping is assumed to be associated with compression in upper plate structures such as folds and/or underlying faults. The relative uplifts of the marine terraces are based on the elevations of 80 ka high-stand deposits, as measured in bay cliff and beach cliff exposures (Table 2).

A vertical deformation index aids in visualizing the heights of the 80 ka high stand deposits binned at approximately 50 kilometer intervals along the coast in the study area (Table 4). Two 50 km intervals are combined in the central part of the study area where the late Pleistocene dune sheet cover is most extensive. Terrace elevation data is summed and averaged for each 50 km interval to calculate mean elevation and standard deviation from the mean. The standard deviation does not represent error about the mean, but instead it represents the square of the measured elevation variance (see further discussion below).

Table 4. Grouped elevation data for lowest marine terrace in along coast intervals from the South Central Cascadia Margin

Along coast Interval UTM-N	Average Position (UTM-N)	Trench Distance (km)	Average Elevation (m)	Standard Deviation (± m 1s)
5159000-5146000	5146000	142	10.0	3.5
5086000-5067000	5076000	119	10.3	3.0
5050000-5008000	5029000	110	9.5	3.4
5008000-4964000	4986000	108	8.0	4.0
4955000-4918000	4936000	101	9.5	3.4
4908000-4812000	4860000	95	10.5	7.4
4800000-4763000	4781000	75	17	11
4750000-4733000	4741000	56	32	23

See Table 2 for lowest terrace elevation Data.

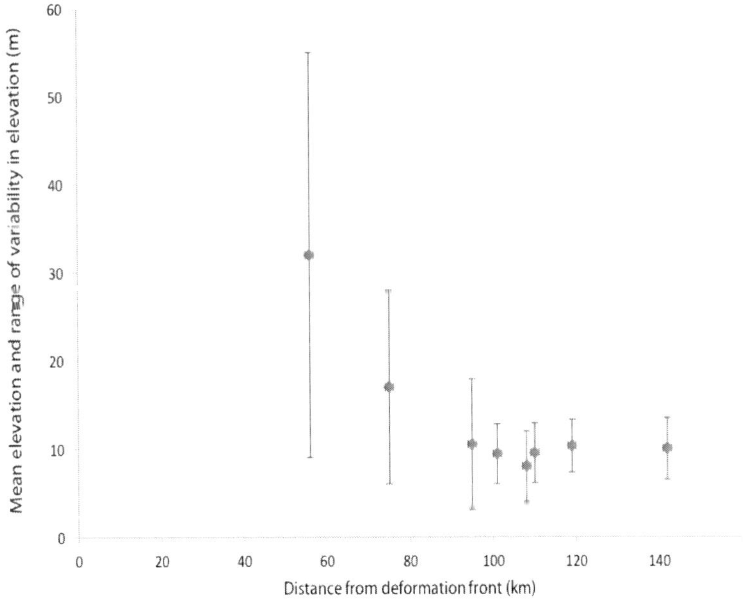

Figure 14. Plot of lowest marine high stand deposit (80 120 ka) elevations including mean and range of variability of elevations in meters (m) NAVD88 for each 50 km alongshore interval (data from Table 2).

The midpoint of each 50 km interval is measured for its distance from the buried trench or deformation front (x-axis in Figure 14). The mean elevation is a simple average of measured sea cliff sections, with no intent to bias sample selection on the basis of terrace height maxima, minima or apparent offsets. The range in elevation variability is taken as the square of elevation variance. It is not an error measurement, but rather a measure of variability of terrace deposit height within each 50 km distance interval (Figure 14).

The ranges of variability of the high stand deposit elevations are shown as ± the square of variance about the mean. The range of elevation variability changes from ± 3–4 m at distances greater than 100 km from the deformation front to ± 10–20 m at 50–70 km distance from the deformation front. The increase in range of variability of terrace elevation reflects broad warping that is likely associated with shore oblique folds and or buried thrust faults in the deformation belt. Localized terrace vertical offsets are not thought to represent plate segmentation or to impact megathrust rupture lengths as has been suggested by other investigators [Kelsey et al., 2002]. However, an increase in the variability of terrace elevations from north to south in the study area is thought to reflect the intersection of the coast with the fold and fault deformation belt. A modest regional uplift of ~ 10 m in the northern coastal terraces occurs landward of the deformation belt. This regional uplift is measured at approximately 10 m per 100,000 years or ~ 0.1 mm yr^{-1}. This regional uplift, possibly reflecting broad underplating in the subduction zone, is differentiated from compressive shortening in the deformation belt. The averaged elevations (~ 30 m) of 80 ka high stand deposits from coastal terraces that occur well within the deformation belt could represent crustal shortening by combined shallow folds, thrusts, and/or more complex structures at depth.

The local terrace warping reaching 80-90 km distance landward of the deformation front in the south central part of the study area is associated with compressive strain in the upper plate. The substantial terrace warping either directly overlies the zone of strong interplate coupling or it extends a very short distance landward of the strongly coupled zone.

Late Holocene Coastal Deformation

Coastal deformation is recorded by alternating interseismic uplift and coseismic subsidence in late Holocene tidal marshes of the South Central Cascadia Margin. One small tidal basin, Netarts Bay, is particularly sensitive to episodic subsidence events in the South Central Cascadia Margin. Two weak subsidence events at 1.3 ka and 2.8 ka were initially found in the Netarts wetlands [Darienzo, 1987; Darienzo and Peterson, 1990] before their subsequent recognition in larger tidal basins [Atwater and Hemphill-Haley, 1996]. The interseismic intervals recorded in Netarts Bay are quite variable, ranging from 200 to 900 years during the last 3.2 ka (Figure 15). The mean recurrence interval for 7 subsidence events in Netarts Bay between 3.2 and 0.3 ka is 430 years. The long–term trends of paleotidal level indicators in Netarts Bay (3.2 ka long record) coincide with eustatic sea level rise of ~ 1.0 m per thousand years. The alternating episodes of interseismic uplift and coseismic subsidence in Netarts Bay are fully elastic.

The ranges of interseismic uplift values are well constrained by the paleotidal level indicators in Netarts Bay, as uplift does not reach supratidal elevations. Maximum uplift ranges between 0.75 and 1.25 m pMSL (Figure 15). A problem exists with regards to the apparent rates of interseismic uplift between subsidence events in Netarts Bay. The longer interseismic intervals (>500 years) do not reflect the total gain in paleotidal elevation that would be expected from the higher rates of uplift demonstrated by the shorter interseismic cycles and the present interseismic cycle (0.3 ka to the present). It is not known whether 1) lower strain rates lead to longer recurrence intervals or 2) whether early rates of uplift are followed by decreasing rates of vertical strain prior to megathrust rupture. These questions relate directly to predictions of the timing of the next great earthquake in the study area. Will the next great earthquake follow a short strain cycle or long strain cycle? Will a change in modern strain rates (Figure 4) signify a precursor to the next megathrust rupture?

In the southern part of the study area the locally enhanced subsidence, such as in South Slough, Coos Bay (Figure 11) is possibly related to upper plate structures (Figure 3). It is not known to what degree the localized strain is elastic, or whether the localized strain reflects stress release or stress accumulation during the megathust ruptures. Such localities of locally enhanced strain, including the Sixes River wetlands (Figures 11 and 12) deserve further investigations, by seismic profiling, to establish the geometry and scale of possible underlying upper plate structures. The focus of this study is on the distribution of regional elastic strain in the upper plate (as discussed below).

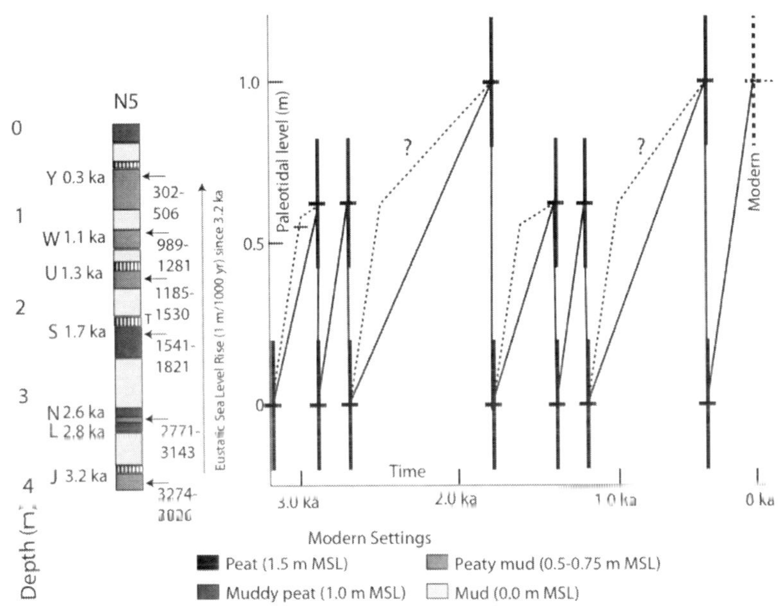

Figure 15. Core log and radiocarbon dates from core site N5 in Netarts Bay (redrafted from Figure 8). Great earthquake designations (letters), and ages (ka), follow those from the Niaw core site in the Niawiakum tidal creek of Willapa Bay (Figure 7) [Atwater and Hemphill-Haley, 1996; Atwater et al., 2004]. Paleotidal levels are shown at the ends of interseismic uplift and coseismic subsidence for each interval in site N5 from Netarts Bay. Two plots of interseismic uplift for Netarts Bay are shown including uplift rates fitted for each earthquake cycle (solid line) and early uplift rates (dashed line) taken from the most recent interseismic uplift (0.3 to 0.0 ka) followed by late uplift rates fitted to the measured paleotidal level datum. The modern marsh elevation has already reached the maximum interseismic uplift that has occurred prior to previous megathrust ruptures. Little or no additional vertical strain accumulation is expected to occur prior to the next megathrust rupture. Data and figure modified from Darienzo and Peterson [1990].

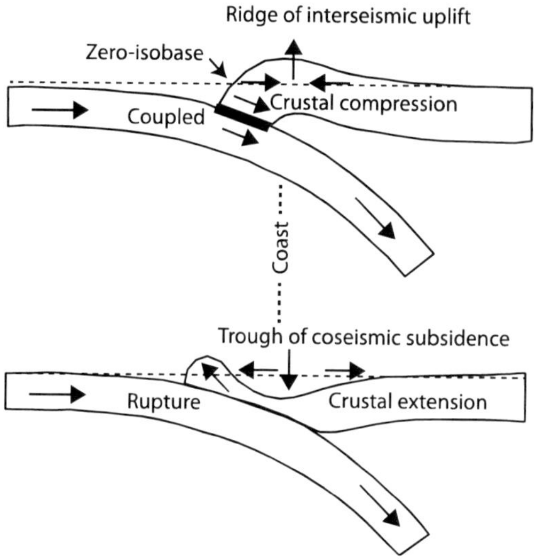

Figure 16. Cross section diagram of interseismic coastal uplift during interplate coupling (upper) and coseismic subsidence during megathrust rupture (lower). Figure is redrafted from several sources. Vertical scales are greatly exaggerated to highlight the trough of coseismic subsidence, reaching a maximum value at the position of the vertical arrow. A zero-isobase occurs between the trough of coseismic subsidence and the seaward area of coseismic uplift. Coseismic subsidence should approach 0 m near the zero-isobase, which approximates the landward extent of the strongly coupled zone.

The regional trough of coseismic subsidence coincides with the ridge of interseismic uplift in the upper plate (Figure 16). A zero-isobase occurs between the trough of coseismic subsidence and the ridge of coseismic uplift.

Coseismic uplift events might be difficult to recognize in wetland settings where plant roots grow or descend well below marsh surfaces. However, the regional zero-isobase should be apparent where mapped subsidence values diminish to zero. Equally important in constraining the geometry of upper plate elastic strain is the trough of maximum coseismic subsidence (see below).

The full width of the trough of coseismic subsidence is contained within the tidal reach of the lower Columbia River (Figure 7). The most landward extents of the last three dated coseismic subsidence events (0.3, 1.1, and 1.3 ka) are observed in site CRCF1, at a distance of 185 km from the deformation front. The greatest landward extent of apparent episodic subsidence in the lower Columbia River at CRLV2 in Longview, Washington, is about 200 km from the deformation front. The trough of maximum coseismic subsidence (1.5 m subsidence) is bracketed between sites JohnD and BlindS, at distances of 145 and 155 km from the deformation front, respectively. These core sites contain at least two recorded subsidence events each of maximum subsidence (~ 1.5 m) based on plant macrofossil indicators of paleotidal level. The lower Columbia River is the only tidal system setting in the South Central Cascadia Margin that spans the full width of the trough of coseismic subsidence.

Coseismic subsidence events in Netarts and Tillamook Bays are modest (~ 1.0 m) but several cores sites in Tillamook Bay show about 1.5 m of subsidence from the last Cascadia Margin earthquake, now correlated to Japanese tsunami records at AD 1700 (Atwater et al., 2005) (Figure 8). The core records in Tillamook Bay do not demonstrate at least two events of 1.5 m subsidence, thereby placing the trough of maximum subsidence landward of the eastern shoreline of Tillamook Bay.

Coseismic subsidence events in the Siuslaw estuary range from less than 0.5 m to 1.0–1.5 m over a landward distance of about 20 km (Figure 9). Sites of maximum subsidence (218 and 219) recorded near Mapleton, Oregon, occur at a distance of 120 km from the deformation front. The trough of maximum subsidence likely occurs within 10 km distance landward of sites 218 and 219. The trough of maximum subsidence in the Siuslaw locality is estimated to be located at distance of 130 km from the deformation front.

Multiple sites lacking evidence of episodic subsidence, as based on plant marcrofossils and microfossils [Briggs, 1994] occur in the lowermost reaches of the Siuslaw and Umpqua estuaries (Figures 9 and 10). These sites occur at distances of 100 ± 10 km from the deformation front. These core sites represent the zero-isobase between coseismic uplift and coseismic subsidence in the central part of the study area (Figure 16). The zero–isobase likely extends through the lowermost reaches of Coos Bay, but it is partially obscured by localized subsidence in South Slough and Pony Slough (Figure 11). The sources of the localized subsidence in Coos Bay are not known, but the South Slough occurs in a north–south syncline axis, possibly associated with underlying fault(s) [Madin et al., 1995].

Coseismic subsidence in the Coquille estuary ranges from 0.5-1.0 m at the coast to approximately 1.5m over a landward distance of about 20 km (Figure 12). The small values of coseismic subsidence in the lowermost reaches of the Coquille estuary reduce the number of expected subsidence events recorded in some wetland sites. The number of coseismic subsidence events increases with distance landward to Coquille, Oregon, where the river turns south. The trough of maximum subsidence likely occurs landward of site 521, at an estimated distance of about 100 km from the deformation front.

Comparison of Geologic Strain Records

A map of the estimated position of the trough of maximum subsidence in the Columbia River, Siuslaw, and Coquille estuary localities suggests a significant narrowing of the coupled zone with distance south in the study area (Figure 17). The distance between the trough of maximum subsidence and the deformation front decreases from 150 to 100 km between the Columbia and Coquille drainages.

Nevertheless, the position of the zero-isobase at 100 km from the deformation front in the south central part of the study area confirms the presence of a relatively wide zone of strong interplate coupling throughout the length of the South Central Cascadia Margin.

The landward edge of the Quaternary deformation belt, based on offshore seismic profiling (Figure 3) and onshore terrace deformation (Table 2 and Figure 6) ranges from about 110 km to about 80 km distance from the deformation front between the Columbia River and Coos Bay, respectively (Figure 17). The landward edge of the Quaternary deformation belt approximately coincides with the late Holocene zero–isobase in the south central part of the study area, though some inferred upper plate structures appear to occur well landward of the zero-isobase in Coos Bay, as based on localized late Holocene subsidence (Figure 11). The landward edge of the late Neogene deformation belt also extends somewhat landward of the the zero–isobase in the south central part of the study area. The similarities in positions of all three strain indicators demonstrates the presence of a relatively wide interplate coupled zone throughout the 500 km length of the South Central Cascadia Margin.

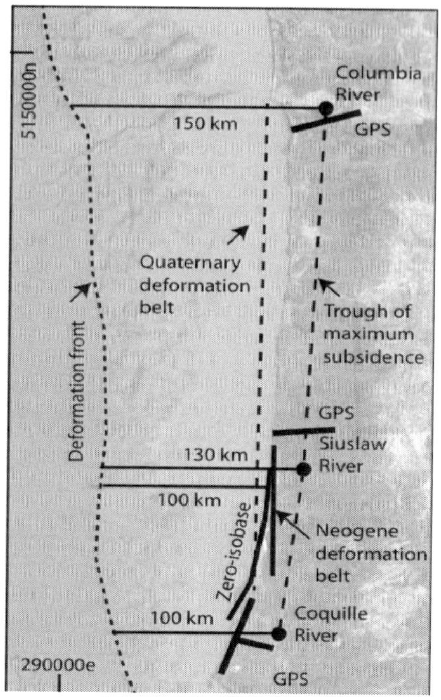

Figure 17. Map of the South Central Cascadia Margin showing deformation front (dotted line), estimated position of the trough of maximum subsidence (bold dashed line) with distances from the deformation front, and the zero-isobase (solid line) with distance from the deformation front in the south central part of the study area. GPS baselines localities are shown with baseline positions (bold lines) The landward edge of the Quaternary deformation belt (medium dashed line) is shown for most of the study area. The landward edge of the late Neogene deformation belt (solid bold line) is shown for the south central part of the study area.

Preliminary analyses of the GPS station baselines (Figure 17) establish modern west–east shortening rates of about 5 mm yr^{-1} normalized to 40 km baseline lengths in the north and central coast. The north and central coast baselines are located within the zone of elastic strain in the upper plate. Baseline shortening rates in the south coast are about twice as large as those in the north and central coast localities. The mid-point distances of the south coast baselines are about 60–70 km from the deformation front. The larger shortening rates in the south coast baselines might reflect their proximity to the interplate coupled zone, which approaches or underlies the south coast baselines. Additional work is underway to analyze many more coastal GPS strain records that have been collected in the South Central Cascadia Margin.

CONCLUSION

Four different time scale records of coastal deformation in the South Central Cascadia Margin demonstrate convergent tectonic strain associated with active subduction of the Juan

De Fuca plate under the North American plate. These strain records refute some previous studies, which suggested that only aseismic creep or decoupled plate subduction occurs in the South Central Cascadia Margin. Two of the coastal deformation records presented here demonstrate long–term inelastic strain during the late Neogene and late Pleistocene. The inelastic strain records extend 90-110 km landward of the buried trench or deformation front in the south central part of the study area, near Florence, Oregon. These records imply strong long term coupling of the interplate megrathrust to distances of 100 km landward of the deformation front. Episodic rupture of the wide coupled zone could lead to large magnitude subduction zone earthquakes and associated nearfield tsunami inundations in the South Central Cascadia Margin.

Coseismic subsidence ranges from at least 1.5 m to less than 0.5 m of abrupt sea level rise in the study area coastal marshes. The maximum trough of coseismic subsidence (> 1.5 m subsidence) occurs at distances of 150 to 100 km from the deformation front, decreasing from north to south in the study area. The zero–isobase (< 0.5 m coseismic subsidence) occurs at a distance of about 100 km from the deformation front, in the south central part of the study area. The position of the zero–isobase is very well constrained in several tidal marsh systems during the last several thousand years. The zero–isobase is assumed to reflect the width of the coupled zone, about 100 km landward distance from the deformation front in the south central part of the study area, near Florence, Oregon. The predicted strengths of seismic shaking, both on the coast and in the more densely populated forearc valleys that are located landward of the coast, depend, in part, on the assumed landward extent of the coupled zone. Building codes are established on the basis of the predicted strengths of seismic shaking, so they are also dependent on the assumed width of the coupled zone. The results presented here will have direct relevance to building code determinations in the South Central Cascadia Margin.

Late Holocene coastal wetlands establish elastic strain cycles, with six interseismic intervals between 3.2 and 0.3 ka averaging 480 years in duration. The two longest strain cycles average 850 years between recorded coseismic subsidence events. The two shortest strain cycles average only 200 years in length. The current interseismic strain cycle, starting at AD1700 is 300 years long. Policymakers are now confronted with indecision about the imminence of the next megrathrust rupture, which could be dependent on whether or not the current strain cycle is part of a cluster of short recurrence intervals. Interseismic uplift between earthquake events is measured in several coastal marsh systems, including Netarts Bay, Oregon. The net amount of measured interseismic uplift is not directly proportional to the length of the interseismic interval. It appears that most of the interseismic uplift occurs within the first couple of millennia following the preceding megathrust rupture. Based on the present high marsh development in most of the study area tidal basins it is expected that little or no additional marsh emergence will occur prior to the next megathrust rupture. The mechanism(s) leading to the observed decline of vertical strain accumulation in the zone of interseismic uplift during the later stages of the longer interseismic intervals are not known.

Modern horizontal strain or shortening is measured perpendicular to the coastline in three GPS baselines in the South Central Cascadia Margin. The measured strain, on the order of at least 0.5 cm per year over ~ 40 km long baselines, demonstrates active strain accumulation in the upper crust. Though small in magnitude the measured strain rates are consistent with a current condition of interplate coupling along the full length of the South Central Cascadia Margin. These are the first direct measurements of GPS baseline shortening made in the Central Cascadia Margin. These initial GPS results suggest declining strain with increasing distance from the coupled zone, and a component of north– south shortening, possibly associated with oblique subduction or regional upperplate compression. Additional monitoring and statistical analyses of many more GPS baselines in the Central Cascadia Margin are underway to resolve the spatial variability of active tectonic strain in the Central Cascadia subduction zone.

ACKNOWLEDGMENTS

We thank Gregg Brigs, Chip Barnett, Charlie Clough, Betsy McLeod, Aaron Wieting, Chris Beard, and Gene Pierson for assisting with field work, laboratory analyses, and study logistics. Brian Atwater provided a useful review of an earlier version of this chapter. This Oregon Sea Grant project was funded by the NOAA Office of Sea Grant and Extramural Programs, US Department of Commerce, under grants number NA36RG0551 and NA36RG0451, project number R/CP-28, and by appropriations made by the Oregon State Legislature. Additional funding was provided through USGS NEHRP 14-08-0001-G1512 and NSF EAR8903903.

REFERENCES

Atwater, B. F. (1987). Evidence for great Holocene earthquakes along the outer coast of Washington State (US). *Science*, 236(4804), 942-944.

Atwater, B. F., and Hemphill-Haley, E. (1996). Preliminary estimates of recurrence intervals for great earthquakes of the past 3500 years at northeastern Willapa Bay, Washington. (*US Geological Survey Open-File Report 96-001*).

Atwater, B. F. (1994). Geology of the Holocene liquefaction features along the lower Columbia River at Marsh, Brush, Price, Hunting, and Wallace Islands, region and Washington (*US Geological Survey Open-File Report* 94-209).

Atwater BF, Tuttle MP, Schweig ES, Rubin CM, Yamaguchi DK, Hemphill-Haley E (2004) Earthquake recurrence, inferred from paleoseismology, in Gillespie, A.R., Porter,S.C., and Atwater, B.F., editors, The Quaternary Period in the United States: Amsterdam, Elvevier, pp. 331-350.

Atwater, B. F., Musumi-Rokkaku, S., Satake, K., Tsuji, Y., Ueda, K., and Yamaguchi, D. K. (2005). The Orphan Tsunami of 1700. Reston, Virginia: US Geological Survey and University of Washington Press.

Atwater, B. F., Nelson, A. R., Clague, J. J., Carver, G. A., Yamaguchi, D. K., Bobrowsky, P. T., et al. (1995). Summary of coastal geologic evidence of past great earthquakes at the Cascadia subduction zone. *Earthquake Spectra*, 11(1), 1-18.

Baldwin, E. M. (1956), Geology map of the lower Siuslaw River area, Oregon, *US Geological Survey*, O. M.-186.

Baldwin, E. M. (1961), Geologic map of the lower Umpqua River area, Oregon, *US Geological Survey*, O. M.-204.

Baldwin, E. M. (1974), Eocene stratigraphy of southwestern Oregon. Oregon Department of Geology and Mineral Industries Bulletin 83, 40 p. 1 map.

Baldwin, E. M., J. D. Beaulieu, L. Ramp, J. Gray, V. C. Jr. Newton, and R. S. Mason, (1973), *Geology and mineral resources of Coos County, Oregon, Bulletin* 80. 82 pp., Oregon Department of Geology and Mineral Industries, Portland, Oregon.

Barnett, E. T. (1997). Potential for coastal flooding due to coseismic subsidence in the central Cascadia margin. Portland State University, Portland, Oregon.

Beaulieu, J. D., and Hughes, P. W. (1975). *Environmental geology of western Coos and Douglas Counties, Oregon* (Bulletin 87 No. 87). Portland, Oregon: Department of Geology and Mineral Industries.

Black, G. L. (Cartographer). (1993). Offshore-onshore geologic cross-section from the Pan American Corporation well (continental shelf) to the Umpqua River (southern Oregon Coast Range) [Open-File Report].

Blakely, R. J., Brocher, T. M., and Wells, R. E. (2005). Subduction-zone magnetic anomalies and implications for hydrated forearc mantle. *Geology*, 33, 445-448.

Bradley, C. C., and Griggs, G. (1976). Form, genesis, and deformation of central California wave-cut platforms. *Geological Society of America Bulletin*, 87, 433-449.

Briggs, G. G. (1994). Coastal crossing of the elastic strain zero-isobase, Cascadia margin, south-central Oregon coast. Portland State University, Portland, Oregon.

Clarke, S. H., and Carver, G. A. (1992). Late Holocene tectonics and paleoseismicity, southern Cascadia subduction zone. *Science*, 255, 188-192.

Clifton, H. E. (1994). Transgressive and early highstand systems tracts: Pleistocene terrace deposits Willapa Bay, Washington. Paper presented at the Field Trip Guide S. E. P. M. Research Conference: Clastic Deposits of the Transgressive Systems Tract, Long Beach, Washington.

CORS. (2011). N. O. A. A. Continuously Operating Reference Stations Data Center. *http://www.ngs.noaa.gov/CORS/*.

Darienzo, M. E., 1987, M. S., Late Holocene geologic history of a Netarts salt marsh, northwest Oregon coast, and its relationship to relative sea level changes. M. S. *Thesis, University of Oregon, Eugene, Oregon*, 94 p.

Darienzo, M. E., 1991, Ph. D., Late Holocene paleoseismicity along the northern Oregon coast. Ph. D. *Thesis, Portland State University, Portland, Oregon*, 167 p.

Darienzo, M. E., and Peterson, C. D. (1987). Netarts Marsh: Subduction zone earthquakes or super-storms. Symposium: Could there be a devastating earthquake in Oregon. Paper presented at the Oregon Academy of Science, Monmouth, Oregon.

Darienzo, M. E., and Peterson, C. D. (1990). Episodic tectonic subsidence of Late Holocene salt marshes, northern Oregon central Cascadia margin. *Tectonics*, 9(1), 1-22.

Darienzo, M. E., Peterson, C. D., and Clough, C. (1994). Stratigraphic evidence for great subduction zone earthquakes at four estuaries in Northern Oregon. *Journal of Coastal Research*, 10(4), 850-876.

Goldfinger, C., Kulm, L. D., Yeates, R. S., Appelgate, B., MacKay, M. E., and Moore, G. F. (1992a). Transverse structural trends along the Oregon convergent margin: Implications for Cascadia earthquake potential and crustal rotations. *Geology*, 20(2), 141-144.

Goldfinger, C., Kulm, L. D., Yeats, R. S., Appelgate, B., MacKay, M. E., and Cochrane, G. R. (1996). Active strike slip faulting and folding of the Cascadia plate boundary and forearc in central and northern Oregon. In A. M. Rogers, W. J. Kochelman, G. R. Priest and T. J. Walsh (Eds.), Assessing and reducing earthquake hazards in the Pacific Northwest (Vol. *US Geological Survey Professional Paper* 1560, pp. 223-256).

Goldfinger, C., Kulm, L. D., Yeats, R. S., Mitchell, C. E., Weldon, R. J., III, Peterson, C. D., et al. (1992b). Neotectonic map of the Oregon continental margin and adjacent abyssal plain (Open File Report 0-92-4). Portland, Oregon: Oregon Department of Geology and Mineral Industries.

Goldfinger, C., Grijalva, K., Bürgmann, R., Morey, A. E., Johnson, J. E., Hans Nelson, C., Gutiérrez-Pastor, J., Ericsson, A., Karabanov, E., Chaytor, J. D., Patton, J., and Gràcia, E. (2008). Late Holocene Rupture of the Northern San Andreas Fault and Possible Stress Linkage to the Cascadia Subduction Zone. *Bulletin of the Seismological Society of America*, 98, 861–889, doi: 10.1785/0120060411.

Hyndman, R. D., and Wang, K. (1995). The rupture zone of the Cascadia great earthquakes from current deformation and thermal regime. *Journal of Geophysical Research*, 100, 22133-22154.

Kelsey, H. M. (1990). Late Quaternary deformation of marine terraces on the Cascadia subduction zone near Cape Blanco, Oregon. *Tectonics*, 9, 983-1014.

Kelsey, H. M., Nelson, A. R., Hemphill-Haley, E., and Witter, R. C. (2005). Tsunami history of an Oregon coastal lake reveals a 4600 yr record of great earthquakes in the Cascadia subduction zone. *Geological Society of America Bulletin*, 117(7/8), 1009-1032.

Kelsey, H. M., Witter, R. C., and Hemphill-Haley, E. (2002). Plate-boundary earthquakes and tsunamis of the past 5500 yr, Sixes River estuary, southern Oregon. *Geological Society of America Bulletin*, 114(3), 298-134.

Kennedy, G. L., Kajoie, K. R., and Wehmiller, J. F. (1982). Aminostratigraphy and faunal correlations of late Quaternary marine terraces, Pacific coast, US. *Nature*, 299, 545-547.

Li, W. (1992). Evidence for late Holocene coseismic subsidence in the lower Eel River valley, Humboldt county, northern California: An application of foraminiferal zonation to indicate tectonic submergence. Humboldt State University, Arcata, California.

Madin, I. P., McInelly, G. W., and Kelsey, H. M. (Cartographer). (1995). Geologic map of Charleston Quadrangle, Coos County, Oregon.

McInelly, G. W., and Kelsey, H. M. (1990). Late Quaternary tectonic deformation in the Cape Arago-Bandon region of coastal Oregon as deduced from wave-cut platforms. *Journal of Geophysical Research*, 95, 6699-6713.

McNeil, L. C., Goldfinger, C., Yeates, R. S., and Kulm, L. D. (1998). The effects of upper plate deformation on records of prehistoric Cascadia subduction zone earthquakes. In I. S. Stewart and C. Vita-Finzi (Eds.), *Coastal Tectonics* (Vol. 146, pp. 321-342). London: Geological Society of London.

Muhs, D. R., Kelsey, H. M., Miller, G. H., Kennedy, G. L., Whelan, J. F., and McInelly, G. W. (1990). Age estimates and uplift rates for late Pleistocene marine terraces: Southern Oregon portion of the Cascadia forearc. *Journal of Geophysical Research*, 95, 6685-6698.

Mulder, R. A. (1992). Regional tectonic deformation of the northern Oregon coast as recorded by Pleistocene marine terraces. Portland State University, Portland, Oregon.

Nelson, A. R. (1987). Apparent gradual rise in relative sea level on the south-central Oregon coast during late Holocene-implications for the great Casdadia earthquake hypothesis. E. O. S., 68, 1240.

Nelson, A. R. (1992). Discordant 14C ages from buried tidal-marsh soils in the Cascadia subduction zone, southern Oregon coast. *Quaternary Research*, 38, 74-90.

Nelson, A. R., Jennings, C. W., and Kashima, A. E. (1996). An earthquake history derived from stratigraphic and microfossil evidence of relative sea-level change at Coos Bay, southern coastal Oregon. *Geological Society of America Bulletin*, 108, 141-154.

Newton, V. C., Jr. (Cartographer). (1980). Prospects for oil and gas in the Coos Basin, Western Coos, Douglas, and Lane Counties, Oregon.

Niem, A. R., and Niem, W. A. (Cartographer). (1985). Oil and Gas investigations of the Astoria Basin, Clatsop and northernmost Tillamook Counties, northwest Oregon [OGI-14].

Personius, S. F. (1993). Age and origin of fluvial terraces in the central Coast Range, Western Oregon (Bulletin 2038): US Geological Survey.

Personius, S. F. (1995). Late Quaternary stream incision and uplift in the forearc of the Cascadia subduction zone, western Oregon. *Journal of Geophysical Research*, 100, 20193-20210.

Peterson, C. D., Barnett, E. T., Briggs, G. G., Carver, G. A., Clague, G., and Darienzo, M. E. (1997). Estimates of coastal subsidence from great earthquakes in the Cascadia subduction zone, Vancouver Island B. C., Washington, Oregon, and Northernmost California (Open-File Report 0-07-5): Oregon Geology and Mineral Industries.

Peterson, C. D., Carver, G., Cruikshank, K. M., Abramson, H., Garrison-Laney, C., Dengler, L., 2011. Evaluation of the use of paleotsunami deposits to reconstruct inundation distance and runup heights associated with prehistoric inundation events, Crescent City, southern Cascadia margin. *Earth Surface Processes and Landforms*. DOI:10.1002/esp.2126.

Peterson, C. D., Cruikshank, K. M., Schlichting, R. B., and Braunsten, S., 2010, Distal Runup Records Of Latest Holocene Paleotsunami Inundation In Alluvial Flood Plains: Neskowin and Beaver Creek, Oregon, Central Cascadia Margin, US, *Journal of Coastal Research*, 26:622-634. Doi 10.2112/08-1147.1

Peterson, C. D., Cruikshank, K. M., Jol, H. M., Schilichting, R. B., 2008, Minimum runup heights of paleotsunami from evidence of sand ridge overtopping at Cannon Beach, Oregon, Central Cascadia Margin, US, *Journal of Sedimentary Research*, v. 78, p. 390-409.

Peterson, C. D., and Darienzo, M. E. (1997). Discrimination of flood, storm and tectonic subsidence events in coastal marsh records of Alsea Bay, Central Cascadia Margin, US. In A. M. Rogers, T. J. Walsh, W. J. Kockelman and G. R. Priest (Eds.), Assessing and Reducing Earthquake Hazards in the Pacific Northwest (Vol. *U. S. G. S. Professional Paper* 1560, pp. 115-146).

Peterson, C. D., Darienzo, M. E., Doyle, D. L., and Barnett, E. T. (1995). Evidence for coseismic subsidence and tsunamis inundation during the past 3000 years at Siletz Bay, Oregon. In G. R. Priest (Ed.), *Explanation of mapping methods and use of the tsunami hazard map of the Siletz Bay area, Lincoln County, Oregon* (Vol. O-95-05, pp. 45-69). Portland, Oregon: Oregon Department of Geology and Mineral Industries.

Peterson, C. D., Darienzo, M. E., Hamilton, D., Pettit, D. J., Yeager, R. K., Jackson, P. L., et al. (1994). Cascadia Beach-Shoreline Data Base, Pacific Northwest Region, US (Open-File Report 0-94-2): Oregon Department of Geology and Mineral Industries.

Peterson, C. D., Doyle, D. L., and Barnett, E. T. (2000). Coastal flooding and beach retreat form coseismic subsidence in the central Cascadia Margin, US. *Environmental and Engineering Geoscience*, 6(3), 255-269.

Peterson, C. D., Komar, P. D., and Scheidegger, K. F. (1986). Distribution, geometry, and origin of heavy minearl placer deposits on Oregon beaches. *Journal of Sedimentary Petrology*, 56(1), 67-77.

Peterson, C. D., and Priest, G. R. (1995). Preliminary reconnaissance survey of Cascadia paleotsunami deposits in Yaquina Bay, Oregon. *Oregon Geology,* 57, 33-40.

Peterson, C. D., Stock, E., Cloyd, C., Beckstrand, D., Clough, C., Erlandson, J., et al. (2006). Dating and morphostratigraphy of coastal dune sheets from the central west coast of North America. Corvallis, Oregon: Oregon Sea Grant Publications.

Peterson, C. D. and K. F. Scheidegger, 1984. Holocene depositional evolution in a small active- margin estuary of the northwestern United States. *Mar. Geol.*, 59:51-83.

Peterson, C. P., and Gray, J. J. (Cartographer). (1986). Geologic map of the ocean floor off Oregon and adjacent continental margin [GMS-42].

PANGA. (2011). Pacific Northwest Geodetic Array Data Center. *http://www. geodesy.cwu.edu/.*

Priest, G. R., Myers, E., Baptista, A. M., Fleuck, P., Wang, K., Kamphaus, R. A., et al. (1997). Cascadia subduction zone tsunamis: Hazard mapping at Yaquina Bay, Oregon (Open-File Report No. O-97-34). Portland, Oregon: Oregon Department of Geology and Mineral Insustries.

Priest, G. R., Myers, E., Baptista, A. M., Fleuck, P., Wang, K., and Peterson, C. D. (2000). Source simulation for tsunamis: Lessons learned from fault rupture modeling of the Cascadia subduction zone, North America. Science of Tsunami Hazards.

R. Development Core Team (2011). R: A language and environment for statistical computing. R Foundation for Statistical Computing, Vienna, Austria. ISBN 3-900051-07-0, *URL http://www.R-project.org/.*

Saltus, R. W., Blakely, R. J., Hauessler, P. J., and Wells, R. E. (2005). Utility of aeromagnetic studies for mapping potentially active faults in two forearc basins: Puget Sound, Washington, and Cook Inlet, Alaska. *Earth, Planets, Space*, 57, 781-793.

Satake, K., Wang, K., and F., A. B. (2003). Fault slip and seismic moment of the 1700 Cascadia earthquake inferred from Japanese tsunami descriptions. *Journal of Geophysical Research*, 108(11), ESE 7-1 - ESE 7-17.

Schlicker, H. G., and Deacon, R. J. (1974). Environmental Geology of Coastal Lane County, Oregon (Bulletin 85 No. 85). Portland, Oregon: Department of Geology and Mineral Industries.

Shennan, I., Long, A. J., Rutherford, M. M., Innes, F. M., Green, F. M., and Walker, K. J. (1998). Tidal marsh stratigraphy, sea-level change and large earthquakes: Submergence events during the last 3500 years at Netarts Bay, Oregon, US. *Quaternary Science Reviews*, 17, 365-393.

Snavely, P. D. Jr., N. S., MacLeod, and H. C., Wagner (1972a), Preliminary bedrock geologic map of the Yaquina and Toledo quadrangles, Oregon. *US Geological Survey open-file map.*

Snavely, P. D. Jr., N. S., MacLeod, and H. C., Wagner (1972b), Preliminary bedrock geologic map of the Cape Foulweather and Euchre Mountain quadrangles, Oregon. *US Geological Survey open-file map.*

Stuiver, M., Reimer, P. J., and Reimer, R. W. (2005). C. A. L. I. B. 5.0: *http://calib.qub.ac.uk/calib/.*

Ticknor, R. L. (1993). Late Quaternary crustal deformation on the central Oregon coast as deduced from uplifted wave-cut platforms. Western Washington University, Bellingham, Washington.

Tolman, B., Harris, R. B., Gaussiran, T., Munton, D., Little, J., Mach, R., et al. (2004). *The GPS Toolkit -- Open Source GPS Software*. Paper presented at the Proceedings of the 17th International Technical Meeting of the Satellite Division of the Institute of Navigation (ION GNSS 2004).

Valentine, D. W. (1992). Late Holocene stratigraphy, Humboldt Bay, California, Evidence for late Holocene paleoseismicity of the southern Cascadia subduction zone. Humboldt Sate University, Arcadia, California.

Venables, W. N. and Ripley, B. D. (2002) *Modern Applied Statistics with S. Fourth Edition*. Springer, New York. ISBN 0-387-95457-0.

Wells, R. E. (1990). Paleomagnetic roations and the Cenozoic tectonics of the Cascade arc, Washington, Oregon, and California. Journal *of Geophysical Research*, 95(B12), 19409-19417.

West, D. O., and McCrumb, D. R. (1988). Coastline uplift in Oregon and Washington and the nature of Cascadia subduction zone tectonics. *Geology*, 16(2), 169-172.

Witter, R. C., Kelsey, H. M., and Hemphill-Haley, E. (2003). Great Cascadia earthquakes and tsunamis of the past 6700 years, Coquille River estuary, southern coastal Oregon. *Geological Society of America Bulletin*, 115, 1289-1306.

In: Earthquakes
Editor: Kostas Konstantinou

ISBN: 978-1-62081-883-1

Chapter 2

THE SEISMIC HAZARD OF THE VILLAVICENCIO CITY, COLOMBIA

Germán Chicangana[1,2*], Carlos A. Vargas-Jiménez[2†], Alexander Caneva[2,3**], Claudia Mojica-Sánchez[1], Tulio Aymerich Hernández-Hernández[1], Johan Ardila-Escobar[5‡] and Alejandra Bernal-Jiménez[4±]

[1]Corporación Universitaria del Meta, Villavicencio, Meta, Colombia
[2]Universidad Nacional de Colombia, Bogotá, Colombia
[3]Universidad Antonio Nariño, Bogotá, Colombia
[4]Fundación Nueva Herencia, Villavicencio, Meta, Colombia
[5]Fundación Kinkaju, Villavicencio, Meta, Colombia

ABSTRACT

Geological aspects, as well as some historical antecedents related to seismic hazard of the Villavicencio City [Colombia, South America] are shown. This city, with a population of 430,000 inhabitants, is located along the Eastern Frontal Fault System (EFFS), in a faulted zone, with the Andes Cordillera region at the west, and the South American Craton at the east. The analysis of historical and instrumental earthquake records, associated to strong motion effects in cities like Villavicencio, Bogotá, Neiva, and Bucaramanga indicates that instrumental networks should be improved in order to evaluate the seismic regime at local and regional scales. The systematic monitoring of

* Centro de Investigaciones en Ingeniería y Arquitectura "Rogelio Salmona", Corporación Universitaria del Meta, Villavicencio, Meta, Colombia. E-mail: *german.chicangana@unimeta.edu.co. claudia.mojica @unimeta.edu.co. ingenierotulio@gmail.com.*

† Grupo de Geofísica, Departamento de Geociencias, Universidad Nacional de Colombia, Bogotá, Colombia. E-mail: *cavargasj@unal.edu.co.*

** Grupo de Geofísica, Centro de Investigaciones, Universidad Antonio Nariño, Bogotá, Colombia. E-mail: *alexander.caneva@uan.edu.co.*

‡ Fundación Kinkaju, Villavicencio, Meta, Colombia. E-mail: *jonar12@hotmail.com.*

± Fundación Nueva Herencia, Villavicencio, Meta, Colombia. E-mail: *fundacionnuevaherencia@gmail.com.*

seismic activity along Algeciras, Guaicáramo, and Servitá faults would represent an important chance for understanding the deformation front related to the continental collision and the basinal geometry of the Colombian Llanos Foothills. On the other hand, field data suggest the presence of large areas with active tectonic evidence that are interrupted by quiet segments, and that could represent a reorientation of the local stress field or its accumulation for promoting scenarios of strong earthquakes. Large tectonic structures passing near Villavicencio, as the Algeciras Fault, have a long history of strong earthquakes on this city (e.g. August 31, 1917). Similarly, the earthquake data corresponding to the Servitá Fault present several events that affected Villavicencio, Bogotá [1743 to Present – day], as well as other small towns. The Guaicáramo Fault also has presented strong activity in several segments near to Villavicencio [e.g. 1923, 1924, and 1995 earthquakes].

The vulnerability of Villavicencio city regarding the seismic hazard is analyzed, as well as the role carried out by local institutions. A description of a possible post-seismic scenario and certain recommendations to mitigate the effects of an earthquake are presented.

INTRODUCTION

Villavicencio is an administrative and economic hub of the Eastern Colombian Orinoquía region, but many of its 430,000 inhabitants are greatly unaware of the seismic hazard in this region, despite being considered an area of high seismic activity by expert seismologists in Colombia [AIS – Ingeominas, 1996; Alfaro and Ramos, 2001].

Since Villavicencio region really began to be colonized in the mid 1700s, the historical seismic data is available but is scarce; so, there is uncertainty about the location of the earthquakes. The epicenters were most likely located in the Orinoquía region. The October 18, 1743 earthquake stands out as the most important one based on the historical record. This earthquake was the first one in the Villavicencio region to have been absolutely confirmed.

The earthquake that struck Villavicencio on August 31, 1917 is documented in its recent history. According to Cifuentes et al. [2006], this earthquake was originated in a fault plane south of the city that is linked to a system of faults known as the East Frontal Fault System (EFFS). However, this authors do not report which fault, among the ones in the EFFS, originated the earthquake; whereas, París et al. [2000] did indicate it in their study.

Villavicencio (including its neighboring area) was inspected for seismic activity on the three main faults associated to the EFFS using methods and tools such as: remote sensing, field work, and the analysis of seismological information taken from Ramirez 1975 historical catalogue (considered by the Colombian seismological community as the main source of historical seismic information in Colombia, Prieto et al. 2004), and the instrumental data recorded between 1993 and 2009 by the National Seismological Network of Colombia (NSNC).

The Algeciras - Uribe Fault System (AUFS), Guaicáramo Fault, and Servitá Fault are the three main faults. All of them are great thrust faults located on the eastern side of the Eastern Cordillera also known as the Llanos Foothills or Piedemonte Llanero. The surfaces of these faults are characterized by a high angle, while in depth have a low angle that extends to the West under the Eastern Cordillera, and even extend to the lower lithosphere near the Moho [Chicangana et al., 2007]. These faults have a long geological history. They originated toward the end of the Mesoproterozoic Era as a result of the formation of the Rodinia supercontinent [Ramos, 2010].

Pondering over Villavicencio´s socioeconomic characteristics, we checked its vulnerability in the event of an earthquake by assessing the roles of the administrative authorities and the scholars, who as a group are responsible for making the decisions and communicating the information regarding the seismic risk the city is facing.

A discussion session can be found at the end of this chapter. In it, a possible post-earthquake scenario is described; recommendations are also given in order to mitigate the effects that an earthquake of a magnitude equal or greater than 6.5 might pose.

Geotectonics

Unlike its southern block, which limits south of 3.5°N, and the northern block that stands north of 6°N, the central block of the Eastern Cordillera of Colombia is characterized by the absence of rocks that belong to the Andean Mesoproterozoic Basement (AMB) [Chicangana et al., 2007]. The oldest rocks found in this block belong to the Quetame complex; these rocks are mainly sedimentary and metamorphic. The latter ones vary their metamorphic grade from very low to low and their ages belong to the lapse that corresponds to the late Neoproterozoic (Cryogenian or Ediacaran?) - Ordovician [Maya, 2001; Chicangana and Kammer, 2011a].

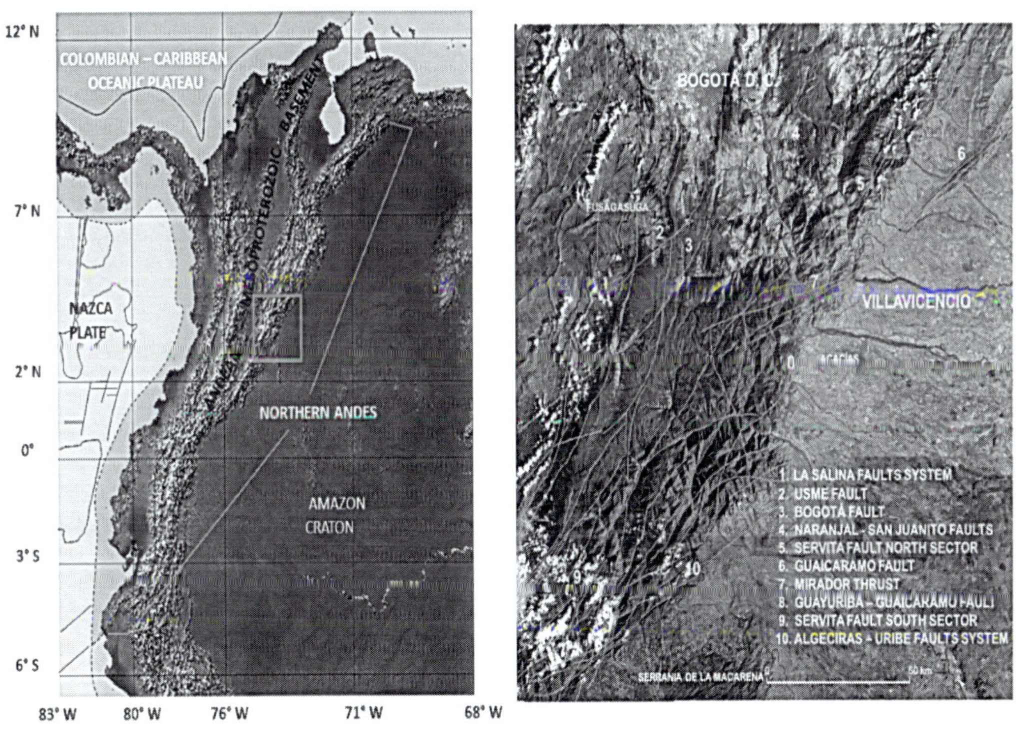

Figure 1. Left, The study area in the square. This is located in the center of Colombia between the Eastern cordillera and the Llanos foothills. Right, Area that correspond to square of the left where faults related to the EFFS are shown and the central part of the Eastern Cordillera of Colombia.

Mora et al. [2006] point out a relative proximity to the AMB based on seismic information. They locate the basement at a depth just a few hundred meters under the surface in different sites east of the Department of Cundinamarca (where the capital city, Bogotá, is located).

A Phanerozoic layer of sedimentary rocks whose age corresponds to the late Paleozoic - Cenozoic lies on this basement. This structure as a whole presents a varying width that ranges from 0.5 to more than 5 kilometers at both the Llanos Foothills and the Eastern Llanos [Mora et al., 2006; INGEOMINAS, 2007].

These rocks have been altered by ongoing tectonic activity that has occurred since the end of the Proterozoic until the Present time. The degree of tectonic alteration in the rocky packages is directly proportional to its age. Chicangana and Kammer [2011 a, b] argue that the Mesoproterozoic basement has had different tectonic deformation events that initially correspond to the Neoproterozoic - Permic lapse as evidenced by the multiple deformed area for its fabric in several sites. These tectonic events produced the inherited fault planes that would control the tectonic reaction of the basement throughout its tectonic history.

The oldest structures that exist in the basement rocks are the planes of the faults and the structures formed after the Mesoproterozoic collision between the Baltic shield, Laurentia, and Gondwana. This collision originated the Rodinia supercontinent during the late Mesoproterozoic in this area of Gondwana, changing the geometry of these rocks, and in some cases varying their polarity as a result of the successive tectonic events that the basement experienced since its origin [Keppie et al., 2001; Ibañez-Mejía, 2010; Ramos, 2010; Chicangana and Kammer, 2011b].

Hence, we can conclude that the great faults of the EFFS have this history at the very least and the areas of weakness on the basement are related to the initial geometry of the collision stemming from the features of the prisms with their corresponding contacts developed during the first accretion. This is why it is proposed that the current thrust front that marks the Guaicáramo Fault and other structures like the Servitá and Guayuriba faults, together with the Saliente de Guape, which is delineated by the AUFS [Hernández, 1996; París et al., 2000], mark the limit of the Mesoproterozoic accretion between the AMB and the Amazon Craton.

In general, the geometry of these structures is like the one of the great faults that have a westerly oriented dip slope. They have a high angle toward the surface but as they get deeper, their dip slope angle decreases so much that they almost become flat and listric [Mora et al., 2006; Chicangana et al., 2007; Parra, 2008]. The continuous reactivations that the main planes of the accretion experienced throughout the Early Neoproterozoic – Mesozoic can be checked thanks to the discordances that are found between the stratigraphic units that affect these faults.

The discordance between the basement and the metasedimentites of the Quetame Complex is the older. This discordance probably developed during the late Mesoproterozoic - Cryogenian as a consequence of the orogeny that resulted from the collision between the terranes or microcontinents that gave origin to Rodinia. It is later seen that there is an evident discordance that occurs between the rocks of the Farallones Group (included in the Late Devonic -Carboniferous time frame) and the rocks of the Quetame Complex, which developed as a result of a strong collision derived from a compressive regime caused by the reactivation of the subduction on this margin of Gondwana during the Mid Ordovician - Lower Silurian interval [Chicangana and Kammer, 2011b].

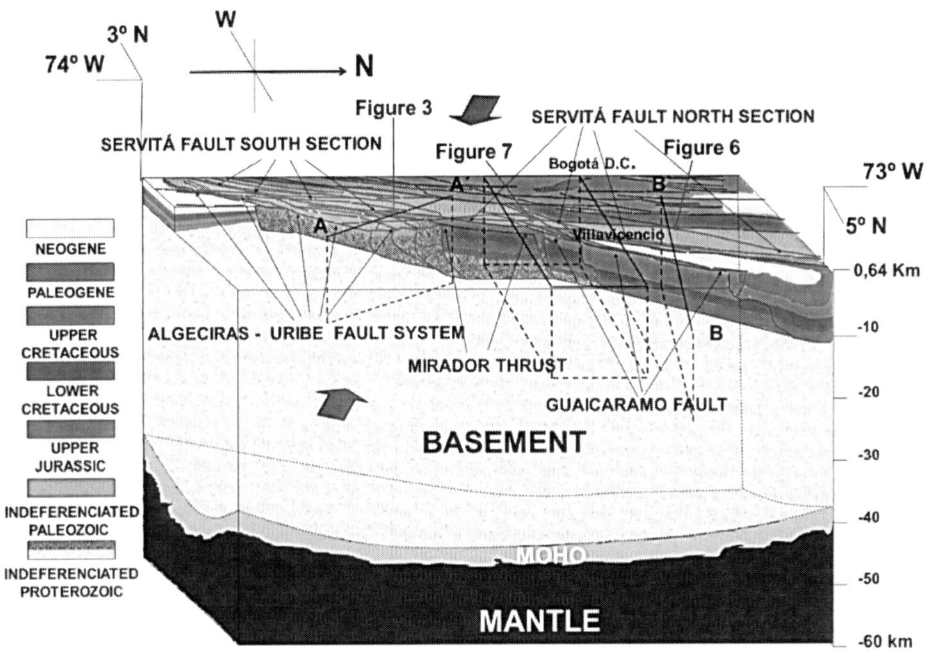

Figure 2. 3-D model of the tecto-stratigraphic display of the central sector of the Eastern Foothills of the Eastern Cordillera. The model shows the convergence between the Mesoproterozoic Andean Basement and the Amazon Craton. The thick arrows represent the stress fields. The compressive force σ1 has a NW - SE direction. The profiles and the cube show us the site of the respective figures.

Finally, we can observe the discordance that occurs between the rocks of the Farallones Group and the sedimentary rocks of the Late Jurassic - Lower Cretaceous lapse. This last discordance was caused by the orogeny developed during the amalgamation of the terranes that allowed the formation of Pangea during the Late Carboniferous - Permic interval. The sedimentary rocks of the Late Jurassic - Lower Cretaceous lapse were formed later. These sedimentary rocks arose as a result of an extensive phase [Horton et al., 2010]. This phase was been of a syn-orogenic effect might have occurred after a slab of the subsiding Farallon plate broke off. This sublithospheric event caused the migration of the asthenospheric wedge toward the lithospheric trench that had previously thinned [Barragán and Baby, 2004].

The reactivation of most of the main basement planes was originated in this extensive phase. On these planes there is evidence of an event that was characterized by a negative tectonic inversion that produced new faults with normal tendency in the Phanerozoic sheets. During the Cenozoic Era the superficial structures that originated in the Lower Cretaceous times, as well as those that affected the older and bigger basement, went through a positive tectonic reversal that accumulated during the two main phases of lithospheric shortening of the last geologic Era.

The first event that produced the Mid - Eocene discordance took place in the Paleogene period of this last Era. Then came a second one during the Late Miocene - Present day time interval of the Neogene. This event originated the structural style that can be seen nowadays in that area of the Eastern Cordillera in Colombia [Chicangana et al., 2007; Parra et al., 2009]. The last event set the current geotectonic arrangement of the rocks that are included in the area considering in this study, and which is also delineated by the thrust front of the Guaicáramo and Servitá faults westerly of the area, bordering the Eastern Footplains.

These thrust faults (Figures 2, 3, 6, and 7) outline the folded structures in this ductile environment. Their propagation is followed along the plane of the fault. The exhumation of Paleozoic and Neoproterozoic lithologies toward the north side of the central block of the Eastern Cordillera along the Servitá Fault [Mora et al., 2006], and the Mesoproterozoic platform in the southern block along the AUFS show a lithospheric contraction in which the faults in the basement constitute the fragile backstop. That is why their structural style is very similar to that of the thrust faults found in extensive basins that are undergoing a tectonic inversion phase [McKlay, 1997]. The southern Guaicáramo Fault, Servitá Fault, and the AUFS are the main structural elements in this area of the Llanos Foothills. The recorded intraplate seismicity is caused by the motion of these structures deep below.

METHODOLOGY

The LANDSAT - 5 TM, 7 ETM+ type images, and a previous field work provided technical support for the analysis of the faults by determining their tectonic limits [Chicangana et al., 2007].

We carried out field visits for the following reasons: to get the stratigraphic control, to get a better understanding of the tectonic geomorphology, and to be able to find evidence of either an active tectonic or recent movement in the traces of the faults related to the Guaicáramo and Servitá faults, and the Algeciras – Uribe Fault System. These considerations, in conjunction with the geotectonic model of each of these faults, helped us to make a seismotectonic model that was used to determine the extent of the seismic potential for these faults in terms of magnitude.

The seismologic information used in this study takes into account historical and instrumental data. The information on the historical seismicity comes from the catalogue put together by Ramirez [1975]. This catalogue was reviewed and updated by Prieto et al. [2004]. We also used the SISRA project list [CERESIS, 2011], and the macroseismic analysis done by Gómez and Salcedo [2000]; and Espinosa [2004].

The instrumental data analyzed in this work is based on the records taken from the National Seismological Network of Colombia (RSNC) for the 1993-2009 period [INGEOMINAS, 2010]. The RSNC seismological information is given with local magnitude (ML) for ML ≥ 2.1.

We took the seismicity records of the Llanos foothills logged by the RSNC keeping in mind the technical handicap, and knowing that it is inaccurate to give a solid result for the seismicity of the region, especially for the motion of the structure for a 16-year period.

In any case, the size of the structures that are analyzed in this study makes us consider the historical seismicity support and its geotectonic style. These elements will help us improve the description of its seismotectonic characteristics.

We used the inductive-deductive method considering the city´s socio-economic characteristics in order to determine Villavicencio´s vulnerability in the event of an earthquake. This method is based on the description of the technical and the social aspects; the former are related to the characteristics of the city´s ground, and the latter relate to the understanding of the seismic risk by either the general public or the authorities.

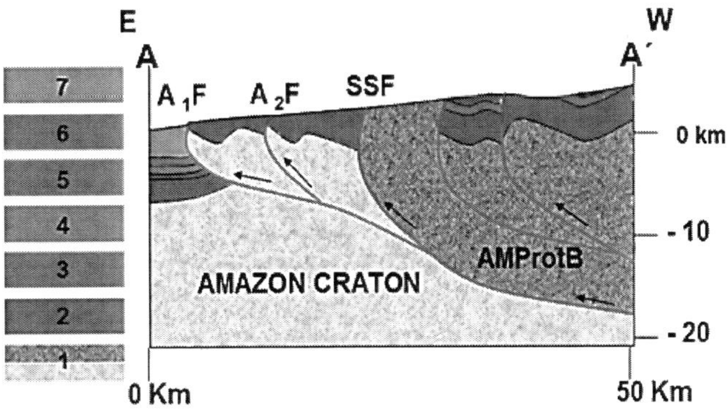

Figure 3. Above. A-A´ profile that is show in figures 2 and 4. Area of the Municipality of Uribe (130 km South of Villavicencio). A_1F: Segment 1 of Algeciras Fault. A_2F: Segment 2 of Algeciras Fault. SSF: southern section of Servitá Fault. AMProtB: Andean Mesoproterozoic Basement. Black and White tones related to the stratigraphy: 1. Proterozoic, undifferentiated. 2. Late Paleozoic. 3. Late Jurassic. 4. Early Cretaceous. 5. Late Cretaceous. 6. Paleogene. 7. Neogene. Below, photograph that shows with the arrows the fault trace related to Segment 1 of the Algeciras Fault. The fault trace crosses the Municipality of La Uribe, Department of Meta.

In the second part we include aspects like the vulnerability analysis of the population´s socio-economic characteristics. The results are used to suggest a possible post-disaster scenario that aims to mitigate the effect of a big earthquake in this region of Colombia.

Results

A. The Faults

1. Algeciras - Uribe Fault System

The Algeciras – Uribe Fault System is a group of thrusts with a N45°E strike, located at the northern tip of the Algeciras Fault as Vergara [1996] or Velandia et al. [2005] had previously described. This fault system is located west of the northern branch of the Serranía de La Macarena (130 km southwest of Villavicencio).

Located in the Municipality of La Uribe in the Department of Meta, this place has a series of parallel fault traces that contour a series of thrusts. Their northward extension under the high plain up to the Ariari River Basin near the Municipality of Lejanías in the Department of Meta can be inferred.

Long ago the Algeciras Fault was described as having a strong seismicity. Ramirez [1975], Espinosa [2004], Prieto et al. [2004], and CERESIS [2011] point out that this fault system is responsible for the 1785, 1827, and 1967 earthquakes. These earthquakes typically have magnitudes much greater than 7.0 according to the previous analysis of their macroseismic effects. Places such as Neiva City, the southern area of the Department of Huila, the Municipality of Timaná, and above all the zone that lies between the Municipalities of Algeciras and Pitalito on the Colombian Southwest, have been affected the most.

Sarabia et al. [2006] maintain that for the 1785 earthquake:

> "...Its effects covered a large area of the country (Neiva, Popayán, Tunja, Ibagué) but most of the information refers to Bogota, where there were both structural damage and victims. The macroseismic parameters are estimated based on both the spatial distribution of the damages and the superficial deformation. The observed intensities show a great range of damage in the area. These can be deduced by the large area covered by the isoseismic lines...The estimated Maximum Epicentral Intensity was VIII, it extends from Bogota to Neiva; Honda and Mariquita registered lower intensities. Even though, there is an evident gap in the historical data of the eastern and northwestern areas of the country because that area was undeveloped, the data on the intensity suggest that the location of the epicenter could be north of the Department of Huila, which might match with the northern segment of the Algeciras – Garzón Fault. This fault has numerous signs of a wide neotectonic and microseismic activity."

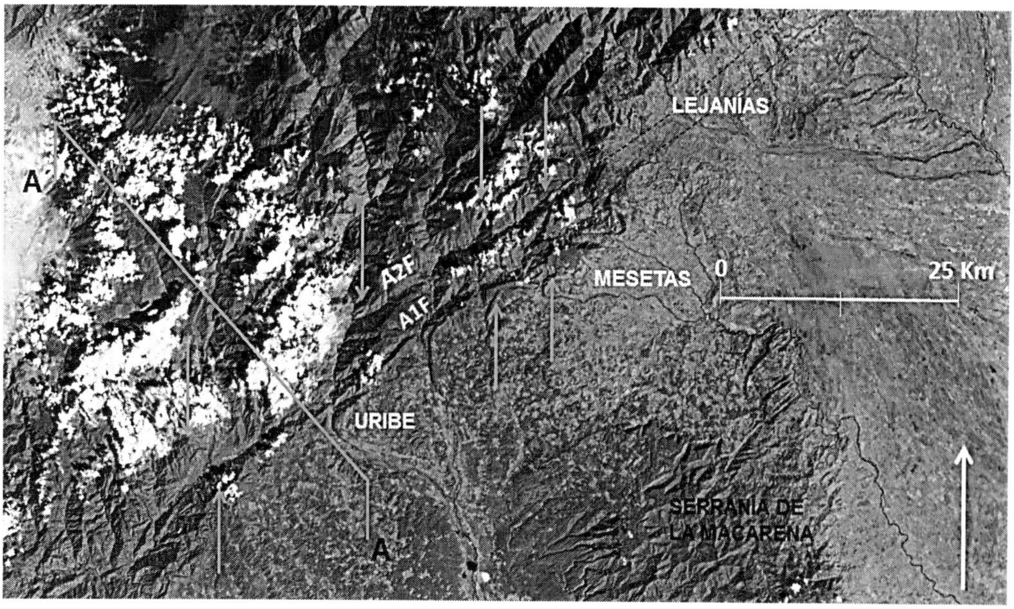

Figure 4. LANDSAT TM - 5 image that corresponds to the northern area of the Serranía de La Macarena, in the Department of Meta, a panoramic view of the segments of the Algeciras Fault System shown in the profile on Figures 1 and 3. The broken lines show the planes of the fault projected to the NE and are indicated by the arrows.

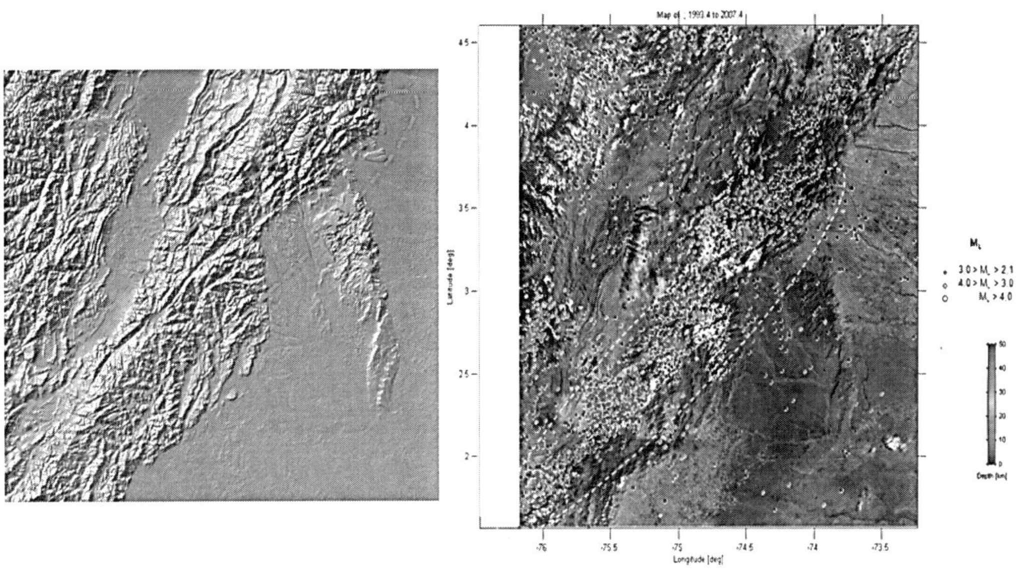

Figure 5. Relief Map taken from the topographic base of GTOPO30 that corresponds to the area covered by the Algeciras Fault System located on the southern block of the Eastern Cordillera. Right, seismicity - fault system relation. The seismic activity was registered for the 1993-2007 period by the RSNC, and lain on the LANDSAT - TM 5 image.

After estimating an intensity of X in the Mercalli Modified scale for the November 16, 1827 earthquake, and its epicenter in Timaná in the Department of Huila, Espinosa [2004] linked this earthquake to the Algeciras Fault. The historical data show that it was felt, like the one in 1785, both in Bogotá and also in a vast area of the country.

Regarding this earthquake, Espinosa maintains that:

> "This event is special because it is the one that affected the area with the greatest population in Colombia´s history…The intensity zones show a clear stretched tendency in the NNE direction, parallel to the division of the Colombian Andes, controlled by both the great NNE direction of fault systems, and by the great lithologic units."

Finally, the Global Seismological Network (IRIS, 2011a), and the Seismological Network of the Instituto Geofísico de los Andes Colombianos at Universidad Javeriana [Ramírez, 1975; Prieto et al., 2004; CERESIS, 2011] helped establish the focus of the M_W = 7.2 February 9, 1967 earthquake at 2.89 N and 74.80 E (IRIS, 2011a). This location is correlated to the Algeciras Fault, and its aftermath is described by Bohórquez and Alfaro [2008].

The following is their comment based on Ramirez [1975]:

> "…It was considered a big tremor because it was felt from Caracas to Iquitos in Peru and from Buenaventura in the Pacific Coast to Mitú in the border line between Brazil and Colombia. It was registered in over 500 seismological stations located around the planet…According to the locals, five aftershocks were felt between 10 a.m. and 4 p.m., and up to twenty aftershocks during the first three days. The seismological station in Bogotá registered a total of 350 aftershocks in the month that followed the big earthquake. The focus was located at 58 km

underneath the Eastern Cordillera and the coordinates of the epicenter (2.9ºN and 74.8ºW) correspond to a point on the surface of the Eastern Cordillera…"

The great focus depth is explained because the Global Seismological Network installed at that time only had four remote stations in Northwestern South America (IRIS, 2011b). Ramírez [1975], who was quoted by Prieto et al. [2004], points out that the M_S was 6.3 and the depth of the focus was established at 36 km according to the records of the Seismological Network of the Instituto Geofísico de los Andes Colombianos. The epicentral value of the earthquake was the location estimated by the Global Seismological Network. The Instituto Geofísico de los Andes Colombianos only had four stations at that time located in Bogotá, Chinchiná, Galerazamba and Laguna de Fúquene [Ramírez, 1975].

The seismic activity in the area where the February 9, 1967 earthquake had place, was continuously registered by the RSNC during 1993-2009 lapse (INGEOMINAS, 2010). The RSNC gives information about the shallow seismicity for the section of the Algeciras Fault located North of the area analyzed by Velandia et al. [2005].

From the above we can conclude that the AUFS near the administrative and governmental headquarters of the Municipality of La Uribe, Department of Meta, is potentially seismic, having a high risk for Villavicencio and its surrounding areas. However, a controversy has arisen. A northwestern direction was established for these faults due to the uplift of the orogeny instead of a southwestern direction as many authors have suggested.

The outcrop of the oldest lithologies toward the west side of the trace of the fault was checked in the field on segment 1 of the Algeciras Fault (Figures 3 and 4). Lodolites alternating with Sandstones are observed in correlation with the Ambicá Sandstone Unit from the Devonian [Fuquen and Osorno, 2002]; but to the East, there is a sequence of Lodolites and Chert that is possibly correlated to the Lomagorda Formation of Albian age. That thrust contours the Llanos Foothills near the northern branch of Serranía de la Macarena together with the northwesterly thrusts that have allowed its recent protrusion.

The current development of that uplift produced the superficial seismicity. It was registered in the region by the RSNC along the Algeciras Fault System as is shown in Figure 5. According to this seismic map, we could think that it is possible to experience an $M \geq 7.0$ earthquake along these faults within 150 years or less. This suggests that the M 7.1 August 31, 1917 earthquake that struck Bogotá and Villavicencio [Dimaté et al. 2005] was probably produced by a segment of this fault system in the area near the Municipality of La Uribe, Department of Meta.

2. Guaicáramo Fault

The thrust in Villavicencio´s urban area developed since the Late Pliocene until the present time. That thrust marks a sudden change in the topography of the city´s central-eastern zone. This change can be inferred by observing on the images the changes of the riverbeds toward the northwest of the urban area.

This structure, locally referred to as the Coladepato Fault, is connected to the San Pedro and Cumaral faults in the boundaries of the Municipality of Cumaral in the Department of Meta and the municipalities of Medina and Paratebueno in the Department of Cundinamarca [París et al., 2000]. This fault had a last motion in the Holocene according to París et al. [2000]. Thirty kilometers northeast of Villavicencio, the San Pedro Fault, which is the base of the regional thrust front, contours the Serranía de Las Palomas. The Serranía de Las Palomas is the Foothills´ westernmost geomorphological feature of this deformation in the region.

The Guaicáramo Fault appears in this zone as a result of the eastern structural propagation of a component of the Mirador Fault due to the spreading of the fault´s plane through the Cenozoic units when the Servitá Fault reactivated during the Pliocene - Pleistocene lapse. The lithospheric shortening, which developed during the last 2 million years, set the structural style of the Llanos foothills in the central region of Colombia near Villavicencio [Mora, 2007; Mora et al., 2010].

Toward north, the Guaicáramo Fault cuts the basement at 4.5°N where the fault spreading its plane into AMB [Bayona et al., 2008]. With the seismic information applied for oil exploration at the 4.5°N latitude, Mora [2007] shows that the fault was activated considerably during the Late Pliocene - Pleistocene lapse. This activation set the structural style of the Llanos Foothills.

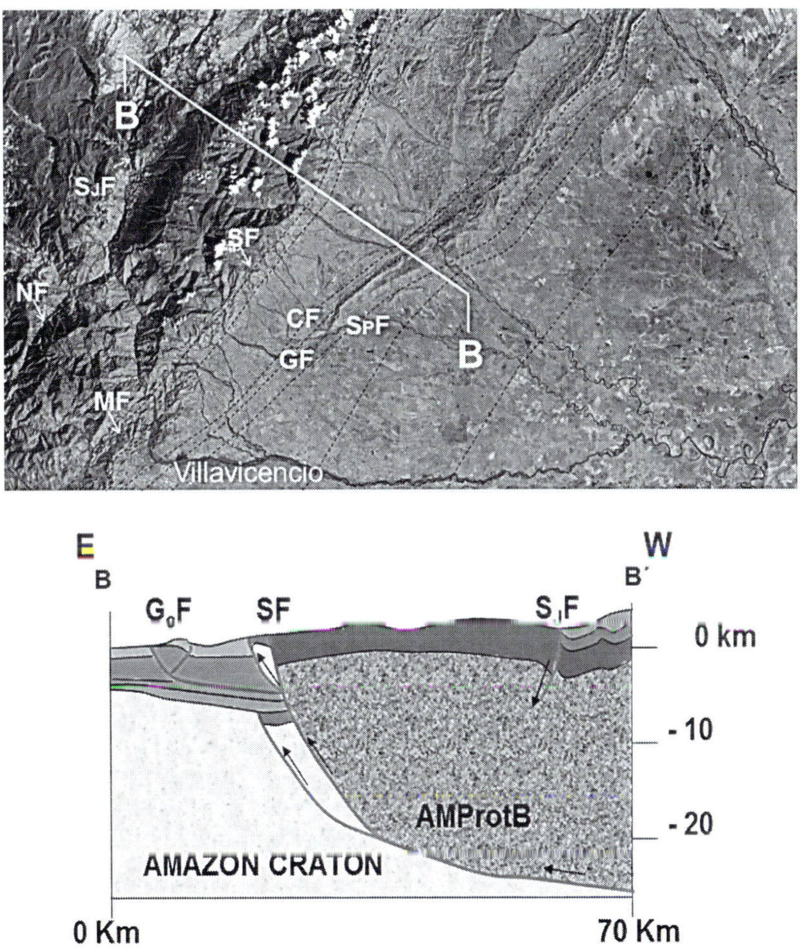

Figure 6. Above, LANDSAT IMAGE - TM 5 of the north region of Villavicencio. Below, B – B´ profile of the area of the Municipality of Paratebueno in the Department of Cundinamarca, 45 km north of Villavicencio. CF: Coladepato – Cumaral Fault; GF: Guaicáramo Fault; MF: Mirador Fault; NF: Naranjal Fault; SF: Servitá Fault; SJF: San Juanito Fault; SPF: San Pedro Fault; AMProtB: AMB. The stratigraphy in black and white tones are shown in Figure 3.

The RSNC records show a regional seismicity correlated to this fault 90 km north of Villavicencio (latitude 5°N, Figure 8). The most recent earthquake related to this fault occurred on January 19, 1995; it had a magnitude of 6.5 in the Tauramena Municipality, Department of Casanare [Dimaté et al., 2003].

Considering its morphotectonic characteristics, it is proposed here that in order to generate a $M \geq 6.5$ earthquake, the rupture model in this fault would be similar to the one stated for the La Laja Fault located in the Serranía de Villicum, San Juan Province, Argentina [Perucca and Paredes, 2003; Meigs et al., 2006]. This idea is supported by the results obtained with the seismic information of the region that is next to Villavicencio [Mora, 2007; Parra et al., 2009; Mora et al., 2010].

There is an activation of a complex structural system in the Argentinean tectonic scenario that consists of the fold of superficial plastic layers prompted by a deep, active and brittle basement that is affected by thrusts of the basement. This geotectonic situation is very similar to the one of the faults in the areas around the municipalities of Cumaral and Medina. There are three great thrust faults in this area that:

1) Delineate the Serranía de Las Palomas.
2) Affect the uppermost folded Neogene tectonic packages, and
3) Leave clear evidence of active tectonics. These faults, located correspondingly from West to East, are Cumaral, Guaicáramo and San Pedro (Figure 6).

The historical records show two fault-related earthquakes East of the Department of Cundinamarca between 1923 and 1924. However, it is suggested here that these earthquakes were due more to the trace of the Servitá Fault than to the Guaicáramo Fault. Nevertheless, considering the great features of active tectonics in this fault, Chicangana et al. [2010] recommended the immediate installation and operation of a regional seismological network that would detect the microseismicity correlated to the tectonic movement of these and other faults near Villavicencio, in order to establish the seismic hazard range.

The active tectonic features seen in the thrusts and marked by the traces of the Cumaral, Guaicáramo, and San Pedro faults respectively, clearly show that the orogenic structure in the Serranía de Las Palomas is still growing. This situation is probably due to a deep movement of the Servitá Fault as a result of the pressure of the AMB on the Amazon Craton (Figures 6 and 7) originating a movement of the thrusts exposed in the easterly oriented surface. Due to its technical constraints, the region´s seismicity reported by the RSNC cannot indicate the relation between the microseismicity and the planes of the fault, and will probably show the periodic activity of the Servitá Fault inaccurately.

3. Servitá Fault

The Servitá Fault is a great thrust fault whose plane extends down to a depth 30 km with a 45° dip slope toward the Northwest. It has the features of a listric fault at a deep crust level [Mora et al., 2006; Mora 2007; Chicangana et al., 2007; Parra, 2008].

This fault describes a great physiographic feature of several tenths of kilometers long, which is easily seen in the satellite images. It is located toward the limits between the Cordillera and the high plain in the eastern side of the Eastern Cordillera in the middle of the country (Figures 1 and 7).

This great physiographic feature is evidence of its recent movement and of its status as a regional landmark. París et al. [2000] point out that this fault not only has a big geomorphologic feature but also belongs to the Servitá – Santa María Fault Group. This description concurs with Page [1986] who points out that the Servitá Fault has had a great movement even though it had little time for doing so.

Neotectonic and tectonic studies aiming to provide a scientific background to its movement. The Servitá Fault is correlated with historic earthquakes that have affected Bogotá; for example the ones that occurred on October 18, 1743; September 4, 1966; March 19, 1988; and the most recent one on May 24, 2008 [Chicangana et al., 2007, 2010].

Lots of rock falls and mudslides occurred in the steep epicentral zone (Figure 8). The landslides found in the epicentral region characterized by a rainfall of over 4000 mm per year, and an average temperature of 24°C.

There is a direct relation between the magnitude of the earthquake and the size of the region that was affected by the landslides phenomena due to the harsh climatic conditions and to the nature of the rocks. In this region the rocks are mainly sedimentary and metamorphic of low to very low grade.

There have been earthquake-related landslides phenomena in the past. Ramírez [1975] and Prieto et al. [2004] include the reports on the earthquake that occurred on October 18, 1743 given by the priest in the town of Fómeque, Fray José Trellera y Guiluz, who said:

> "…There was, after a short while, a small earthquake following the first one; and then after the second one, within fifteen minutes there was another one that matched the first one, and each of the bigger earthquakes lasted more than a Miserere; and many more followed, small ones, many people reported them until the evening, and after the people arrived, the images of saints and virgins were taken out of the church, and we found an intact image of Nuestra Señora, the original of the bigger altar, without the smallest lesion, only its crown had fallen, even being with screws in the head."

Then he adds:

> "At the moment of prayer there was a small earthquake, after that one came a big one, and at 7 p.m. a tougher one. At first, there was noise during all of them with somewhat of a din; since then, there was a series of earthquakes, especially the one at five in the morning and at night; this lasted one month, but their number is unknown because some were felt and others were not."

The aftershocks process of the main earthquake is confirmed in the following description of Fray José Trellera y Guiluz:

> "The rivers grew forcefully and the ponds broke their mouths; the ground split up everywhere more in some surroundings than in others; there are countless volcanoes and many covered paths. Simultaneously, there were deafening thunders and all the huts were damaged…the volcanoes covered the paths. Several people were buried within cliffs. A volcano covered a girl in Tengavitá…The ground split open right where Felipe Rubio was threading sugar cane in Guachavitá, meanwhile…the only place that did not split open was where he was standing, although he had already seen himself swallowed into the depths. Let Cáqueza tell us about the ruins of its church and Ubaque about theirs; and Choachí, Une and Chipaque the ones that they suffered, that single I tell for bigger those of Fomeque."

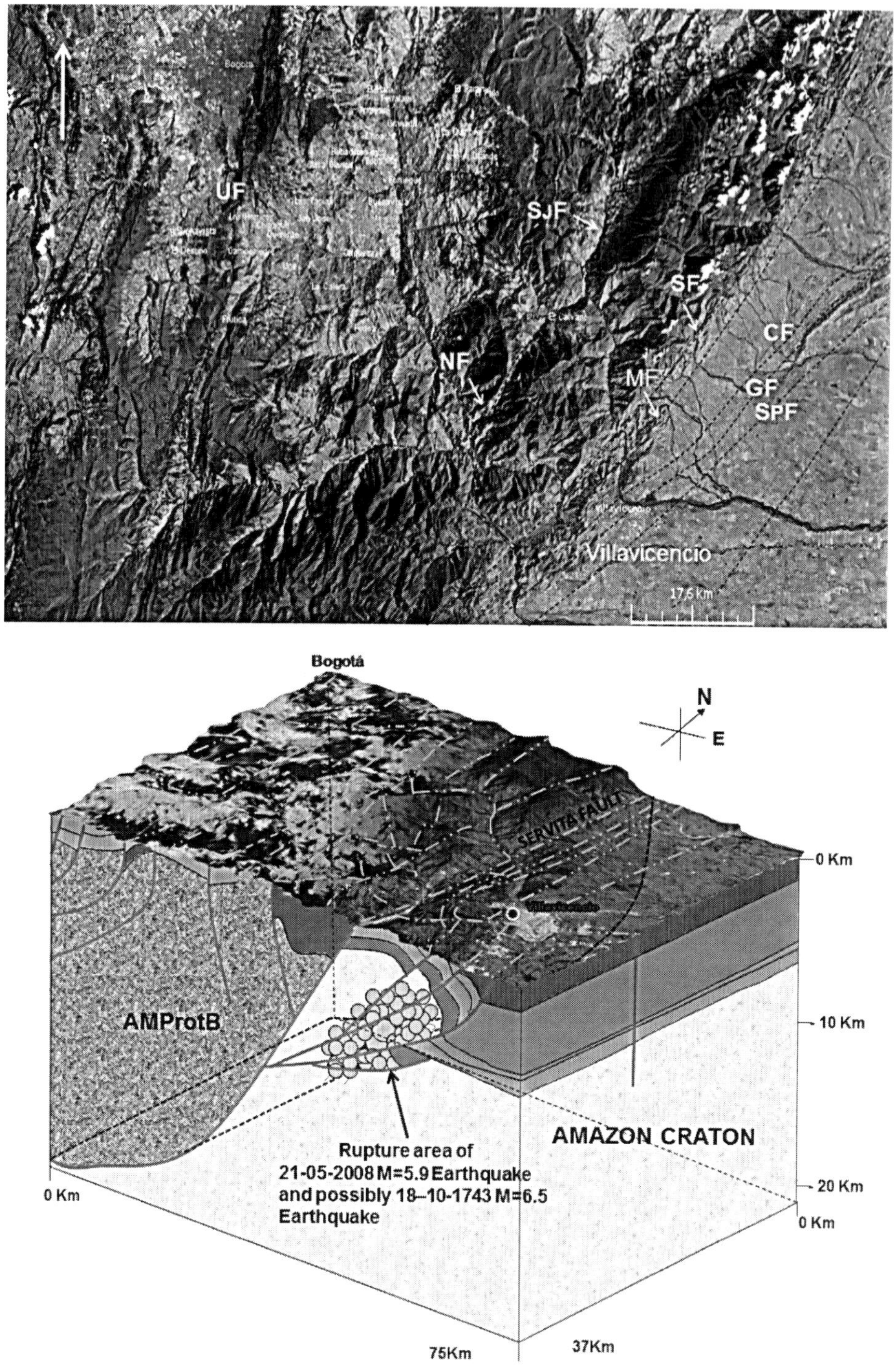

Figure 7. Above, LANDSAT - TM 5 image of western Villavicencio. Below, 3-D block corresponding to the LANDSAT - TM 5 image above and that also is show in figure 2. It shows from a hypothetical point of view, the proposed geometry of the Servitá Fault with the May 24, 2008 Meta Quetame earthquake rupture area. CF: Coladepato – Cumaral Fault; GF: Guaicáramo Fault; MF: Mirador Fault; NF: Naranjal Fault; SF: Servitá Fault; SJF: San Juanito Fault; SPF: San Pedro Fault; UF: Usme Fault. Refer to Figure 3 to check the stratigraphic black and white tones.

Figure 8. Two photos that show the landslides correlated to the coseismic effect of M 5.9 Quetame earthquake, 2008 along the Río Negro Canyon in the Municipality of Quetame in the Department of Cundinamarca. This is a region located next to the epicentral zone with a steep topographic slope. The effect of the earthquake caused simultaneous rock fall in many places.

In this eyewitness' narration we can see that the Eastern region of the Department of Cundinamarca had, in many places, the coseismic type mass removal phenomenon because in those days the term "volcano" meant landslide.

According to this, when compared to the area affected by the 2008 Quetame earthquake, the effect after the 1743 earthquake was greater. It is possible that many of the big ancient landslides that are seen nowadays, and that are partially or completely covered by vegetation (Figure 9), correlate with the same historic earthquake or older earthquakes that had a similar or greater magnitude than the one occurred in 1743.

Figure 9. Photos of old landslides in the Eastern Department of Cundinamarca and the Northwestern Department of Meta regions. 1. Vereda Servitá, Municipality of Villavicencio. 2. Río Negro Canyon, Municipality of Guayabetal, Cundinamarca. 3. Cerro Buenavista, Municipality of Villavicencio. 4. Río Negro Canyon, Municipality of Quetame, Cundinamarca. Some of these landslides remain active although partially compared to their original size. Note that albeit their close location to the epicentral areas, neither these slides as photo 4 was modified by M 5.9 Quetame earthquake, 2008.

A permanent microseismicity for the period between 1993 - 2009 is observed in the tectonic activity of the Servitá Fault recorded by the RSNC. The recorded pre-seismic and post-seismic processes of the May 24, 2008 Quetame earthquake (INGEOMINAS, 2009b) confirm that the size of the rupture zone that the aforementioned earthquake generated is due more to the kinematic mechanism of a great structure that sets a cortical boundary of a great thrust block (Figure 10). This in turn would correspond to the Servitá Fault according to Mora´s [2007] and Parra´s [2008] sense in geological and structural terms.

The size of the rupture zone that provoked that earthquake easily exceeds the activation of a course or normal fault like the Naranjal or San Juanito faults [check Mora et al., 2006; Mora, 2007 or Parra, 2008]. The aftershock process of the 2008 earthquake, whose foci exceeded the 10 km depth, was due more to the adjustment of a great thrust block that commands the growth of the Eastern Cordillera in the middle of Colombia. That block limits La Salina Fault to the West, and the Servitá Fault to the East. The pressure that the Nazca plate and the Panama Arc have on northwestern South America produces the movement of this block, and this movement is evident in the seismicity registered by the RSNC both in the La Salina Fault zone and in the Servitá Fault.

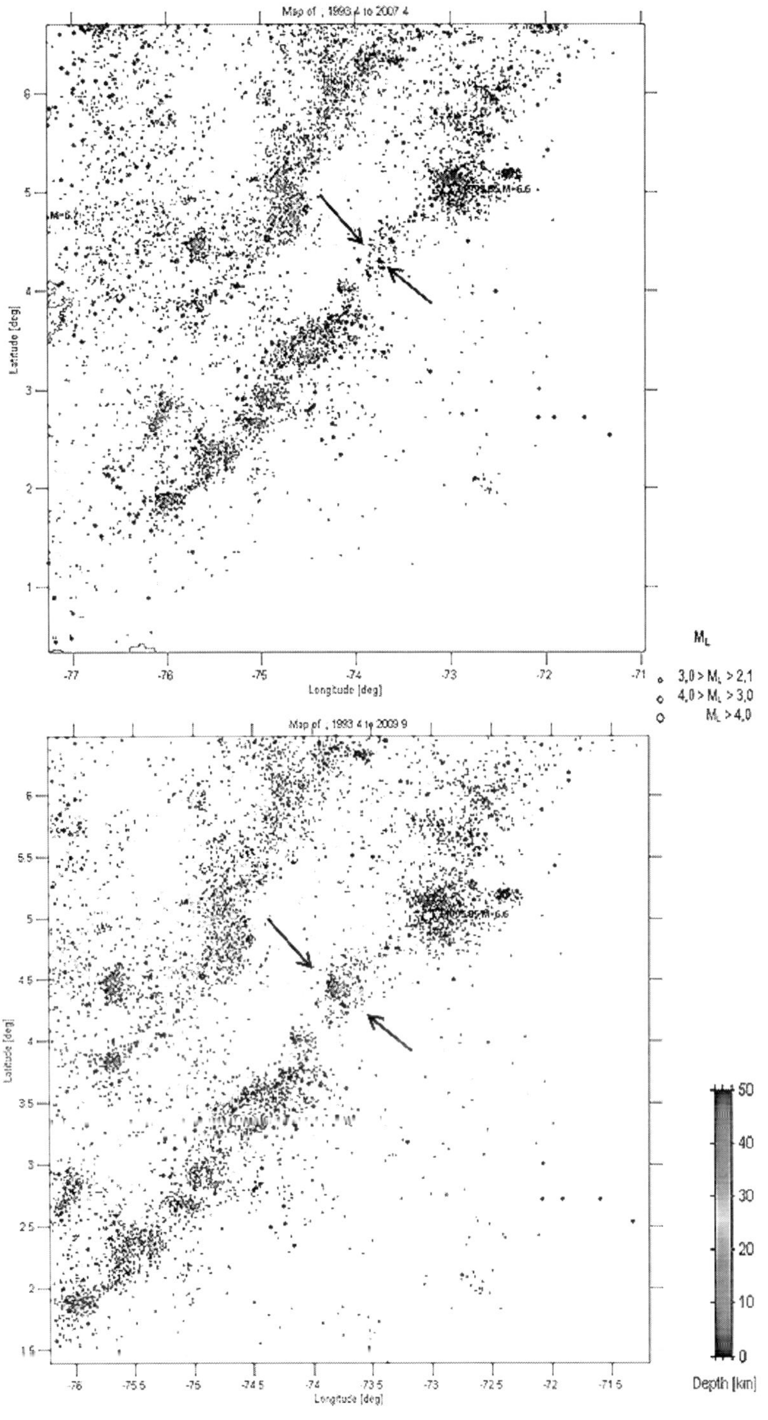

Figure 10. Above, the RSNC seismic events for 1993 – 2007 period. Below, To the previous catalog we add the earthquakes of the lapse 2008 - 2009 (INGEOMINAS, 2008a, 2010). The region where the earthquake struck is compared in both cases. The arrows show the location of the May 24, 2008 Quetame earthquake; the pre-seismic phase, recorded until April, 2007, and the post-seismic phase, which corresponds to the seismicity occurred between May, 2007 and August 2009 can be seen here.

B. The Soils of the City

In addition to the proximity of the faults, Villavicencio also increases its vulnerability in the seismic hazard because is built on soils and substrata recently shaped.

According to the first phase of the indicative seismological and geotechnical zoning of Villavicencio, carried out by INGEOMINAS [Ojeda and Alvarado, 2002], the city lies on a great alluvial block with a thickness of about 100 meters average in more of 60 % of urban area. It originated by the dragging and the accretion of materials from the different rivers in the area during the last million years [Chicangana, 2009].

The geotechnical zoning from INGEOMINAS divided the urban area into five zones, and it determined that more than 70% of the soil in the urban area exceeded a spectral acceleration of 1.6 g. The city´s downtown stands out at a zone called by INGEOMINAS as Piedemonte Fluviotorrencial. By earthquake occurrence this sector of the urban area can have a spectral acceleration up to 2.0 g.

The seismological and geothecnical study carried out by INGEOMINAS recorded 0.20 and 0.75 seconds for the main periods in Villavicencio´s urban area. These match the periods in which earthquakes were registered with a portable network. The moderate variation in the periods shows the relative heterogeneity and rigidity of the soils in this city [Ojeda and Alvarado, 2002].

Additionally, the types of soils that are able to present liquefaction i.e. alluvial and unconsolidated sands and silts found in the stream-banks of the Arenoso and Pendejo Brooks and the Ocoa River are located in the urban area of the Foothills´ southwest, an area that has a high density of population.

It is forbidden by law to build houses along the stream-banks of brooks, rivers, and in wetlands; but this regulation is ignored due to historical events like the 1950-1960 migration, and the displacement caused by different reasons, such the country´s political violence and the war caused by drug trafficking (1980s-2011).

C. Socioeconomic Factors

1. The Population´S Awareness over an Earthquake

Villavicencio has a great historical disadvantage in terms of social perception about earthquakes. Its recent founding (1850) provides a historical amnesia regarding its vulnerability over an earthquake. This is a great disadvantage for the region because nobody knows about earthquake periodicity let alone the knowledge on interseismic lapses correlated to the East Frontal Fault System – EFFS near Villavicencio.

The society lacks experience in dealing with this type of phenomenon thanks to the confidence perceived collectively by the expression: "there are infrequent earthquakes in the region".

This is supported by the fact that earthquakes are rarely felt and that people have not been living in the city for an extended time. Obviously there have been two earthquakes that proved people wrong: the one that occurred in the Department of Huila in 1967 and the other one that struck Tauramena in 1995. These events were strong enough to topple buildings and wreak havoc among the population.

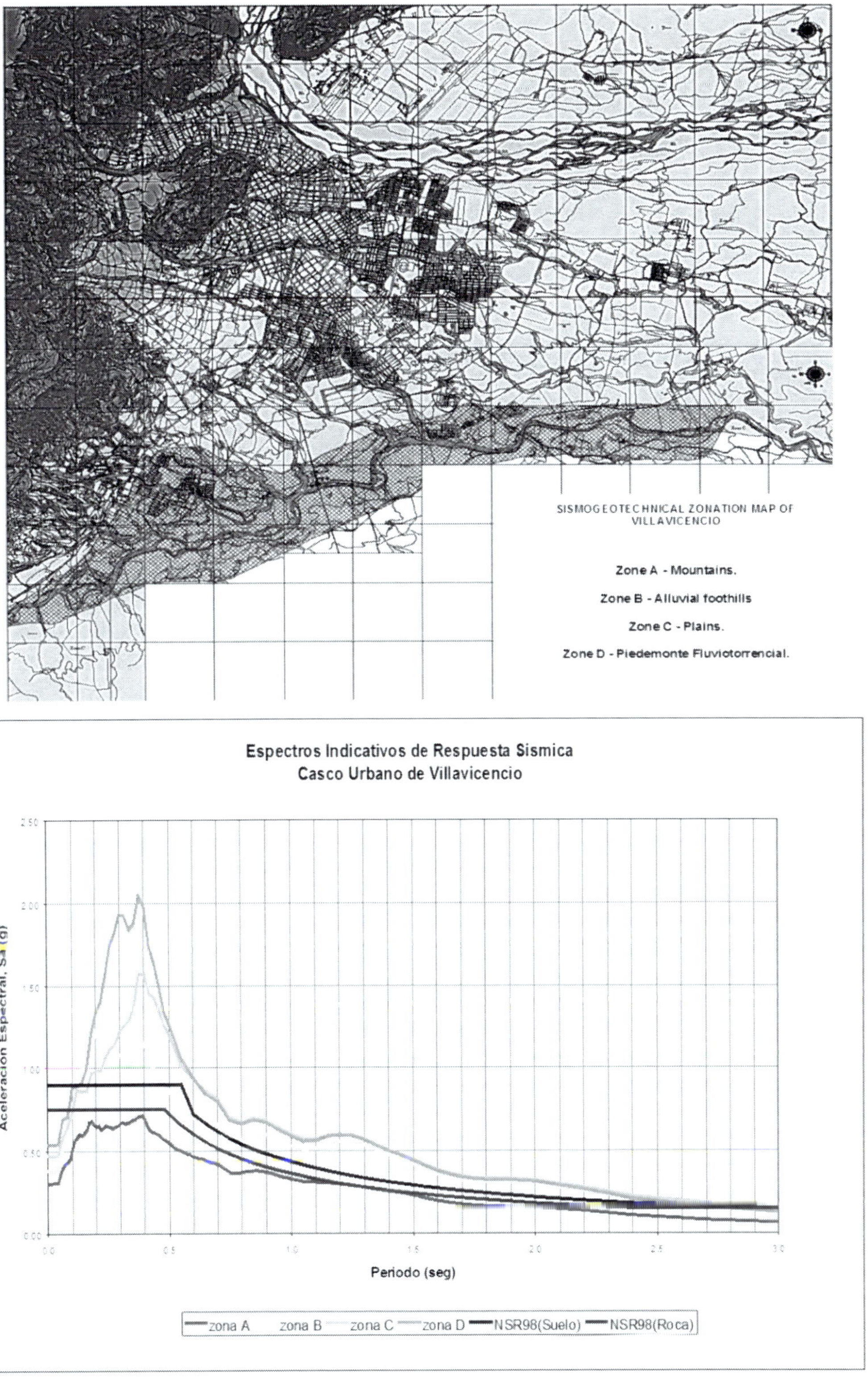

Figure 11. Seismological and geotechnical zoning map of Villavicencio. After Ojeda and Alvarado [2002].

Villavicencio´s population has grown rapidly over the last 40 years; going from 90,000 in 1973 [Tursktra, 1998] to the current official count of 430,000 people [DANE, 2008]. This rapid growth is a conspicuous problem that is related to its inhabitants´ historical amnesia, and is explained by the constant inflow of people that caused a strong growth of this Colombian city during all this time.

The people´s immigration to Villavicencio has occurred due to factors like theprocesses related to the political violence experienced in Colombia during the 1950s and 1960s, the illegal drug trafficking, the armed conflict in this region (since 1990 until these days); together with the economic growth brought by oil drilling and the development of the farming industry (since 1990 until today). It is estimated that close to 70% of the city´s population has only lived in it no longer than the time equal to that of two generations.

August 17, 1917 was a day in which Villavicencio was rocked by an earthquake. At that time, the city barely had over 3,000 inhabitants [Tursktra, 1998], and the effects of this earthquake left 7 fatal victims in Villavicencio and 4 in San Martín. That earthquake caused great financial loss and severe damage to buildings and houses in Bogotá. The serious effects that this earthquake had in Villavicencio and San Martín, according to the historical records, led Cifuentes et al. [2006] to point out that the epicentral area was located between these cities of the Llanos Foothills.

Bogotá, the most important urban center in the country, has a vast historical record of more than 470 years old that makes this city a leading seismological witness and this region the location of several earthquakes. There are less than 75 km between Bogotá and Villavicencio; and Bogotá has been a historical reference to analyze the historical seismicity in the central area of the country.

Espinosa [2004] points out that there have been at least 14 earthquakes throughout the history of the plateau in which Bogotá is nestled. That history starts with some uncertainty since 1614. However, it can be said historically speaking, that the earthquakes that have been felt in Bogotá might come from different areas of the country and not only from the Llanos Foothills region.

The historical seismicity that has taken place in Bogotá might have come from different sources; these can be: the subduction zone located in Western Colombia, the Northeastern Bucaramanga seismogenic nest, or it is related to the Algeciras Fault; which is the case of the great earthquake of 1827 [Velandia et al., 2005; Espinosa, 2004].

Villavicencio and the Department of Meta have a very young and scarce seismic history except for two towns that were founded during the Spanish colonization period in the Llanos foothills: Medina and San Martín.

Located 30 km north of Villavicencio in the area of the Llanos Foothills that corresponds to the Department of Cundinamarca, the Municipality of Medina experienced an earthquake on December 22, 1923 with an intensity of VIII on the Mercalli Modified scale [Espinosa, 2004]. This earthquake itself together with the one that struck Villavicencio in 1917 are evidence of the seismic activity that takes place at a relative close distance from Villavicencio in the Llanos foothills.

Espinosa measured the intensity at the epicenter of an earthquake based on the range of the damage in the buildings and the historical descriptions in the documents; but he neither measures the magnitude itself nor gives a good estimate of the location of the epicentral focus. This reckoning has obvious limitations but is valid in seismology.

The first seismograph was installed in Colombia right after the earthquake that occurred in Medina in 1923. This marked the beginning of operations of the Instituto Geofísico de los Andes Colombianos [Caneva, 2002]. Only then it was possible to estimate with certain scientific accuracy both a reliable approach to the seismogenic zone that caused the earthquake and its real regional magnitude.

2. Population´S Economic Situation

The communities in Villavicencio that live in hazard zones are constantly affected by phenomena that stem from the tectonic activity in the region, such as the mass removal events correlated to it.

Regarding this matter, the habit prevails have always attended the emergency right after the disaster took place. However, the prevention measures, that include avoiding the establishment of the poorest communities (generally the displaced from other regions people) in those dangerous areas, are not enough because the lack of action of the authorities in that sense.

The term vulnerability is understood as the conditions determined by the physical, social, economic, and environmental factors or processes that increase the community´s sensitivity to the impact of the hazards.

Tursktra pointed out in his 1998 study that given its recent urban development, in 1994, 25% of the population in Villavicencio had a low socioeconomic level (Figure 12). DANE showed that for 2008, 17.77% of the city´s population did not meet their basic needs and 25.7% of the homes had a standard housing deficit. The Public Services Office established that until 2009, 49.1% of the population belonged to classes 1 and 2 according to the Colombian socioeconomic scale, which correspond to the poorest classes; this population earned less than 320 American dollars per month based on the 2011 exchange rate. Within the two population groups, the people categorized in Class 1 represented 17.4% of the total population. They earned less than 300 American dollars per month based on the 2011 exchange rate (CENAC, 2009). This value has not changed for Villavicencio since 1994 when Tursktra [1998] finished his work on the city´s urban development.

Since Tursktra´s work, which supplied the population´s vulnerability toward natural hazards like floods and landslides, is few what the Government has done in order to reduce the effects of those natural hazards that will potentially affect the most vulnerable population.

Since the second half of the 1950s, the low income immigrants took over inadequate lands and started to build on it after being displaced from their original regions due to the political violence. The occupation lands usually are in active landslides, unstable zones, stream banks, and areas that are affected by floods.

Similarly, more and more wetlands have been taken over since the 1980s after an illegal settlement was established by a new group of people who was displaced by force as a consequence of the political violence and illegal drug trafficking [Tursktra, 1998]. The population´s density in zones of high vulnerability is increased by the displacement that feeds the city´s current population growth.

Taking into account that 25% of the population belongs to the most vulnerable type, here we have to consider a DANE [2008] nationwide estimate, which states that 10% of the population does not have access to schooling. This situation easily increases the vulnerability by creating new delinquent groups in post-disaster situations.

Figure 12. Left, map of Villavicencio that shows the distribution of the socioeconomic classes. Right, the quality of the condition of the buildings in the urban area. Both indicators correspond to 1994 and they were taken from Turkstra (1998).

Villavicencio is considered an average Colombian city regarding the number of vulnerable people. We have to keep in mind that the vulnerability in case of an earthquake is much higher in the population that has a standard housing deficit; which also means that they have more than one basic need unfulfilled. In a strict sense, a population of 430,000 has more than 100,000 people who belong to Class 2 based on the Colombian socioeconomic scale. This does not include another 100,000 people that have constant financial needs. As a final remark, if we consider the unconventional job market, 67.6% of the population has an unconventional job compared to 32.4% of the people who have a conventional one [CENAC 2009].

3. The Authorities Role over Risk Promotion

The royalties from oil drilling and the taxes that are collected after the development of a growing *agriculture* during the last decade make us think about the feasibility of starting a local seismological network for both the Department of Meta and for the municipalities that have enough money to materialize this idea for the Colombian Orinoquía region.

The cost of installation and maintenance of a local seismological network by the local government is low compared to the administrative and infrastructural expenses of the Department of Meta and of the region.

The seismological network will sooner or later help mitigate the effects of a disaster caused by an earthquake. The justification for its deployment is established in the Colombian law as an obligation of the authorities in the Ordinance 919 of 1989, which puts into effect the National Prevention and Attention Disaster System.

This regulation enacts that the department and municipal authorities have the obligation of participating in beefing up such a system throughout the country; this is precisely the objective of the seismological network.

Figure 13. Bad structural and seismic resistance conditions in some downtown buildings; they include among others: lack of propping, building adjacent and irregular geometry with staggering or regression. These are examples of the infringement of the Seismic Resistant Standard and show clear evidence that the vulnerability in case of an earthquake in Villavicencio increases dramatically.

The financial resources which the Department of Meta and some of its municipalities supposedly have, clearly surpass that of other regions of Colombia, including the coffee district, which comprises the Departments of Caldas, Risaralda and Quindío. This region was affected by the 1995 and 1999 strong earthquakes in Pereira and Quindío respectively.

INGEOMINAS has already done a seismological and geotechnical study but the authorities yet have not put into practice these parameters for the benefit of the city and their population.

The lack of a Land Use Plan adequate for the city and not carrying out the seismological and geotechnical study have made increase the seismic vulnerability because there is neither a regulation for the takeover of land by the people nor for the implementation of the Seismic Resistant Standard for construction.

The construction of commercial space that does not comply with the Seismic Resistant Standard has increased due to the generalized administrative corruption during the last decades in Villavicencio (Figure 13). This factor give as a result this situation and increase of poor housing quality for the most vulnerable sectors of the population, raise the hardship of an earthquake in the urban area.

Despite having the Ordinance 93 of 1998 that both regulates the National Prevention and Attention Disaster Plan and clarifies the obligation of the authorities to look after the well-being and economy of the region and its inhabitants, there has not been any appropriate support from the Government in Villavicencio.

The agencies like Red Cross, civil defense, volunteers firemen, among others, (being part of the National Prevention and Attention Disaster Plan System) do not have gotten any support from the authorities. The recent warning of a seismic hazard given to the community has been treated only by some scholars in the community.

This warning is the one that has driven people locally and nationwide over the technical and scientific aspects and its consequences. It also advises the different agencies that make up the National Prevention and Attention Disaster System.

Likewise, the scholars in their role of service to the community and working with research groups at the Universities in Villavicencio, have been knocking on the doors of the Government for over a decade in order to get support but the authorities have given just little support without understanding the benefits, which a seismological and geotechnical zoning may give to the city.

The operability lack of the seismological and geotechnical zoning can be seen in two ways; the first one is the inability of the authorities to take measures to prevent, anticipate, mitigate or to have a quick response in case of an earthquake; and the second one is the lack of attention by factors like the concentration of power, the lack of long term forecasting, government weakness, and corruption, among others. These factors increase the vulnerability to deal with an earthquake and the hardship of the expected post-disaster scenario.

4. The Post-Disaster Scenario

Villavicencio´s present-day population stands at over 430,000. It is possible that in less than two decades it will reach over 1 million due to the high population and economic growth rate that have taken place during the last three decades. This projection is based on studies that development and urban planning experts, like IGAC, which is the National Geographic Survey of Colombia, and the International Institute for Geo-Information Science and Earth Observation (ITC) of Netherlands [Turkstra, 1998; Gobernación del Meta – IGAC, 2004] have done during the last 20 years.

This growth projection could lead to the following situation if corrective measures will not implemented in the short term of a post-disaster scenario:

In case of an earthquake that has a magnitude greater or equal to 6.5, whose epicenter is similar to any of the ones occurred in 1917 or 1923, and keeping in mind what has been observed regarding the geology, the tectonics, the seismicity, the geotechnics, and the vulnerable population, such an event would cause at least 1,000 fatal victims in the city and 1,500 in the region during the quake; and 2,000 more would die in the rubble of the collapsed buildings if they are not rescued soon enough. There would be at least 50,000 injured people and 300,000 people would be affected in Villavicencio and its surrounding region.

The city's operations would be paralyzed due to a shortage of the utility services for at least three months given the poor technical conditions of their infrastructure around the city. Gas leaks would produce local fires injuring people and increasing the material losses.

The road that connects Bogotá to Villavicencio would be blocked because of the numerous mass removal phenomena, and possible collapsing of road sections. This would delay the quick third party reaction from Bogotá and the center of the country. The road problem would cause an increase in number of victims.

According to the prior indicative seismogeotechnical zoning of the city, and keeping in mind the people who make the 25.7% of the population that has a housing deficit, 30% of the buildings would be totally ruined because of the high (70%) vulnerability of the city's constructions in terms of the quality needed to withstand a natural hazard.

The financial losses would be at least three times that of the January 25, 1999 Quindío earthquake. The local farming industry would be hindered and the closing of the main communication route between the Orinoquía region with the rest of the country (the Bogotá – Villavicencio road) would cause financial losses and would substantially raise the price of products in the region and in the whole country for several months while the road is fixed. The earthquake would cause losses that may surpass 3 points in the GDP.

It is possible that the city's commercial capability at the time of the quake might be recovered in less than five years; but its urban recovery could last over a decade considering that more than half of the city is in a low socioeconomic echelon that lacks a good Land Use Plan designed according to a long term reaction plan.

This situation has already been observed in other Colombian cities like Armenia and Popayán, in which the hardship of the earthquakes that struck them could have been lower compared to the hypothetical case presented here for Villavicencio. In this case, a lack of investment in preventing an earthquake would be more expensive in the long term than the potential same investment that would have been done to increase the quality of the lifestyles and the sustainable development of its inhabitants.

RECOMMENDATIONS

Since there is a high vulnerability in the city in case of an earthquake and considering the lack of interest by the authorities regarding its seismic hazard (mainly due to the lack of forecasting and long term planning), the scholars in conjunction with other social entities, like private companies and the mass media, should make the local Government put into practice the 1989 Ordinance 919 and the 1998 Bill 83 in order to forcefully demand the compliance of what the National System for the Prevention and Attention of Disasters requires regarding this type of natural hazard for the city.

This will only be realized by constantly demanding the compliance of the regulations by the regional and local authorities in order to take measures to prevent and mitigate the seismic hazard in the city and in the Llanos Foothills in the Department of Meta.

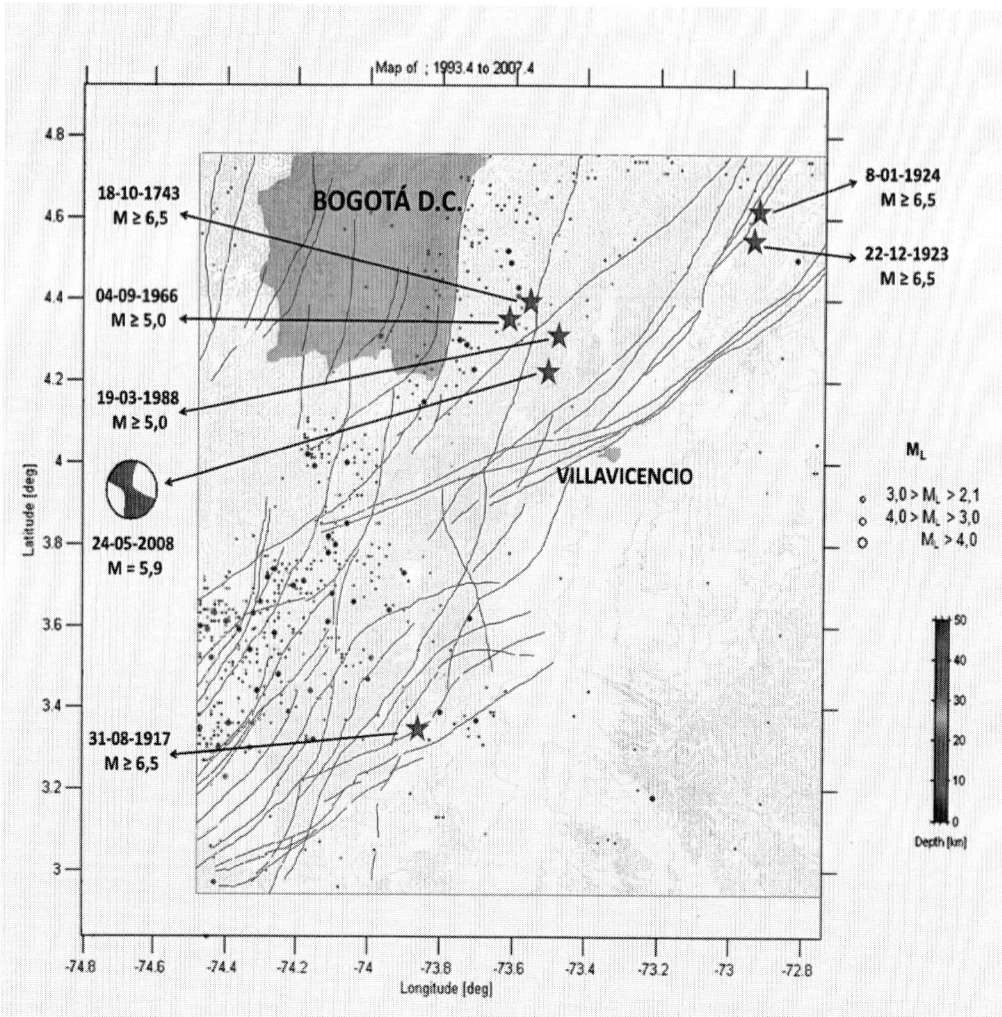

Figure 14. Great historical earthquakes and RSNC instrumental seismicity records for the period between 1993 and 2007 in the Villavicencio and Llanos foothills region in Central Colombia.

Similarly, the Government should demand starting the operation of a local seismological network; such a network will perform an important role in the prevention and mitigation of the seismic hazard in the short and long term.

According to the regulations, the local Government must give support in organizing educational events and workshops for the most vulnerable communities in order to prevent and mitigate the undesirable effects of an earthquake in Villavicencio´s urban area. The idea is to develop a safe city in the short term aiming to have a Land Use Plan that suits its seismic hazard and that follows the updated Seismic Resistance Standards. In this way, there should be a promotion of studies that complement the seismological and geotechnical zonification of the city; this promotion should be adapted to the Land Use Plan of Villavicencio, which in turn should be adjusted to its seismological and geotechnical zonification.

These actions will be led by the scholars and Government agencies such as IGAC and INGEOMINAS.

CONCLUSION

We show in this chapter that the faults correlated to the Algeciras – Uribe Fault System, the Guaicáramo Fault, and the Servitá Fault in the Llanos Foothills in Central Colombia can generate earthquakes as confirmed by both the historical and instrumental seismicity records (Figure 14).

However, it is clear that not having a local seismicity monitoring system for this region, in which the Llanos Foothills and Villavicencio are included, increases the uncertainty on the range of its seismic hazard. This poses a higher vulnerability over this phenomenon because the lack of a local seismic network leads to not knowing about the movement of these faults. That is why, to this day, it is not possible to visualize the state of the stresses that will allow estimating with certainty the place where a spontaneous release of these stresses might take place within the short or middle term.

Regional morphotectonic aspects and active tectonic evidence at a local level have been identified in these structures. They are related to a recent tectonic movement that have led us to conclude that a high seismic hazard underlies in this region of the Llanos Foothills [París et al., 2000; Chicangana et al., 2007].

Based on the aforementioned, we point out that in the event of an earthquake, with a magnitude equal or greater than 6.5 and with an epicenter in a similar location to the ones of the 1743, 1917, 1923, or 1924 earthquakes, there could be a post-disaster scenario of great damage and many fatal victims in Villavicencio and its neighboring municipalities. Similarly, Bogotá and the 10 million people who live in its metropolitan area could experience serious damage and even fatal victims.

The economy in Villavicencio´s region would lose a total of 3% of GDP [Chicangana et al., 2010] and more than 25% of GDP if the center of the country it is affected, a region right where Bogotá is located. This forecast model, which summarizes the results shown here, is based on economic and sociocultural features on a layout of a seismic vulnerability previously done [Chicangana et al., 2010] and following the post-disaster results taken from the UN´s human development index (IDH) established for Colombia in a historical context after the 1983 earthquake in Popayán and the 1999 earthquake in Quindío [UNDP, 1998, 2011a, 2011b; DANE, 2008].

It is impossible to avoid in Villavicencio an earthquake of the characteristics explained in this chapter; but if the recommendations are followed strictly in the short term, and if the earthquake were to strike within 10 years, its lethal effects could be lowered to 75% or even less.

ACKNOWLEDGMENTS

This work is the result of the research and actions taken to spread the word about the range of the seismic hazard in Villavicencio and the Llanos Foothills, in the Department of Meta, in Colombia, to social entities like the scholars, the civilians, and the Government of Villavicencio and the Department of Meta.

Our deep appreciation to: the members of the Corporación Universitaria del Meta in Villavicencio; Research Groups "Estudios sobre Riesgo Sísmico y Amenazas Naturales del

Piedemonte Llanero", The Geophysics Group of the Department of Geosciences of Universidad Nacional de Colombia, Bogotá; the Geophysics Group of Centro de Investigaciones of Universidad Antonio Nariño, Bogotá; GEOSLAC - Asociacion Latinoamericana y del Caribe en Geociencias; and the Fundations "Kinkaju" and "Nueva Herencia" in Villavicencio; Omar Agudelo in CEPREVE at Universidad Nacional de Colombia, Bogotá; Andrés Alfaro Castillo at Universidad de la Salle, Bogotá; the students and professors at Vicerrectoría de Investigaciones, Universidad del Meta; and all the people in Villavicencio that took part in completing this work that began in 2006 and continues until today. Equally a special acknowledgement to Jaime Andrés Ochoa García for his contribution in the improvement of the text.

REFERENCES

A. I. S. – INGEOMINAS (1996). Estudio General de Amenaza Sísmica de Colombia. Asociación Colombiana de Ingeniería Sísmica – *A. I. S.* 300, Bogotá, 252 p.

Alfaro, A. and Ramos, A. (2001). Fuentes Sismogénicas y deslizamientos en Villavicencio para la Microzonificación Sísmica. *Revista Ingeniería y Universidad, Pontificia Universidad Javeriana,* Bogotá, 4 (2), pp. 42-51.

Barragán, R. and Baby, P. (2004). Magmatismo alcalino intra – placa en la Cuenca Cretácica Oriente, Ecuador: Evidencias Geoquímicas, geocronológicas y tectónicas. En: La Cuenca Oriente: Geología y Petróleo. P. Baby, M. Rivadeneira and R. Barragán, (eds), *I. F. E. A. – I. R. D. – P. E. T. R. O. E. C. U. A. D. O. R.*, Quito, Ecuador, pp. 69-91.

Bayona, G., Cortes, M., Jaramillo, C., Ojeda, G., Aristizabal, J. J. and Reyes-Harker, A. (2008). An integrated analysis of an orogen–sedimentary basin pair: Latest Cretaceous–Cenozoic evolution of the linked Eastern Cordillera orogeny and the Llanos foreland basin of Colombia. *G. S. A. Bulletin*, 120 (9/10). pp. 1171-1197.

Bohórquez, M. and Alfaro, A. (2008). Análisis probabilístico de la amenaza sísmica de Neiva-Colombia. 40 años del macrosismo del Huila (1967 – 2007). *Revista Épsilon*, 11, pp. 41-48.

Caneva, A. (2002). Magnitud representativa del catálogo de sismos en Colombia. *Revista Ingeniería y Universidad*, Pontificia Universidad Javeriana, Bogotá, 6 (1), pp. 9-18.

CENAC (2009). Boletín Estadístico, Contexto Sectorial Villavicencio – Meta. Bogotá D. C. Centro de Estudios de la Construcción y el Desarrollo Urbano y Regional- C. E. N. A. C. *http://www.cenac.org.co* (last access 25/08/2011).

CERESIS (2011). Consulta al Catálogo de Hipocentros (1520-1991). Centro Regional de Sismología para América del Sur – CERESIS, Lima, Perú. *http://www.ceresis.org/portal/catal_hipo.php*(last access 25/08/2011).

Chicangana, G. (2009). *La Amenaza sísmica de Villavicencio y el centro de Colombia. Villavicencio, Colombia.* Corporación Universitaria del Meta, Punto Gráfico, Villavicencio, Colombia, 110 p.

Chicangana, G. and Kammer, A. (2011a). Evolución tectónica de la cordillera Oriental de Colombia. Desde la apertura del océano Iapeto hasta la conformación de la Pangea: Una visión preliminar. Primera parte: Aspectos geológicos. *Revista Ingeniería, Investigación y Desarrollo*, (in press).

Chicangana, G. and Kammer, A. (2011b). Evolución tectónica de la cordillera Oriental de Colombia. Desde la apertura del océano Iapeto hasta la conformación de la Pangea: Una visión preliminar. Segunda parte: Evolución geodinámica. *Revista Ingeniería, Investigación y Desarrollo*, (in press).

Chicangana, G., Vargas, C., Kammer, A., Hernández, T. and Ochoa, L. (2007). Caracterización Sismotectónica Regional Preliminar de un sector del Piedemonte Llanero colombiano: Corredor San Juan de Arama – Cumaral, Meta. *Boletín de Geología – U. I. S.*, 29 (1), pp. 61-74.

Chicangana, G., Vargas, C., Caneva, A., Mojica, L., Hernández, T., Ardila, J. and Bernal, A. (2010). La sociedad frente a la gestión del riesgo: Caso sobre la amenaza sísmica en la ciudad de Villavicencio. *Boletín de Geología – U. I. S.*, 32 (1), pp. 125-141.

Cifuentes, H.; Sarabia, A.; Robertson, K. and Dimaté, C. (2006). Parámetros Macrosísmicos del Sismo de 1917 en Colombia. II Congreso Latinoamericano de Sismología, Memorias (CD-ROM), *GEOSLAC*, Bogotá.

DANE (2008). Censo General 2005, Nivel Nacional. Departamento Administrativo Nacional de Estadística. Bogotá D. C. *http://www.dane.gov.co/censo/files /libroCenso2005nacional.pdf* (last access 25/08/2011).

Dimaté, C., Rivera, L. and Cisternas, A. (2005). Re-visiting large historical earthquakes in the Colombian Eastern Cordillera. *Journal of Seismology*, 9 (1), pp.: 1-22.

Dimaté, C., Taboada, A., Rivera L., Delouis, B, Osorio, A., Jiménez E., Fuenzalida, A., Cisternas, A., Gómez, I., 2003. The 19 January 1995 Tauramena (Colombia) earthquake: geometry and stress regime. *Tectonophysics* , 363, 159-180.

Espinosa, A. (2004). *Historia Sísmica de Colombia* (1550 – 1830). (CD-ROM), Gedes – Uniquindío, Armenia, Colombia.

Fuquen, J. and Osorno, J. (2002). Geología de la Plancha 303 - Colombia, Departamentos de Huila, Tolima y Meta, Escala 1: 100000. Memoria explicativa. *INGEOMINAS*, Bogotá, 90 p.

Gobernación del Meta – I. G. A. C. (2004). *El Meta: Un territorio de oportunidades.* Instituto Geográfico "Agustín Codazzi". Bogotá. (CD-ROM).

Gómez, A. and Salcedo, E. (2000). Determinación de la profundidad focal de algunos sismos ocurridos en Colombia a partir de datos macrosísmicos. *Revista Red Sismológica del Eje Cafetero*, Viejo Caldas y Tolima. 5, pp. 12-16.

Hernández, G. (1996). Bases georeferenciadas para el análisis estructural del Piedemonte Llanero. VII Congreso Colombiano de Geología, Sociedad Colombiana de Geología, *Bogotá,* V. 2, pp. 391-399.

Horton, B., Saylor, J., Nie, J., Mora, A., Parra, M., Reyes-Harker, A. and Stockli, D. (2010). Linking sedimentation in the northern Andes to basement configuration, Mesozoic extension, and Cenozoic shortening: Evidence from detrital zircon U-Pb ages, Eastern Cordillera, Colombia. *G. S. A. Bulletin*, 122 (9/10), pp. 1423-1442. doi: 10.1130/B30118.1.

Ibañez-Mejía, M. (2010). New U-Pb geochronological insights into the Proterozoic tectonic evolution of northwestern South America: The Meso - to Neoproterozoic Putumayo Orogen of Amazonia and its implications for Rodinia reconstructions. (MSc thesis). 69 p. Department of Geosciences, The University of Arizona, Tucson, Arizona,US *http://www.geo.arizona.edu/Antevs/Theses/IbanezMejiaMSc2010.pdf* (last access 25/08/2011).

INGEOMINAS (2007). Mapa Geológico de Colombia. *Escala*: 1: 1000000. INGEOMINAS, Bogotá.

INGEOMINAS (2008a). Boletín de Sismos 1993-2007. INGEOMINAS – Red Sismológica Nacional de Colombia, Bogotá.

INGEOMINAS (2008b). El Sismo de Quetame del 24 de mayo de 2008. Aspectos Sismológicos y Evaluación preliminar de daños. Informe Preliminar N° 2. INGEOMINAS - Red Sismológica Nacional de Colombia, Bogotá. *http://seisan.ingeominas.gov.co/RSNC/pdfs/Info_quetame.pdf* (last access 25/08/2011).

INGEOMINAS (2010). *Boletín de Sismos 1993-2009*. INGEOMINAS – Red Sismológica Nacional de Colombia, Bogotá.

IRIS (2011a). 1967 Huila (Colombia) Earthquake Archive. Incorporated Research Institutions for Seismology, Acoustic and Seismics Laboratory, Institute of Science and Technology, The University of Michigan, Ann Harbor, Michigan, US *http://www.iris.edu/seismo/quakes/1967huila/* (last access 25/08/2011).

IRIS (2011b). *Handbook: Word – Wide Standard Seismograph Network*, April 1964. 500 p. Incorporated Research Institutions for Seismology, Acoustic and Seismics Laboratory, Institute of Science and Technology, The University of Michigan, Ann Harbor, Michigan, US *http://www.iris.edu/seismo/info/stations/WWSSN1964.pdf* (last access 25/08/2011).

Keppie, J. D., Dostal. J., Ortega-Gutiérrez, F. and López, R. (2001). A Grenvillian arc on the margin of Amazonia: evidence from the southern Oaxacan Complex, southern Mexico. *Precambrian Research*, 112 (3 -4), pp. 165-181.

Maya, M. (2001). Distribución, facies y edad de las rocas metamórficas de Colombia. I - 2426, *INGEOMINAS*, Bogotá, 57 p.

Meigs, A., Krugh, W. C., Schiffman, C., Verges, J. M. and Ramos, V. A. (2006). Refolding of Thin - Skinned Thrust Sheets by Active Basement-Involved Thrust Faults in the Eastern Precordillera Of Western Argentina. Revista de la Asociación Geológica Argentina, 61 (4), pp. 589-603. *http://www.scielo.org.ar/pdf/raga/v61n4/v61n4a13.pdf* (last access 20/6/2011).

Mora, A. (2007). *Inversion tectonics and exhumation processes in the Eastern Cordillera of Colombia* [Dissertation zur Erlangung des akademischen Grades Doktor der Naturwissenschaften (Dr.rer.nat)]. 133 p. Wissenschaftsdisziplin Geologie eingereicht an der Mathematisch-Naturwissenschaftlichen Fakultät, Universität Potsdam, Postdam, Germany.

Mora, A.; Parra, M.; Strecker, M. R.; Kammer, A.; Dimaté, C. and Rodríguez, F. (2006). Cenozoic contractional reactivation of Mesozoic extensional structures in the Eastern Cordillera of Colombia. *Tectonics*, 25 (2): TC2010 doi:10.1029/2005TC001854.

Mora, A., Parra, M., Strecker, M. R., Sobel, E. R., Zeilinger, G., Jaramillo, C., Ferreira Da Silva, S. and Blanco, M. (2010). The eastern foothills of the Eastern Cordillera of Colombia: An example of multiple factors controlling structural styles and active tectonics. *G. S. A. Bulletin*, 122 (11/12), pp. 1846-1864.

Ojeda, J. and Alvarado, C. (2002). Zonificacion Sismogeotécnica Indicativa de la Ciudad de Villavicencio. In: Zonificación integral por amenazas naturales para la ciudad de Villavicencio - Meta. *INGEOMINAS Alcaldía de Villavicencio, Convenio interadministrativo* 009/ 2000. V. II, 178 p.

Page, W. D., 1986. Seismic geology and seismicity of northwestern Colombia. San Francisco. Woodard - Clyde Consultants *Report for I. S. A. and Integral Ltda.* Medellín, Colombia. 200 p.

París, G., Machette, R., Dart, R. L. and Haller, K. M. (2000). Database and Map of Quaternary faults and folds of Colombia and its offshore regions, Open – File Report 00 – 0284. *http://pubs.usgs.gov/of/2000/ofr-00-0284/ofr-00-0284.pdf* (last access 17/6/2011).

Parra, M. (2008). Cenozoic foreland-basin evolution in the northern Andes: insights from thermochronology and basin analysis in the Eastern Cordillera, Colombia. [Dissertation zur Erlangung des akademischen Grades Doktor der Naturwissenschaften (Dr. rer. nat.)]. 121 p. Wissenschaftsdisziplin Geologie, Institut für Geowissenschaften, Mathematisch-Naturwissenschaftlichen Fakultät, Universität Potsdam, Postdam, Germany. *http://opus.kobv.de/ubp/volltexte/2009/2933/pdf/parra_diss.pdf* (last access 17/6/2011).

Parra, M., Mora, A., Jaramillo, C., Strecker, M. R., Sobel, E. R., Quiroz, L., Rueda, M. and Torres, V. (2009). Orogenic wedge advance in the northern Andes: Evidence from the Oligocene-Miocene sedimentary record of the Medina Basin, Eastern Cordillera, Colombia. *G. S. A. Bulletin*, 121 (5 – 6), pp. 780- 800. doi:10.1130/B26257.1.

Perucca, L. P. and Paredes, J. (2003). Fallamiento Cuaternario en la zona de La Laja y su relación con el terremoto de 1944, Departamento de Albardón, San Juan, Argentina. *Revista Mexicana de Ciencias Geológicas*, 20 (1), pp. 20-26.

Prieto, J., Briceño, L. A., Caneva, A. and Ramos, A. (2004). Actualización de la historia de los terremotos de Colombia de Jesús Emilio Ramírez,S. J. Instituto Geofísico de la Universidad Javeriana, Pontificia Universidad Javeriana, *Bogotá*, 186 p.

Ramírez, J. E. (1975). La Historia de los Terremotos en Colombia. Instituto Geográfico "Agustín Codazzi", *Bogotá D.E.* 250 p.

Ramos, V. A. (2010). The Grenville - age basement of the Andes. *Journal of South American Earth Sciences*, 29 (1), pp. 77-91.

Sarabia, A., Cifuentes, H., Dimaté, C. and Robertson, G. (2006). Análisis histórico y parámetros macrosísmicos del sismo de julio 12 de 1785 en el centro de Colombia. II Congreso Latinoamericano de Sismología, Memorias (CD-ROM), GEOSLAC, Bogotá.

Turkstra, J. (1998). Urban Development and Geographical Information, Spatial and temporal patterns of urban development and land values using integrated geo – data, Villavicencio, Colombia. *I. T. C. Publication Series*. Enschede, Netherlands. 60, 268 p.

U. N. D. P. (1998). The State of Human Development. United Nations Development Programme, 1998. *http://hdr.undp.org/en/media/hdr_1998_en_chap1.pdf* (last access 25/08/2011).

U. N. D. P. (2011a). Informe del Desarrollo Humano para Colombia en 1998. *http://hdr.undp.org/en/reports/national/latinamericathecaribbean/colombia/colombia_1998_sp.pdf* (last access 25/08/2011).

U. N. D. P. (2011b). H. D. I. 2010 Index. *http://hdr.undp.org/en/media/Lets-Talk-HD-HDI_2010.pdf* (last access 25/08/2011).

Velandia, P., Acosta, J., Terraza, M. R. and Villegas, V. H. (2005). The current tectonic motion of the Northern Andes along the Algeciras Fault System in SW Colombia. *Tectonophysics*, 399 (1-4), pp. 313-329.

Vergara, S. H. (1996). Rasgos y actividad neotectónica de la Falla Algeciras. VII Congreso Colombiano de Geología. *Memorias, Sociedad Colombiana de Geología, Bogotá*, V. 1, pp. 491-500.

In: Earthquakes
Editor: Kostas Konstantinou
ISBN: 978-1-62081-883-1

Chapter 3

THE 9 JULY 1956 (MW 7.6) AMORGOS EARTHQUAKE, SOUTHERN AEGEAN: SEISMOTECTONICS OF A LARGE TSUNAMIGENIC EVENT

K. I. Konstantinou [*] ***and S. Rontogianni***
Institute of Geophysics, National Central University, Taiwan

ABSTRACT

On 9 July 1956 a large earthquake occurred near the island of Amorgos, southern Aegean, causing extensive damage while it was followed by a tsunami that inundated the coasts of nearby islands also reaching the Turkish coast. This chapter aims first at providing a thorough review of previous studies about this earthquake and then developing a seismotectonic model for its occurrence by combining all available information. Analysis of historical seismograms of the Amorgos event shows that its focal mechanism is a purely normal fault and that the fault plane exhibits a low-angle (< 30 degrees) dip. This analysis also revealed a relatively deep centroid (45 km) placing the high slip area inside the upper mantle, unlike previous estimates which suggested a shallow focus (~10 km). This relatively large depth seems to be supported by numerical modeling of the tsunami generation and propagation that points to submarine landslides rather than co-seismic rupture of the seafloor as its cause. In addition, the crustal rheology of the area implies that the Amorgos earthquake could have nucleated near the depth of peak strength (33 km) rupturing up to a depth of 7 km without breaking the seafloor. However, these considerations are at odds with geological observations of high-angle faulting near the epicentral area of the earthquake. The geological history of the southern Aegean can offer clues so as to resolve this apparent controversy. Amorgos lies at the eastern edge of a metamorphic core complex that started being exhumed 23-19 Ma due to the rollback of the subducting African lithosphere. A direct consequence of this exhumation is the development of large scale detachment faults consisting of upper crust high-angle and lower crust shallow-angle normal faults. The existence of these faults is

[*] Institute of Geophysics, National Central University, 300 Jhongda Rd, Jhongli, 320 Taiwan. E-mail: *kkonst@ncu.edu.tw*.

based on geological observations of exhumed metamorphic rocks and thermo-mechanical numerical models of the exhumation process. Such a detachment fault can explain the geometry and nucleation depth of the Amorgos event thus resolving all the aforementioned discrepancies.

INTRODUCTION

The Hellenic subduction system was created by the northward movement of the African plate at a rate of 1 cm/yr and its subduction beneath the Aegean micro-plate which moves towards SW at a rate 3.5 cm/yr [McClusky et al., 2000; Le Pichon and Kreemer, 2010] (Figure 1). This difference in geodetic velocity rates implies the retreat of the subduction system which is commonly referred to in the literature as “slab rollback”. Such a rollback may result from the interaction (draping) of the subducting slab with the upper/lower mantle discontinuity at 660 km depth which tends to slow down slab descent into the lower mantle [e.g. Fiuniciello et al., 2003]. Seismic tomography seems to support this interpretation since tomographic images show a high-velocity anomaly of the descending slab down to 1800 km [Spakman et al., 1993; Bijwaard et al., 1998]. The geometry of the high-velocity anomaly changes distinctly across the 660 km discontinuity providing evidence for slab draping. Slab rollback also results in a major pulse of horizontal extension in the overriding plate, as it is actually reported by GPS studies in the Aegean region [Nyst and Thatcher, 2004; Hollenstein et al., 2008; Rontogianni, 2010]. Large scale continental extension itself is responsible for the subsequent exhumation of high-pressure metamorphosed rocks and the formation of a metamorphic core complex in the area of the Cyclades islands in the southern Aegean.

Seismicity in the southern Aegean takes the form of both intermediate depth and shallow crustal earthquakes with magnitudes that range from small to moderate/large events. The deformation of the subducting slab causes intermediate depth earthquakes with hypocentral depths between 40-180 km that form a well-defined Wadati-Benioff zone of amphitheatrical shape [Papazachos and Nolet, 1997; Papazachos et al., 2000; Konstantinou and Melis, 2008]. Various deformation modes inside the slab may be responsible for this type of seismicity, such as tearing of the slab, slab-pull or along-strike compression [Rontogianni et al., 2011]. Shallow seismicity is usually restricted within the first 15-20 km and is associated with either interplate events around the Hellenic trench, or intraplate ones that occur along numerous faults. Piper and Perissoratis [2003] studied the Quaternary neotectonics of southern Aegean by interpreting a large number of seismic reflection cross-sections. The authors mapped active faults in eastern and southern Peloponnese, Crete, Rhodes and Kos islands while faults inside the Cretan Sea basin and the Cyclades were considered inactive. These conclusions are also supported by historical [Papazachos and Papazachou, 1997] and instrumental observations of seismicity (e.g. the global relocation database of Engdahl et al., 1998 and its subsequent updates).

The area of the southern Aegean has been struck several times in the past by large earthquakes of shallow origin, some of them also followed by tsunamis. A catalog of tsunamis that followed historical or instrumentally observed large earthquakes is also compiled by Papazachos and Papazachou [1997] for the southern Aegean. Most of them have intensities between II-III according to the Ambraseys-Sieberg scale [Ambraseys, 1962], however, a few have intensities that exceed IV. These larger intensities are observed for tsunamis affecting the coasts of western Crete and Rhodes. It can be concluded that the hazard stemming from tsunami activity in the southern Aegean is not as great as in other areas worldwide (e.g. Japan, south America), but it cannot be considered negligible either. On the other hand, intermediate depth earthquakes are usually too small (Mw < 4.0) to cause any significant damage even though there are notable exceptions such as the 8 January 2006 (Mw 6.7) earthquake near Kythira island [Konstantinou et al., 2006; 2009].

On 9 July 1956 a large earthquake occurred near the Cyclades island of Amorgos causing extensive damage to the nearby area. The event was followed by a tsunami which started from SE Amorgos and inundated the coasts of neighbouring islands also reaching the Turkish coast. This chapter aims first at providing a thorough review of previous studies about this earthquake in terms of its seismological characteristics, tsunami observations and modeling that are viewed in connection with other geological observations around Amorgos. All this information is then utilized in order to develop a seismotectonic interpretation for the occurrence of this event. First, we start with an introduction to the local geology and neotectonics of the area around Amorgos and then review the source parameters and macroseismic observations of the mainshock. This is followed by a description of the tsunami effects, such as run-up heights and its modeling which has been only recently published.

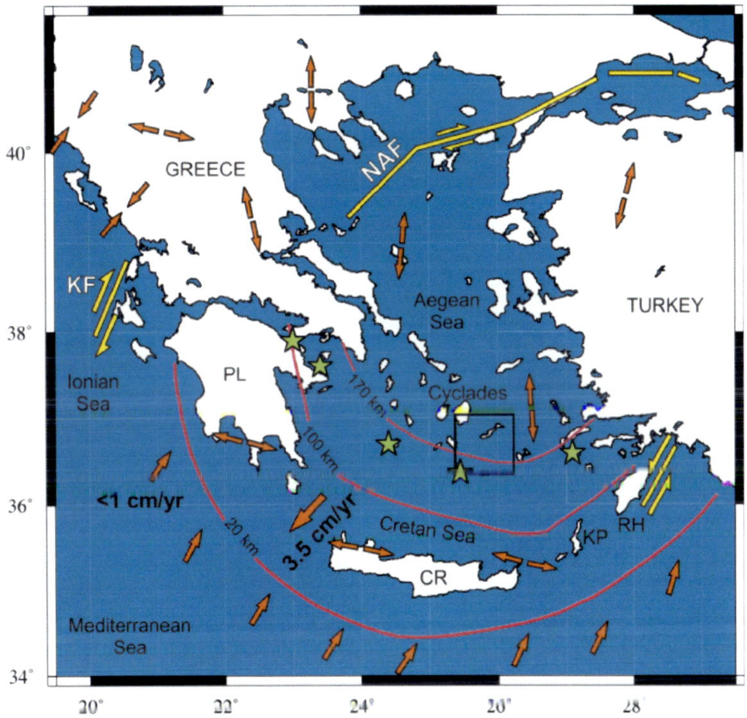

Figure 1. Map of the Greek region showing the main tectonic characteristics adopted from Papazachos and Papazachou [1997]. Single arrows indicate the plate motion direction along with average rate of convergence after McClusky et al. [2000]. Double arrows indicate areas where either extension or compression prevails. The curves for 20, 100, 170 km are isodepth curves showing the hypocentral distribution of earthquakes along the Wadati-Benioff zone. The green stars are locations of volcanic centers. The square shows the area of interest between Amorgos and Santorini islands. PL: Peloponnese, CR: Crete, KP: Karpathos, RH: Rhodes, KF: Kefalonia Fault, NAF: North Anatolian Fault.

We then examine the role played by crustal rheology in the area and how it might have influenced the nucleation depth of the Amorgos earthquake. These points give us enough hints to related its occurrence with pre-existing zones of weakness that were generated by the exhumation of the metamorphic core complex. Finally, we close this chapter with a conclusion regarding the seismotectonic setting of the Amorgos event.

AMORGOS ISLAND NEOTECTONICS

The geology of Amorgos island is characterized by high-pressure metamorphosed rocks (e.g. schists, quartzites, marbles etc), post-Triassic to Eocene carbonate rocks and middle-Eocene to Oligocene flysch. Pleistocene and alluvial deposits can be found only in very few locations. Papadopoulos and Pavlides [1992] have studied the brittle deformation structures on the island through field mapping and slickenside measurements (Figure 2). The results of their study show a master neotectonic fault with a strike N20°-30°E which coincides with the shoreline of the eastern part of Amorgos. Coastal morphology there reveals that this is the most active side of the island with the largest displacement and very steep shorelines. Such characteristics make this part of Amorgos inaccessible and completely uninhabited. This neotectonic fault has also the effect of causing a pronounced asymmetry in the river network, since the main branches of streams are nearly parallel to one another and directed to the northwest. Stress inversion of the slickenside measurements suggests that the mean stress axes correspond to a NW-SE trending main extensional stress.

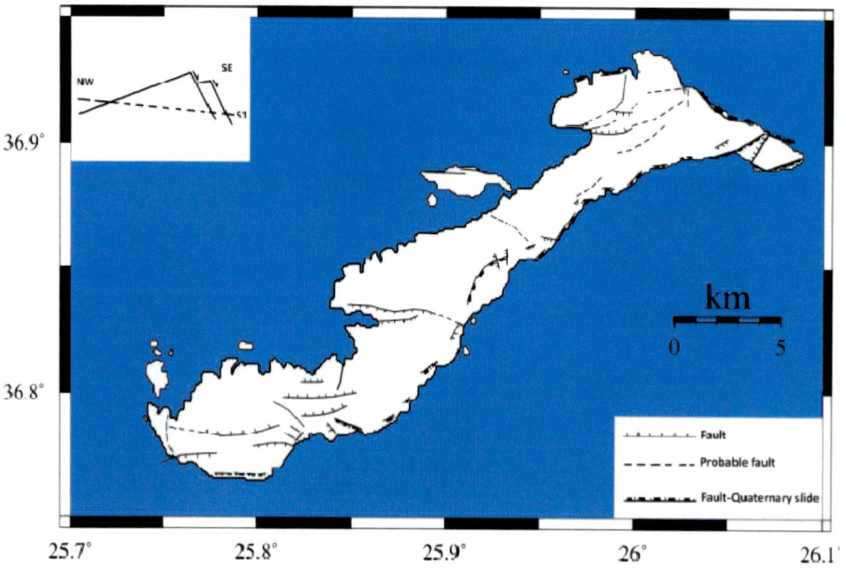

Figure 2. Map of the Amorgos island showing the main structural features observed by Papadopoulos and Pavlides [1992]. Faults were deduced from analysis of aerial photographs and neotectonic field observations. The inset at the upper left hand corner shows a cross-section along NW-SE direction highlighting the observed tilting of the island towards NW and the steep fault surfaces along the SE.

The aforementioned results imply that Amorgos can be regarded as a tilted and uplifted block of this master normal fault which controls the high-gradient relief of the eastern coast of the island (cf Figure 2 inset). In this case, the questions that are put forward is whether this uplift is a long-term feature and if so, what is the rate of uplift over the last few thousand years. Stiros et al. (1994) investigated these issues by using geomorphic and biological evidence along with archaeological findings. Coastal uplift is clearly demonstrated by the identification of marine biogenic detritus at a height of 200 m from the present sea level at sites along the SW coast. The stratigraphy of these sediments points to the possibility that their age is 1-2 Ma (middle to lower Pleistocene) which in turn corresponds to an uplift rate of 0.1-0.2 mm/yr. Additionally, buildings of Roman age can be found at several sites along the NW coast of Amorgos which have been submerged below the sea by more than 0.5 m. The authors assumed that the floors of these buidlings were constructed at least 1 m above the water level and inferred a minimum of 1.5 m relative subsidence in less than 2000 years, or a minimum subsidence rate of 0.75 mm/yr. From the description given so far, it is evident that the present coastline of Amorgos has recorded the occurrence of several episodes of tectonic activity over the Holocene with the generation of the 1956 earthquake being one of them.

SEISMOLOGICAL CHARACTERISTICS OF THE MAINSHOCK

The Amorgos earthquake occurred at a time period when the global seismic network was at its infancy, therefore information regarding its source parameters and focal mechanism was until recently rather ambiguous. The earthquake took place at 03:11 (GMT) or 05:11 local time which means that the event was felt at all sites in daylight since sunrise time was 04:55. Previously there were three published location estimates given in the catalogs of Galanopoulos [1982], Comninakis and Papazachos [1986] and Makropoulos et al. [1989]. As expected, the hypocentral depth is the most poorly determined parameter and the three catalogs report depths 10-15 km. Surface wave magnitude was calculated by the aforementioned authors and was found between 7.4-7.5 while magnitude determinations can also be found in Bath and Duda [1979] (Ms = 7.8) and Abe (1981) (Ms = 7.7). There were three published versions of the focal mechanism based on P-wave first motions by Papazachos and Delibasis [1969], Ritsema [1974] and Shirokova [1972] (Figure 3). The first two solutions exhibit strike-slip faulting with a small normal component and one nodal plane striking NE-SW which was identified as the fault plane. On the other hand, the third solution shows a purely normal fault with nodal planes oriented along the NE-SW direction contradicting the other two solutions. Papadopoulos and Pavlides [1992] argue that Shirokova's solution is the most compatible one with their field neotectonic observations.

In order to clarify these issues Okal et al. [2009] have undertook a careful study of the existing primary data related to the Amorgos event, such as travel times listed in the International Seismological Summary (ISS), the Bureau Central International de Seismologie (BCIS) and analog seismograms recorded at stations worldwide. In addition, travel times of the largest aftershocks and one alleged foreshock were also retrieved and processed. The authors relocated the mainshock and the other events using an interactive, iterative method. This method calculates for each event a 95% confidence ellipse through a Monte Carlo algorithm by injecting into the data randomly generated Gaussian noise. In this way, the

Figure 3. Focal mechanism solutions obtained from P-wave first motions for the 9 July 1956 Amorgos earthquake. Only the solution of Shirokova [1972] is consistent with the strike and sense of slip (normal) observed on the island by Papadopoulos and Pavlides [1992].

mainshock and 21 aftershocks were relocated producing refined epicentral locations (Figure 4). It can be seen that the relocated epicenter of the mainshock was shifted several kilometers towards the SW coast of Amorgos island when compared to previous locations. Focal mechanism solution was determined by utilizing digitized versions of available historical seismograms and using a method which inverts the spectral amplitudes of mantle Rayleigh and Love waves. The resulting focal mechanism shows a purely normal fault with two nodal planes dipping 25° and 67° respectively. Okal et al. [2009] note that this asymmetry of the focal mechanism (not apparent in Shirokova's solution) is a feature necessary in order to match the low amplitude of Love waves relative to the stronger amplitude of the Rayleigh waves. The inversion also revealed a seismic moment 3.9x1027 dyn-cm (corresponding to Mw = 7.4) and a centroid depth of 45 km, implying a deeper than expected source.

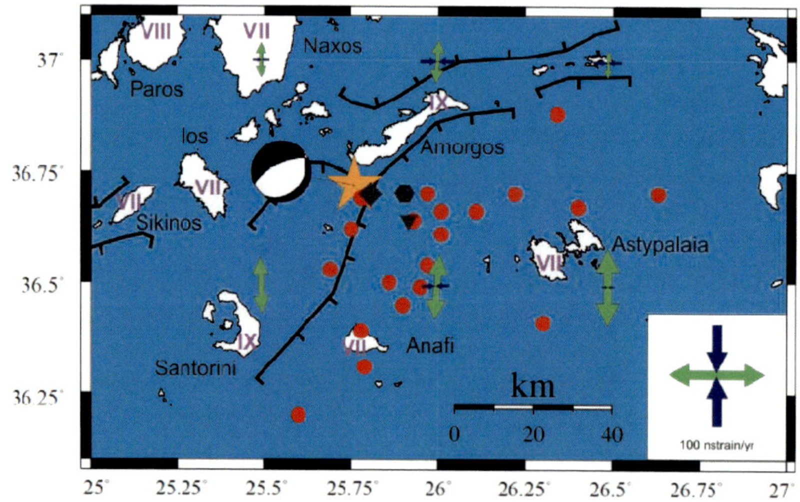

Figure 4. Map of the area where the 1956 Amorgos earthquake occurred. The star indicates the epicenter of the Amorgos mainshock and the red circles aftershocks epicenters determined by Okal et al. [2009]. The hexagon indicates the location given by Galanopoulos [1982], the diamond is the location reported by Comninakis and Papazachos [1986] and the inverted triangle is the location found in Makropoulos et al. [1989]. The beach ball represents the focal mechanism of the mainshock obtained after inversion of historical seismograms by Okal et al. [2009]. Roman numerals shown on the nearby islands are intensity estimates for the mainshock according to the Modified Mercalli intensity scale [from Schenkova et al., 2005]. Black lines indicate possible detachment faults adopted from Bohnhoff et al. [2006]. Double arrows represent principal axes of the strain rate tensor for this area determined from GPS data [Hollenstein et al., 2008].

The main body of information regarding the macroseismic effects of the Amorgos earthquake comes from observations collected by the National Observatory of Athens (NOA). Since 1893 NOA follows generally the same procedure in processing such observations which has two steps: first, printed questionnaires are directed to local authorities of towns and villages that are selected using as criteria their spatial distribution and density of population; second, the completed questionnaires are graded according to the Modified Mercalli intensity scale. As can be seen in Figure 4, the largest intensities were observed in northern Amorgos (IX), Santorini (IX) and Paros (VIII) while other nearby islands registered smaller values. These intensity values are derived from the database of Schenkova et al. [2005] and are consistent with those published by Papadopoulos and Pavlides [1992]. It is especially worthy to mention that Santorini sustained heavy damages and casualties in most of its towns as a result of (a) the occurrence of a large aftershock (Ms = 7.2) very near the island about 13 minutes after the mainshock, and (b) possible site effects owing to the soft sedimentary material of volcanic origin that most buildings were founded on.

TSUNAMI OBSERVATIONS AND MODELING

Few descriptions of the tsunami that followed the earthquake were previously available and they were mostly focused on second-hand rather than eye-witness accounts [e.g. Ambraseys, 1960; Papazachos et al., 1985]. Okal et al. [2009] performed the first systematic study of the Amorgos tsunami in terms of accumulating eye-witness observations from many different locations in the southern Aegean. The authors obtained testimonies of elderly witnesses and in this way they were able to map the exact extent of penetration of the wave using topographic methods in order to find inundation (i.e. horizontal extent of maximum penetration of the wave) at each location. It was also possible to obtain run-up heights where "run-up" is considered as the altitude above unperturbed sea level at the point of maximum penetration. Figure 5 shows the complete dataset that was accumulated for run-up heights of the waves at Amorgos itself and several other Greek islands [Astypalaia, Naxos, Folegandros, Santorini, Anafi, Nisyros, Tilos, Crete, Rhodes, Symi, Patmos, Leros, Kalymnos] as well as three locations in the SW Turkish coast. The largest values of run-up height are observed in the southern part of Amorgos (~20 m), Astypalaia (~10 m) and Folegandros (~15 m) islands while in other locations run-up varies between 1-4 m.

This pattern of observed run-up was utilized in order to constrain the results of hydrodynamic simulations using initially as seafloor displacement the static displacement of the double-couple source estimated previously. The modeling was carried out by solving the non-linear equations of hydrodynamics under the shallow water approximation using a finite difference algorithm [for more details see Okal et al., 2009]. The results showed that static displacement due to an earthquake source is not compatible with the observed run-up heights. Submarine landslides caused after the earthquake could also be responsible for the generation of the tsunami; Okal et al. [2009] tested this suggestion by assuming different ad hoc landslide models which led to acceptable fit between observed and simulated run-up heights. The submarine landslide scenario is also supported by two other considerations: first, that large run-up amplitudes are expected in the immediate vicinity of the landslide source (in this case in Amorgos, Folegandros and Astypalaia); second, interviews of eye-witnesses from the

Turkish coast sites revealed that the tsunami did not reach the coast there before late in the morning (10-11 am), which is at odds with the earthquake occurrence time (local time 05:11). A consequence of a landslide as the tsunami source is that the Amorgos earthquake probably did not rupture the seafloor. This means that its nucleation depth may be deeper than previously thought, however, it is difficult to assess this as centroid/hypocentral depths are not well-constrained. Lithospheric rheology plays a crucial role in controlling the maximum depth of seismogenesis and in the next section we will examine this factor with respect to the Amorgos event depth.

Crustal Rheology and the Amorgos Earthquake

In recent years, there has been a substantial debate over whether the strength of the continental lithosphere resides mostly in the upper mantle or in the seismogenic part of the crust. The former suggestion that is widely known as the "jelly sandwich (JS) model" predicts that continental lithosphere consists of the upper crust and mantle that are strong, separated by a weak lower crust [Chen and Molnar, 1983; Ranalli and Murphy, 1987; Handy and Brun, 2004; Burov and Watts, 2006]. This model has been used during the last three decades in order to explain the rheological stratification of continental lithosphere and the depth-frequency of earthquakes. The latter suggestion (dubbed as "créme brûlée (CB) model") appeared more recently when a re-assessment of focal depth distribution in southern Iran, Tibet and northern India showed an absence of seismicity in the continental mantle [Maggi et al., 2000a, b; Jackson, 2002; Priestley et al., 2008]. Based on this model, the lack of seismicity in the upper mantle implies that it is actually weak while most of the strength lies in the upper and lower crust that are seismogenic. However, taking into account that different continental lithospheres have different compositions and tectono-thermal history, it is more likely that these two models may be the end members of a whole range of rheological behavior [Afonso and Ranalli, 2004; Burov, 2010].

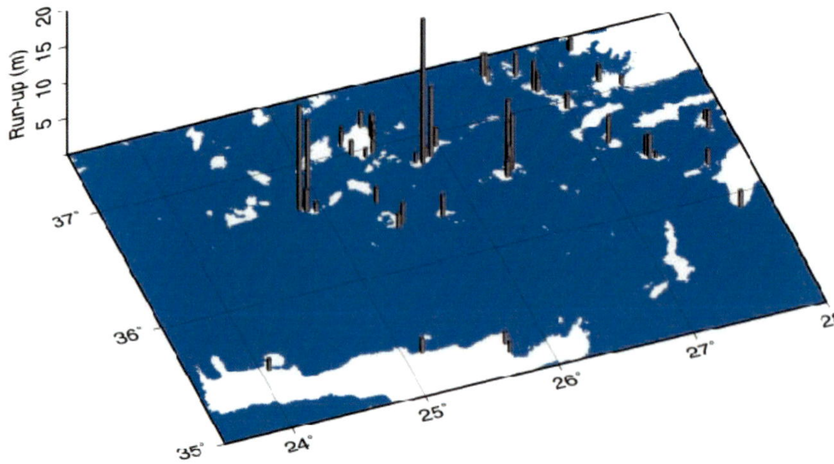

Figure 5. Map showing the spatial distribution of run-up heights for the tsunami that followed the Amorgos earthquake. The co-ordinates (latitude, longitude) and run-up height values are taken from the study of Okal et al. [2009].

The crustal and upper mantle structure of the southern Aegean and the area around Amorgos has been studied by numerous authors [Makris, 1978; Papazachos and Nolet, 1997; Sachpazi et al., 1997; Karagianni et al., 2005; Di Luccio and Pasyanos, 2007; Sodoudi et al., 2006; Endrun et al., 2008]. These studies indicate that the crust has a thickness of about 25 km with a relatively flat Moho topography (thickness differences within 1 km) and it is partitioned into upper and lower crust with their boundary located at 15 km depth. The seismic velocities of these layers and of the upper mantle can be correlated with laboratory determined velocities of known rock types compiled by Christensen and Mooney [1995]. Konstantinou [2010] did this correlation and found that the upper crust can be represented by quartzite, the lower crust by diabase and the upper mantle by pure peridotite (Figure 6).

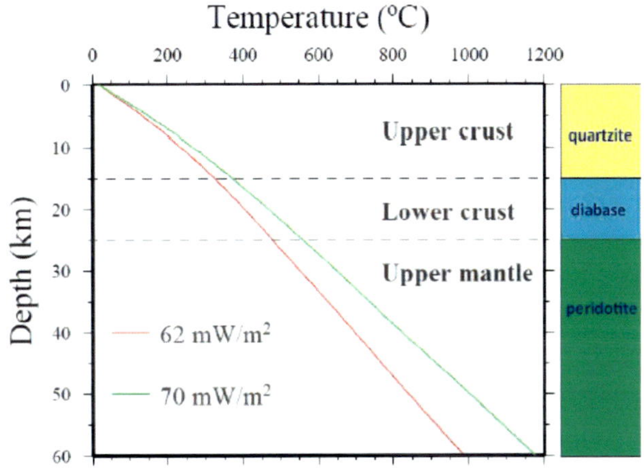

Figure 6. Diagram showing the one-dimensional variation of temperature as a function of depth for the focal area of the Amorgos earthquake. Two values of heat flow have been used in the calculations. The lithosphere was divided into three layers based on previous studies of the lithospheric structure for the area (see text for more details). Seismic velocities reported in these studies were correlated with rock types (quartzite, diabase, peridotite) in order to calculate rheological profiles [after Konstantinou, 2010].

The first step towards developing a thermo-mechanical model for the crust and upper mantle in any area is to have an estimate of how temperature varies as a function of depth. Geotherms are usually calculated by using an analytical solution to the one-dimensional heat conduction equation and taking into account the thermal properties of the crustal and upper mantle rocks [e.g. see Aldersons et al., 2003]. Heat flow is a critical parameter that influences significantly the geotherm curve, however, in most areas worldwide it is also the least constrained parameter. The southern Aegean is no exception to this and Figure 6 shows geotherms that correspond to two different values of heat flow, namely 62 mWm-2 reported by Jongsma [1974] and 70 mWm-2 suggested as a lower bound by Makris and Stobbe [1984]. These geotherms can be used in order to model the rheological behavior of the crustal and upper mantle layers through the estimation of Yield Strength Envelopes (YSEs). A rheological profile or YSE is a curve that depicts the distribution of differential stress as a function of depth and has been utilized extensively as an estimate of the mechanical strength of the lithosphere. Konstantinou [2010] has calculated YSEs for the area of the Amorgos earthquake based on the aforementioned geotherms and by assuming laboratory estimated parameters of brittle and ductile deformation for each rock type (quartzite, diabase, peridotite).

However, ductile deformation depends not only on the geotherm but also on whether the rocks can be considered as dry or wet. The upper crust in the area of the Amorgos earthquake lies underneath the sea therefore assuming a rheology of wet quartzite is the most reasonable choice. The geochemistry of erupted lavas from Santorini also point to a dry lower crust, since the original magma shows contamination of only upper crustal material implying no melting component in the lower crust [Zellmer et al., 2000]. Finally, dehydration reactions due to the subducting African slab suggest a wet upper mantle rather than a dry one. One way to check the consistency of the calculated YSEs is to compare them with the frequency-depth distribution of earthquakes. The only catalog that can provide earthquake locations around Amorgos is the one compiled by the National Observatory of Athens (NOA). The catalog spans several years and can be considered as much representative of the long-term strength of the lithosphere as possible. Konstantinou [2010] selected well-constrained events located around the rupture area of the 1956 earthquake based on number of observations (at least 5 stations), rms residuals of 1 s or smaller and location errors of 5 km or smaller. This yielded a set of 139 events which were augmented by 3 more locations obtained from the global relocation database of Engdahl et al. [1998] so the final number was 142 events (Figure 7).

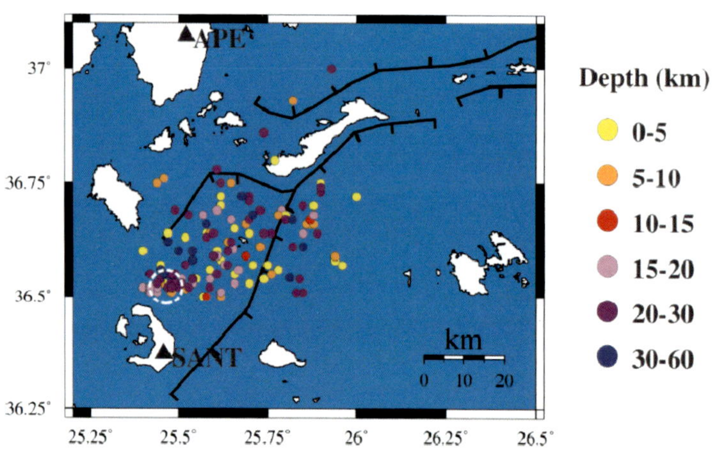

Figure 7. Map showing the distribution of event locations taken from the NOA catalog for the period January 2000 until September 2009. Hypocentral depths follow the color scale shown at the right hand side of the plot. Also shown are the locations (black triangles) of the two closest stations APE and SANT. The white dashed circle indicates the location of Coloumbo submarine volcano. Other symbols are the same as in Figure 4 [after Konstantinou, 2010].

Figure 8 shows the two calculated YSEs (for 62 and 70 mWm-2) along with a histogram describing the frequency-depth distribution of the selected earthquakes. Two peaks at about 5 and 25 km can be seen while there is a substantial decrease in earthquake activity between 7-15 km and below 35 km depth. The shape of the YSE that corresponds to the geotherm with heat flow 62 mWm-2 seems to fit the observed frequency-depth distribution reasonably well, if we take into account the inherent uncertainties involved in such calculations. In this YSE the upper mantle exhibits maximum strength at 33 km becoming more ductile at greater depths. The lower crust retains significant strength, but it is relatively weaker than the upper mantle confirming the JS model for the continental lithosphere. Based on the empirical relationships derived by Wells and Coppersmith [1994] the down-dip rupture width of the Amorgos event is 26 km. Such large earthquakes tend to nucleate near the base of the seismogenic layer and rupturing a large part of it. This means that the rupture may have extended from the depth of peak strength in the upper mantle at 33 km to 7 km upwards. A scenario like this agrees well with the tsunami modeling results where no seafloor rupture should be expected. On the other hand, the centroid depth of 45 km inferred by Okal et al. [2009] falls well within the ductile depth of the upper mantle therefore it appears that it is an overestimation of the true value which may lie at a much shallower depth.

Seismotectonic Interpretation

The 9 July 1956 Amorgos earthquake occurred at the edge of the Cyclades islands group which constitutes a block that exhibits very little internal deformation [Kreemer and Chamot-Rooke, 2004; Hollenstein et al., 2008; Rontogianni, 2010]. The geology of these islands is characterized by high-pressure metamorphism that was the result of underthrusting of continental crust fragments and their later exhumation. Exhumation took place when large-scale lithospheric extension commenced at 23-19 Ma creating metamorphic core complexes and the opening of the Aegean Sea basin [Ring et al., 2010]. Mechanically, exhumation is generally agreed that it occurs in extrusion wedges that are delineated by large-scale, normal faults known as detachments. Amorgos island in particular is also delineated by two of these detachments, one along its western and one along its eastern coast (cf Figure 4). As mentioned earlier, the detachment fault along the eastern coast is still active and causes the observed uplift that tilts the whole island towards the NW direction. This situation implies that this normal fault is high-angle which is at odds with the low-angle focal mechanism obtained by Okal et al. [2009].

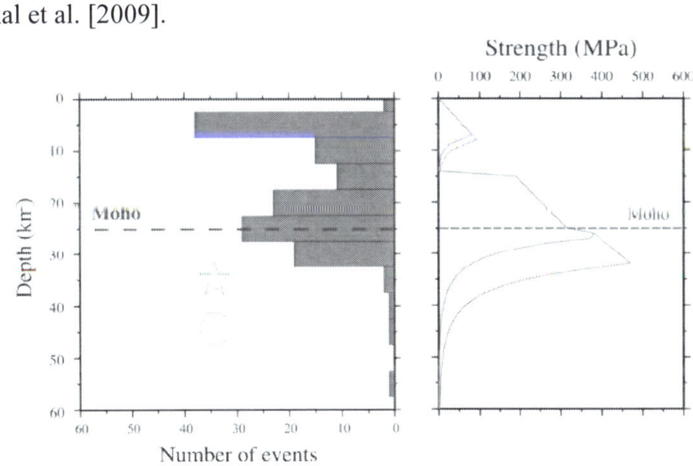

Figure 8. Comparison of rheological modeling results and the depth-frequency of earthquakes in the Amorgos area. Left hand side panel: Histogram showing the distribution of hypocentral depths for 142 events that occurred in the study area. The star indicates the hypocenter of the Amorgos event according to the Engdahl et al. [1998] relocation database (it should be noted that the depth is considered not well-constrained). The hexagon indicates the centroid depth for the 1956 earthquake reported by Okal et al. [2009]. Right hand side panel: Rheological profiles calculated using the lithosphere stratification shown in Figure 6 and two different values of heat flow (red curve: 62 mW m^{-2}, green curve: 70 mW m^{-2}) [after Konstantinou, 2010].

In order to reconcile these seemingly contradicting observations we have to examine more carefully the mechanical processes that created the metamorphic core complexes and the detachments in the first place. Recently, Tirel et al. [2008] used a fully coupled thermo-mechanical numerical simulation code that accounts for elastic-brittle-ductile properties of constituent rocks. The authors assumed a compositional (density) anomaly with the properties of granite, placed centrally in the crust to localize strain at the onset of deformation. During the first stages of the simulations the deformation pattern is a relatively symmetrical dome which is amplified by horizontal flow in the ductile crust. At later stages the dome becomes asymmetric (tilted to the one side) with the localization of a detachment zone along the one dome limb.

The model stemming from these simulations can be used in order to explain two main observations: (a) the high-angle normal fault and the NW tilt of Amorgos, (b) the low-angle normal faulting for the Amorgos earthquake reported by Okal et al. [2009] and its large nucleation depth. Figure 9 shows a sketch of the asymmetric dome and the detachment zones that were formed during the dome widening (exhumation) process. The detachment zone at the side that is uplifted is steeper at a considerable depth range but becomes more flat as it approaches the Moho; on the other side that is subsiding, the detachment zone has approximately an S shape and it flattens at a more gentle rate near the Moho. It should be noted that these detachment zones were not faults created by brittle failure of rock, but rather surfaces that accommodated the flow of lower crustal material upwards. These surfaces however, can function as pre-existing zones of weakness once the exhumation process is over and the rocks have cooled. The scenario that emerges therefore is that similar detachment

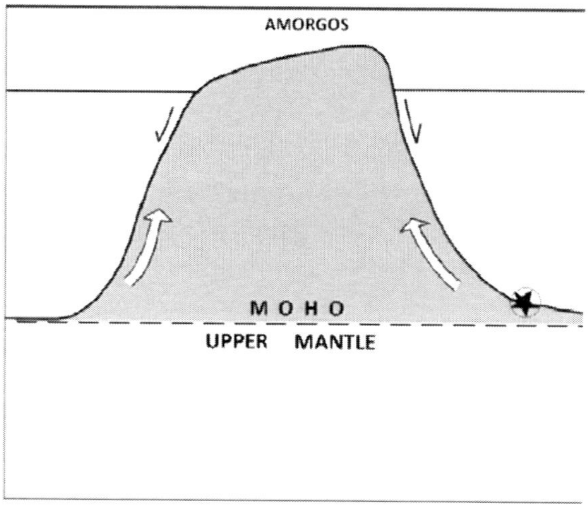

Figure 9. Cartoon illustrating the seismotectonic interpretation for the Amorgos earthquake. The gray area indicates the metamorphic dome that was exhumed 23-19 Ma and exhibits an asymmetric, tilted pattern [after Tirel et al., 2008]. The white arrows represent the flow of material along the two detachment faults at each side. The black arrows show the sense of motion of the two blocks and the black star indicates the possible nucleation location of the Amorgos event.

zones delineate Amorgos island and the 1956 earthquake nucleated near the point where the zone becomes flatter (i.e near or slightly below the Moho consistent with the crustal rheology results). This interpretation is also supported by a recent study of crustal and mantle anisotropy in the southern Aegean that shows that anisotropy in the lower crust is the result of fossil mineral fabric, remnant of the extension that accompanied the formation of the metamorphic core complexes [Endrun et al., 2011]. Such a scenario is consistent with the geological history of the Cyclades and can systematically explain all the aforementioned observations.

CONCLUSION

The southern Aegean as a closed sea basin is particularly susceptible to tsunami activity that may follow a large earthquake. The 1956 Amorgos event is an example of this, since the tsunami generated affected all the nearby islands and the Turkish coast. However, this chapter has highlighted two characteristics of this earthquake in terms of seismic and tsunami hazard: first, that the tsunami was not produced by the co-seismic displacement of the seafloor but it was rather the result of a submarine landslide. Second, that the event itself may have nucleated along a pre-existing zone of weakness (remnant of the metamorphic core complex exhumation) at a depth near or below the Moho. The first characteristic effectively means that it is extremely difficult to give evacuation warning in advance of a tsunami caused by a landslide, as its time of occurrence cannot be predicted. The second one makes the Amorgos earthquake different from usual crustal earthquakes and implies that such events originating in the lithospheric upper mantle may have much different recurrence timescales.

ACKNOWLEDGMENTS

We would like to thank the National Science Council of Taiwan for its financial support of this research in the form of a grant allocated to K.I. Konstantinou and a postdoctoral fellowship awarded to S. Rontogianni.

REFERENCES

Abe, K., 1981. Magnitudes of large shallow earthquakes from 1904 to 1980, *Phys. Earth Planet. In.*, 27, 72-92.

Afonso, J. C., G. Ranalli, 2004. Crustal and mantle strengths in contintental lithosphere: is the jelly sandwich model obsolete?, *Tectonophysics*, 394, 221-232, doi:10.1016/j.tecto.2004.08.006.

Aldersons, F., Z. Ben-Avraham., A. Hofstetter, E. Kissling, T. Al-Yazjeen, 2003. Lower-crustal strength under the Dead Sea basin from local earthquake data and rheological modeling, *Earth Planet Sci. Lett.*, 214, 129-142.

Ambraseys, N. N., 1960. The seismic sea wave of July 9, 1956, in the Greek archipelago, *J. Geophys. Res.*, 65, 1257-1265.

Ambraseys, N. N., 1962. Data for the investigation of the seismic sea waves in the eastern Mediterranean, *Bull. Seism. Soc. Am.*, 52, 895-913.

Bath, M., S. J. Duda, 1979. Some aspects of global seismicity, *Seismol. Inst. Uppsala Rep.*, 1-79.

Bijwaard, H. and W. Spakman, 1998. Closing the gap between regional and global travel time tomography, *J. Geophys. Res.*, 103, 30055– 30078.

Burov, E. B., 2010. The equivalent elastic thickness (Te), seismicity and the long-term rheology of continental lithosphere: Time to burn-out "créme brûlée"? Insights from large-scale geodynamic modeling, *Tectonophysics*, 484, 4-26, doi:10.1016/j.tecto.2009.06.013.

Burov, E. B., A. B. Watts, 2006. The long-term strength of continental lithosphere: `jelly sandwich' or "créme brûlée"? *G. S. A. Today*, 16, doi:10.1130/1052-5173(2006)016.

Chen, W.-P., P. Molnar, 1983. Focal depths of intracontinental and intraplate earthquakes and their implications for the thermal and mechanical properties of the lithosphere, *J. Geophys. Res.*, 88, 4183-4214.

Christensen, N. I., W. D. Mooney, 1995. Seismic velocity structure and composition of the continental crust: A global view, *J. Geophys. Res.*, 100, 9761-9788.

Comninakis, P. E., B. C. Papazachos, 1986. A catalog of earthquakes in Greece and the surrounding area for the period 1901-1985, Univ. Thessaloniki Geophys. Lab. Publ., 1.

Di Luccio, F., M. E. Pasyanos, 2007. Crustal and upper-mantle structure in the Eastern Mediterranean from the analysis of surface wave dispersion curves, *Geophys. J. Int.*, 169, 1139-1152, doi:10.1111/j.1365-246X.2007.03332.x.

Endrun, B., T. Meier, S. Lebedev, M. Bohnhoff, G. Stavrakakis, H.-P. Harjes, 2008. S velocity structure and radial anisotropy in the Aegean region from surface wave dispersion, *Geophys. J. Int.*, 174, 593-616, doi:10.1111/j.1365-246X.2008.03802.x.

Endrun, B., S. Lebedev, T. Meier, C. Tirel, W. Friedrich, 2011. Complex layered deformation within the Aegean crust and mantle revealed by seismic anisotropy, *Nature Geosci.*, 4, 203-207, doi:10.1038/NGEO1065.

Engdahl, E. R., Van der Hilst, R. and R. Buland, 1998. Global teleseismic earthquake relocation with improved travel times and procedures for depth determination, *Bull. Seism. Soc. Am.*, 88, 722-743.

Galanopoulos, A. G., 1982. The damaging shocks and the earthquake potential of Greece, *Ann. Geol. Pays Hellen.*, 30, 648-724.

Fiunicello, F., C. Faccenna, D. Giardini, K. Regenauer-Lieb, 2003. Dynamics of retreating slabs: 2. Insights from three-dimensional laboratory experiments, *J. Geophys. Res.*, 108(B4), 2207.

Handy, M. R., J.-P. Brun, 2004. Seismicity, structure and strength of the continental lithosphere, *Earth Planet Sci. Lett.*, 223, 427-441, doi:10.1016/j.epsl.2004.04.021.

Hollenstein, C., M. D. Müller, A. Geiger, H.-G. Kahle, 2008. Crustal motion and deformation in Greece from a decade of GPS measurements, 1993-2003, *Tectonophysics*, 449, doi:10.1016/j.tecto.2007.12.006.

Jackson, J., 2002. Strength of the continental lithosphere: Time to abandon the jelly sandwich?, *G. S. A. Today*, 12, 4-10, doi:10.1130/1052-5173(2002)012.

Jongsma, D., 1974. Heat flow in the Aegean area, Geophys. *J. R. astr. Soc.*, 37, 337-346.

Karagianni, E. E., C. B. Papazachos, D. G. Panagiotopoulos, P. Suhaldoc, A. Vuan, and G. F. Panza 2005. Shear velocity structure in the Aegean area obtained by inversion of Rayleigh waves, *Geophys. J. Int.*, 160, doi:10.1111/j.1365-246X.2005.02354.x.

Konstantinou, K. I., I. S. Kalogeras, N. S. Melis, M. C. Kourouzidis, and G. N. Stavrakakis 2006. The 8 January 2006 earthquake (Mw 6.7) offshore Kythira island, southern

Greece: Seismological, strong-motion and macroseismic observations of an intermediate depth event, *Seism. Res. Lett.*, 77, 540-549.

Konstantinou, K. I. and N. S. Melis, 2008. High-frequency shear-wave propagation across the Hellenic subduction zone, *Bull. Seismol. Soc. Am.*, 98, 797-803, doi:10.1785/0120060238.

Konstantinou, K. I., S.-J Lee, C. P. Evangelidis, N. S. Melis, 2009. Source process and tectonic implications of the 8 January 2006 (Mw 6.7) Kythira earthquake, southern Greece, *Phys. Earth Planet. In.*, 175, doi:10.1016/j.pepi.2009.03.010.

Konstantinou, K. I., 2010. Crustal rheology of the Santorini–Amorgos zone: Implications for the nucleation depth and rupture extent of the 9 July 1956 Amorgos earthquake, southern Aegean, *J. Geodyn.*, 50, 400-409, doi:10.1016/j.jog.2010.05.002.

Kreemer, C., N. Chamot-Rooke, 2004. Contemporary kinematics of the southern Aegean and the Mediterranean ridge, *Geophys. J. Int.*, 157, 1377-1392, doi:10.1111/j.1365-246X.2004.02270.x.

Le Pichon, X., C. Kreemer, 2010. The Miocene-to-Present kinematic evolution of the Eastern Mediterranean and Middle-East and its implications for dynamics, *Ann. Rev. Earth Planet. Sci.*, 38, 323-351, doi:10.1146/annurev-earth-040809-152419.

Maggi, A., J. A. Jackson, D. MacKenzie, K. Priestley, 2000a. Earthquake focal depths, effective elastic thickness and the strength of the continental lithosphere, *Geology*, 28, 495-498.

Maggi, A., J. A. Jackson, K. Priestley, C. Baker, 2000b. A re-assessment of focal depth distributions in southern Iran, the Tien-Shan and northern India: Do earthquakes really occur in the continental mantle?, *Geophys. J. Int.*, 143, 629-661.

Makropoulos, K. C., J. K. Drakopoulos, J. B. Latousakis, 1989. A revised and extended earthquake catalog for Greece since 1900, *Geophys. J. Int.*, 98, 391-394.

Makris, J., 1978. The crust and upper mantle of the Aegean region from deep seismic soundings, *Tectonophysics*, 46, 269-284.

Makris, J., C. Stobbe, 1984. Physical properties and state of the crust and upper mantle of the eastern Mediterranean Sea deduced from geophysical data, *Mar. Geol.*, 55, 347-363.

McClusky et al., 2000. Global positioning system constraints on plate kinematic and dynamics in the Eastern Mediterranean and Caucasus, *J. Geophys. Res.*, 105, 5695-5719.

Nyst, M., Thatcher, W., 2004. New constraints on the active tectonic deformation of the Aegean, *J. Geophys. Res.*, 109, B11406, doi:10.1029/2003JB002830.

Okal, E. A., C. E. Synolakis, B. Uslu, N. Kalligeris, E. Voukouvalas, 2009. The 1956 earthquake and tsunami in Amorgos, Greece, *Geophys. J. Int.*, 178, 1533-1554, doi:10.1111/j.1365-246X.2009.04237.x.

Papazachos, C., and G. Nolet, 1997. P and S deep velocity structure of the Hellenic area obtained by robust nonlinear inversion of travel times, *J. Geophys. Res.*, 102, 8349-8367.

Papazachos, B. C., N. D. Delibasis, 1969. Tectonic stress field and seismic faulting in the area of Greece, *Tectonophysics*, 7, 231-255.

Papazachos, B. C., C. Koutitas, P. M. Hatzidimitriou, B. G. Karakostas, C. A. Papaioannou, 1985. Source and short-distance propagation of the July 9, 1956 southern Aegean tsunami, *Mar. Geol.*, 65, 343-353.

Papadopoulos, G. A., S. B. Pavlides, 1992. The large 1956 earthquake in the south Aegean: Macroseismic field configuration, faulting, and neotectonics of Amorgos island, *Earth Planet. Sci. Lett.*, 113, 383-396.

Papazachos, B. C., Papazachou, K., 1997. The earthquakes of Greece, Ziti editions, Thessaloniki.

Papazachos, B. C., V. G. Karakostas, C. B. Papazachos, and E. M. Scordilis, 2000. The geometry of the Wadati-Benioff zone and lithospheric kinematics in the Hellenic arc, *Tectonophysics*, 319, 275-300.

Piper, D. J. W., C. Perissoratis, 2003. Quaternary neotectonics of the south Aegean arc, *Mar. Geology*, 198, 259-288, doi:10.1016/S0025-3227(03)00118-X.

Priestley, K., J. Jackson, D. McKenzie, 2008. Lithospheric structure and deep earthquakes beneath India, the Himalaya and southern Tibet, *Geophys. J. Int.*, 172, 345-362, doi:10.1111/j.1365-246X.2007.03636.x.

Ranalli, G., D. C. Murphy, 1987. Rheological stratification of the lithosphere, *Tectonophysics*, 132, 281-295.

Ring, U., J. Glodny, T. Will, S. Thompson, 2010. The Hellenic subduction zone: High-pressure metamorphism, exhumation, normal faulting, and large-scale extension, *Ann. Rev. Earth Planet. Sci.*, 38, 45-76, doi:10.1146/annurev-earth-050708-170910.

Ritsema, A. R., 1974. Earthquake mechanisms of the Balkan region, R. Neth. *Meteorol. Inst. Sci. Rep.*, 74(4), 36 pp.

Rontogianni, S., 2010. Comparison of geodetic and seismic strain rates in Greece by using a uniform processing approach to campaign GPS measurements over the interval 1994–2000. *J. Geodyn.*, 50, 381-399, doi:10.1016/j.jog.2010.04.008.

Rontogianni, S., K. I. Konstantinou, N. S. Melis, C. P. Evangelidis, 2011. Slab stress field in the Hellenic subduction zone as inferred from intermediate-depth earthquakes, *Earth Planets Space*, 63, 139-144, doi:10.5047/eps.2010.11.011.

Sachpazi, M., A. Hirn, A. Narcessian, F. Avedik, J. McBride, M. Loucoyannakis, R. Nicolich, 1997. A first coincident normal-incidence and wide-angle approach to studying the extending Aegean crust, *Tectonophysics*, 270, 301-312.

Schenkova, Z., I. Kalogeras, V. Schenk, R. Pichl, M. Kourouzidis, G. Stavrakakis, 2005. Atlas of isoseismal maps of selected Greek earthquakes (1956-2003), Evonymos Ecological library, Athens.

Shirokova, E. I., 1972. Stress pattern and probable motion in the earthquake foci of the Asia-Mediterranean seismic belt, In: L. M. Balakina et al. (Eds), Elastic strain field of the Earth and mechanisms of earthquake sources, Nauka, Moscow.

Sodoudi, F., Kind, R., Hatzfeld, D., Priestley, K., Hanka, W., Wylegalla, K., Stavrakakis, G., Vafidis, A., Harjes, H.-P., Bohnhoff, M., 2006. Lithospheric structure of the Aegean obtained from P and S receiver fucntions, *J. Geophys. Res.*, 111, B12307, doi:10.1029/2005JB003932.

Spakman, W., S. Van der Lee, R. Van der Hilst, 1993. Travel-time tomography of the European-Mediterranean mantle down to 1400 km, *Phys. Earth Planet. In.*, 79, 3-74.

Stiros, S. C., L. Marangou, M. Arnold, 1994. Quaternary uplift and tilting of Amorgos island (southern Aegean) and the 1956 earthquake, *Earth Planet. Sci. Lett.*, 128, 65-76.

Tirel, C., J.-P. Brun, E. Burov, 2008. Dynamics and structural development of metamorphic core complexes, *J. Geophys. Res.*, 113, B044403, doi:10.1029/2005JB003694.

Wells, D. L., Coppersmith, K. J., 1994. New empirical relationships among magnitude, rupture length, rupture width, rupture area, surface displacement, *Bull. Seism. Soc. Am.*, 84, 974-1002.

Zellmer, G., S. Turner, C. Hawkesworth, 2000. Timescales of destructive plate margin magmatism: new insights from Santorini, Aegean volcanic arc, *Earth Planet Sci. Lett.*,174,265-281.

In: Earthquakes
Editor: Kostas Konstantinou
ISBN: 978-1-62081-883-1

Chapter 4

Characteristics of Interaction between Interplate Earthquakes from the View of Multi-Scale Simulations

Keisuke Ariyoshi and Yoshiyuki Kaneda*
Japan Agency for Marine-Earth Science and Technology, Japan

Abstract

We review previous numerical simulations of interaction between fault segments on a plate interface based on a friction law, comparing coupled events with single events on the amount of coseismic and preseismic slip. We find out the characteristics of coupled event:

(i) In case of multiple fault segments nearby enough to have short delay time of rupturing, coseismic slip is increased nearly in proportion to the size of coupled fault segment while preseismic slip is almost the same as single event.
(ii) In case of those distant enough to have long delay time of rupturing, seismic moment magnitude is almost the same as single event while preseismic slip is significantly increased by several tens percents.

We also investigate the changes in slip rate and recurrence interval of repeating earthquakes using 3-D numerical simulations of a subduction zone involving large and small asperities. Our results reveal that the postseismic slip of a large earthquake can trigger an event at a nearby small asperity, although the event may differ to other repeating earthquakes at the asperity because the slip velocity of the event tends to be slower than that of other repeaters; the slip distribution also differs. We discuss several coupled earthquakes in Sumatra with short-term and long-term delay times, and point out that the slip amount estimated from repeating earthquake analyses tends to be smaller than that estimated from GPS analyses for large amounts of postseismic slip, as observed for postseismic activity following the 2003 Tokachi-oki earthquake. These results may be applied to the postseismic slip off Kanto and off Sanriku after the 2011 off the Pacific

* Earthquake and Tsunami Research Project for Disaster Prevention, Japan Agency for Marine-Earth Science and Technology, Yokohama 236-0001, Japan.

Coast of Tohoku Earthquake, which may trigger megathrust aftershocks in the near future. In southwest Japan, Dense Oceanfloor Network System for Earthquakes and Tsunamis (DONET) may be useful to detect the pre-seismic change of a possible M9 class coupled megathrust earthquake composed of Tokai, Tonankai, Nankai and Hyuga-nada asperities.

INTRODUCTION

On March 11, the 2011 off the Pacific Coast of Tohoku Earthquake occurred in Japan, whose magnitude (Mw 9.0) was one of the greatest mega-thrust earthquakes in world history (Figure 1). As suggested by The Headquarters for Earthquake Research Promotion [2011], it is thought that the area with large slip covers a part of the region along the trench off-shore south of Sanriku and the region along the trench off-shore north of Sanriku through Boso, and it is also thought that the focal area furthermore includes the middle region off-shore of Sanriku and the regions off-shore of Miyagi, Fukushima and Ibaraki prefectures.

On the 2004 Sumatra Andaman earthquake, whose moment magnitude was estimated to be 9.1 by real-time analysis [United States Geological Survey, 2004] and 9.3 by considering duration time as long as 6 to 7 minutes [e.g., Ishii et al., 2005], there are thought to be three divided fault segments (Sumatra, Nicobar and Andaman) ruptured from Sumatra to Andaman [Lay et al., 2005].

As those mega-thrust earthquakes, we refer to the phenomenon that nearby earthquakes occur after time-lag significantly shorter than recurrence interval as "coupled earthquakes".

CHARACTERISTICS OF COUPLED EARTHQUAKES

Considering two models proposed by Ariyoshi et al. [2009, 2011] categorized the characteristics of the coupled earthquakes into two models:

(i) Model of slip proportional to fault size, and
(ii) Characteristic slip model. In the following sections, we review the two proposed model.

MODEL OF SLIP PROPORTIONAL TO FAULT SIZE

Model of slip amount (D) proportional to fault size has two types: "L-model" proportional to fault length (L) [Scholz, 1982], and "W-model" proportional fault width (W) [Romanowicz, 1992]. In case of proportional relation between W and L, Both L-model and W-model are the same relation as similarity model with stress drop ($\Delta\sigma$) constant [Kanamori and Anderson, 1975]. For coupled earthquakes, however, it is possible that the aspect ratio (L/W) of fault does not always keep constant as discussed later. In recent study, it is proposed that $W \propto L^{2/3}$, where $D \propto L^{5/6}$, $D \propto W^{5/4}$, $\Delta\sigma \propto L^{1/6}$, $\Delta\sigma \propto W^{1/4}$ [Leonard, 2010], which can explain the observed relationship between D, W, L and $\Delta\sigma$ better than usual models such as W-model and L-model.

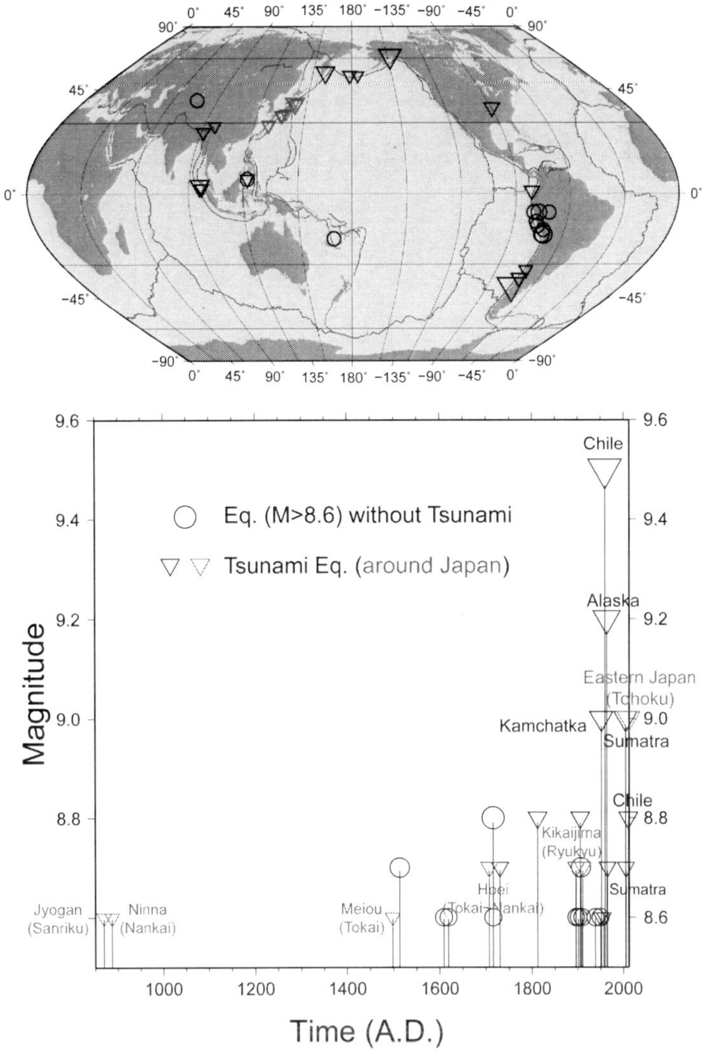

Figure 1. (top) Map of epicenters and (bottom) time series for giant earthquakes (M>8.6) originally revised from Global Significant Earthquake Database [National Geophysical Data Center, 2011] with major plate boundaries (trench: blue, ridge: purple, transform: orange) determined by Coffin et al. [1998]. "▽" represents tsunami earthquakes around (red color) and apart from (black color) Japan, while "□" represents earthquakes without tsunami. The original magnitude is chosen from the available magnitude scales in this order: M_w, M_s, M_b, M_l, and M_{fa}. In these Figures, their database is revised on the magnitude for some earthquakes around Japan [Ariyoshi et al., 2011].

Characteristic Slip Model

In characteristic slip model, coupled earthquake keeps slip amount same as single earthquake (Figure 3). In other words, slip amount is independent of fault size (W and L) even if several earthquakes on the same fault occur simultaneously, which was observed in the North Anatolian fault [Kondo et al., 2005] and the 1992 Landers earthquake [Sieh et al., 1996].

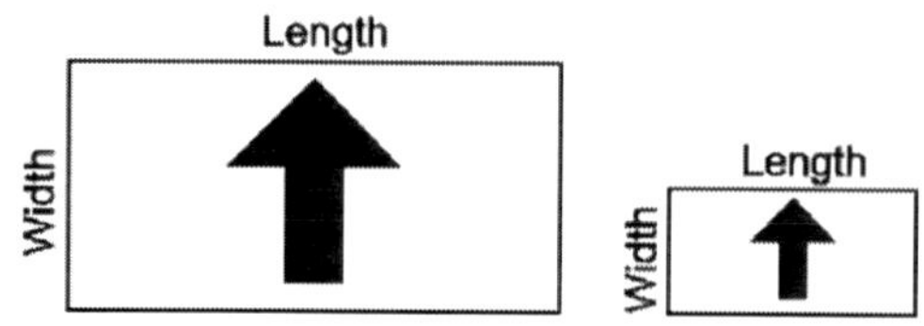

Figure 2. Schematic illustration of "model of slip proportional to fault size" in case of $L = 2W$ [e.g., Kanamori and Anderson, 1975]. Bold arrow size represents slip amount (D).

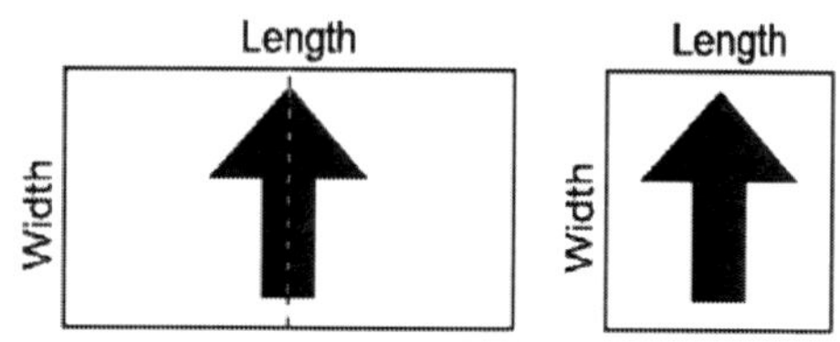

Figure 3. Schematic illustration of characteristic slip model by comparing between coupled earthquake (left) and single earthquake (right).

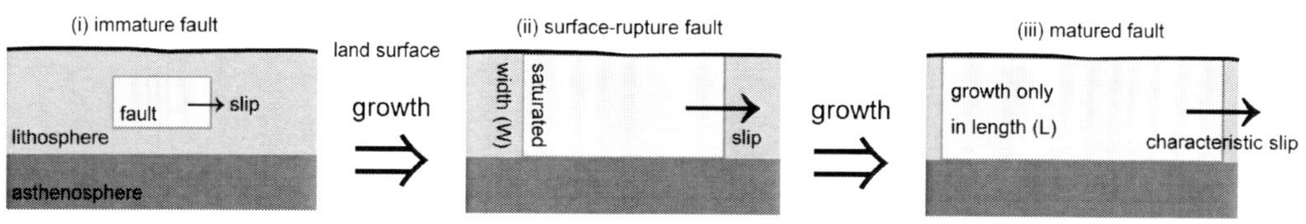

Case ofindicate te that ble toteristic slip model is applicable to (ii) and (iii). matured fault which has the same slip amoun.

Figure 4. Schematic illustration of fault growth process: (i) immature fault, (ii) fault with saturated width, and (iii) matured fault with the same slip amount as (ii). In case of "W-model", characteristic slip model is applicable between (ii) and (iii) because of failure of seismic slip to occur in asthenosphere.

For inland earthquakes with lateral faults and high dip angle, the upper limitation of fault width tends to significantly lower than fault length [Romanowicz, 1992]. This is probably because seismic slip decays abruptly in asthenosphere. It is thought that matured active faults often have deeper part of fault reaching asthenosphere (Figure 4). This means that fault width can not develop for deeper part any more, and keeps constant width in case of large earthquakes. Since keeping constant fault width makes constant slip amount for "W-model", "characteristic slip model" may be applied to matured active faults from the view of "W-model" [e.g., Fujii and Matsu'ura, 2000].

An Important Subject on Coupled Interplate Earthquakes

For trench-type megathrust earthquakes, the upper limitation of fault width may be high enough to avoid saturation because of low dip angle. For examples, fault geometry (width × length) of the 2011 off the Pacific Coast of Tohoku Earthquake is thought to be 500km × 200km [Japan Meteorological Agency, 2011], and sub-fault geometries of the 2004 Sumatra

Andaman earthquake are 420km × 240km for Sumatra segment, 325km × 170km for Nicobar segment and 570km × 160km for Andaman segment, respectively [Lay et al., 2005].

By treating the three sub-faults of the 2004 Sumatra Andaman earthquake as one giant fault, its aspect ratio does not keep constant as obeyed in the "model of slip proportional to fault size". On the other hand, the "characteristic slip model" would not quantitatively explain 20 meters of its maximum slip amount.

In this chapter, we discuss the validity of both the "model of slip proportional to fault size" and "characteristic slip model" in some cases on the basis of numerical simulations, revealing the unknown characteristics and give a road map for earthquake prediction.

A Model of Non-Similar Small Repeating Earthquakes

Since recurrence interval of trench-type megathrust earthquake such as the 2011 off the Pacific Coast of Tohoku Earthquake (about 1000 years) and the 2004 Sumatra Andaman earthquake is so long, it is actually difficult to compare the last event with previous ones in order to understand perturbation or interaction process. Therefore, it is reasonable to focus on small repeating earthquakes as observed off Kamaishi [Igarashi et al., 2003; Uchida et al., 2004]. In this section, we introduce numerical simulation study of small repeating earthquakes and interpret observed phenomena accompanied with nearby large interplate earthquakes on the basis of the simulation results.

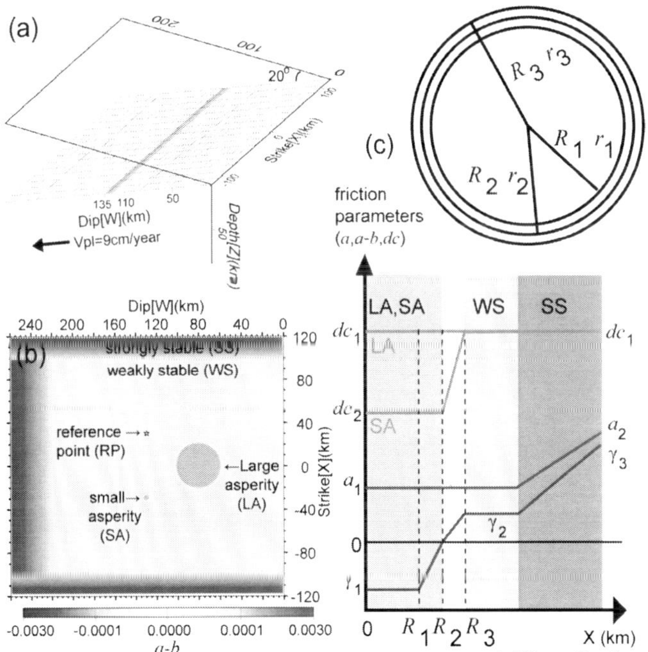

Figure 5. A model of small repeating earthquake perturbed by large earthquakes [Ariyoshi et al., 2007]. (a) Schematic diagram of the 3-D simulation model used in this chapter. (b) Spatial distribution of $\gamma = a - b$. Four regions are shown: LA and SA ($\gamma << 0$), WS ($\gamma \geq 0$), SS ($\gamma >> 0$), and RP ($\gamma \geq 0$), corresponding to the mirror image point of SA with respect to the symmetry axis ($X = 0$). (c) Frictional parameters (a, γ, d_c) as functions of distance from the center of LA (R) or SA (r).

Figure 5 shows a three-dimensional (3-D) subduction plate boundary model of small repeating earthquakes perturbed by large characteristic earthquakes occurring in nearby a large asperity. The model considers a planar plate interface dipping at 20 degrees from the free surface in a half-space homogeneous elastic medium (Figure 5a). The plate interface is divided into 512 (strike) x 380 (dip) cells, the slip of which is assumed to have only a pure dip component and to obey the quasi-static equilibrium between shear and frictional stresses. The equations used in the simulation are as follows.

$$\tau_i^{\text{dislocation}} = \sum_{j=1}^{N} K_{ij}(u_j(t) - V_{\text{pl}}t) - \frac{G}{2\beta}\frac{\mathrm{d}u_i}{\mathrm{d}t}, \quad (1)$$

$$\tau_i^{\text{friction}} = \mu_i \sigma_i, \quad (2)$$

$$\tau_i^{\text{dislocation}} = \tau_i^{\text{friction}}. \quad (3)$$

Here, the subscripts i and j denote the cell locations of an observation and a source, respectively. Equation (1) describes the stress in the i-th cell caused by dislocations, where V_{pl} is the relative speed of the two plates, t denotes time, N is the total number of cells, G is rigidity, β is the shear wave speed, and K_{ij} is the Green's function for shear stress [Okada, 1992] on the i-th cell. K_{ij} is calculated from the quasi-static solution for uniform pure dip-slip u relative to average slip $V_{\text{pl}}t$ [Savage, 1983] over a rectangular dislocation in the j-th cell. The first term of the right-hand side is similar to the form of convolution. The fast Fourier transform can be used for analysis of the along-strike component because the Green's function K_{ij} for the strike component is dependent on the relative distance between the observation and source points [Rice, 1993]. The last term in Eq. (1) is introduced to incorporate radiation damping [Rice, 1993]. Equations (2) and (3) represent the frictional stress and quasi-static equilibrium condition, respectively. In Eq. (2), the effective normal stress σ is given by $\sigma_i(z) = (\rho_{\text{rock}} - \rho_{\text{w}})gz$, where ρ_{rock} and ρ_{w} are the densities of rock and water, respectively, g is acceleration due to gravity, and z is depth. The friction coefficient μ in Eq. (2) is assumed to obey a rate- and state-dependent friction law [Dieterich, 1979; Ruina, 1983], as given by

$$\mu_i = \mu_0 + a_i \log(V_i/V_0) + b_i \log(V_0 \theta_i/d_{ci}), \quad (4)$$

$$\mathrm{d}\theta_i/\mathrm{d}t = 1 - V_i \theta_i/d_{ci}, \quad (5)$$

where a and b are friction coefficient parameters, d_c is the characteristic slip distance associated with b, θ is a state variable for the plate interface, V is slip velocity, and μ_0 is a reference friction coefficient defined at a constant reference slip velocity of V_0. The five equations above are solved using the Runge–Kutta method with adaptive step-size control [Press et al., 1992]. The constant parameters common to all models tested here are as listed in Table 1.

Frictional stability is usually represented by $\gamma = a - b$. If $\gamma > 0$, frictional stress increases with sliding velocity, similar to viscous coupling. If $\gamma < 0$, frictional stress decreases with increasing sliding velocity, resulting in stick-slip behavior [Ruina, 1983]. In the present chapter, an asperity denotes a region with $\gamma < 0$, following *Boatwright and Cocco* [1992]. The plate interface is demarcated into four regions, as shown in Figure 5: (i) large asperity (LA), (ii) small asperity (SA), (iii) weakly stable region (WS), and (iv) strongly stable region (SS). A reference point (RP) is defined in WS, corresponding to the mirror image point of SA with respect to the symmetry axis ($X = 0$). For the parameters shown in Figure 5c are based on rock laboratory results [e.g., Blanpied et al., 1998] and given the following values: Radii for LA, $(R_1, R_2, R_3) = (20, 22, 24)$ [km]; radii for SA, $(r_1, r_2, r_3) = (2, 3, 4)$ [km]; $(a_1, a_2) = (2, 5)$ [$\times10^{-3}$]; frictional stability parameters, $(\gamma_1, \gamma_2, \gamma_3) = (-7.2, 1, 49)$ [$\times10^{-4}$]; and characteristic slip distances, $(d_{c1}, d_{c2}) = (4, 1.35)$ [$\times10^{-2}$ m]. The sizes of the cells h_i in the asperities are set such that all cells in LA and SA are at least 5.7 times smaller than the critical cell size $h^* = \eta G d_c/\sigma(b - a)$, where $\eta = K_{ii}h_i/G$ is a geometrical factor [*Rice*, 1993].

Table 1. The values of constant parameters common to all models

Rigidity.	G	30 GPa	Rock density.	ρ_{rock}	2.75×10^3 kg/m^3
S-wave velocity.	β	3.75 km/s	Water density.	ρ_w	1.0×10^3 kg/m^3
Gravity acceleration.	g	9.8 m/s^2	Reference slip vel.	V_0	1 μm/s
Poisson's ratio.	ε	0.25	Reference friction.	μ_0	0.6

Follows nt parameters common to all models from the view of askes around Japan.

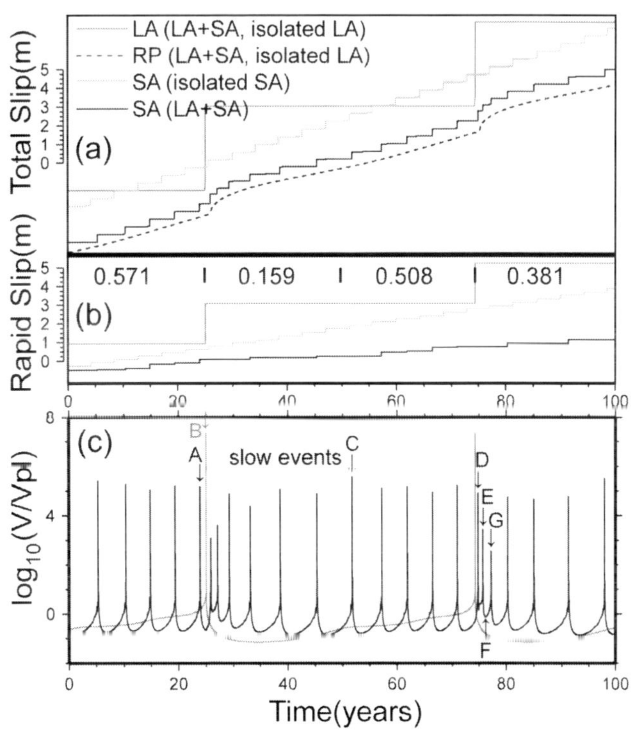

Figure 6. Cyclic temporal history of (a) cumulative slip and (b) rapid slip (>1 mm/s) at the center of LA (cyan), SA in the LA+SA model (black), isolated SA (magenta), and RP (dashed). The numerals represent the amount of rapid slip (meter) for every quarter cycle. (c) Average slip-velocity histories for the areas of LA (cyan) and SA (black). Labels A–G correspond to the snapshots shown in Figure 7. Yellow time windows represent periods during which time large postseismic slip associated with events at LA pass near SA. This figure is derived from Ariyoshi et al. [2007].

Simulation of Non-Similar Small Repeating Earthquakes

We first describe a single-asperity model in which LA or SA exists alone. Hereafter, we refer to this model as the 'isolated LA/SA model'. The cumulative slip histories at the centers of the LA and SA regions are shown in Figure 6a as cyan and magenta lines, respectively. Cumulative slip histories calculated for rapid slip ($V_c > 10^{-3}$ m/s) are plotted in Figure 6b. Recurrence interval (T_r in year) and magnitude (M_w) based on seismic moment calculated from slip with $V > 10V_c$ are constant in both the isolated LA model ($T_r = 49.3$ y and $M_w = 7.3$) and the isolated SA model ($T_r = 4.3$ y and $M_w = 3.9$). The regularity evident in the isolated asperity system has also been reported in previous studies [e.g., Kato, 2004].

Next we investigate the effect of interaction between LA and SA by comparing the model shown in Figure 5b (hereafter termed the 'LA+SA model') with the isolated LA/SA models. The slip history at the center of the LA region in the LA+SA model is found to be the same as that in the isolated LA. This result indicates that the stress change due to SA is negligible at LA. In contrast, the slip histories at the center of SA are clearly different in the isolated SA and LA+SA systems, as shown in Figure 6a and 6b. The recurrence interval and magnitude fluctuate in the LA+SA case ($T_r = 4.9 \pm 0.6$ y and $M_w = 4.0 \pm 0.1$), but the overall pattern shown in Figure 6 repeats with an interval of 98.6 years. Figure 6b shows that cumulative rapid slip at the center of SA for the LA+SA system is substantially smaller than that for the isolated SA system. There is a possibility that this small amount of rapid slip might be due to the difference in the spatial slip distribution. Even if the slip velocity is slow at the center of the asperity, the slip at the edge of the asperity might be fast. In this case, the event generates seismic waves. Therefore, the average slip velocity over the area of LA or SA is introduced in order to ascertain whether slow slip events really occur at the asperity. Figure 6c shows the average slip velocity normalized with respect to V_{pl} on a common-logarithmic scale for the LA+SA case. In addition to seismic events (2 events at LA and 19 at SA), 4 slow events at SA are also recognized, in which the slip velocity does not exceed 10 V_c anywhere (indicated by the yellow background in Figure 6c).

Figure 7 shows snapshots of the slip velocity for LA+SA at the times labeled A–G in Figure 7c. These data indicate that the characteristics of the event at SA vary considerably according to the stage (seismic, postseismic, healing, or preseismic) of LA. Specifically, slow slip events (shots E and G in Figure 7c) are generated when the postseismic slip of the event at LA passes near SA. The locked area in the interseismic stage of SA (shot F) is smaller than usual (shot B), as shown by the high average velocity for SA (close to V_{pl}) in Figure 6c. These results suggest that the recurrence interval of events at SA is reduced when large aseismic slip (such as postseismic slip caused by a nearby large earthquake) passes near SA; the events tend to become slow events because insufficient time is available in which to heal the fault. The seismic area colored with dark red or black in Figure 7, just after the occurrence of LA event (shot D), is different from that of other seismic events (shots A and C) despite the similar peak value of average slip velocity (see Figure 6c). This result indicates that the passage of postseismic slip causes not only slow slip events but also non-similar seismic events with contrasting slip velocity distributions. Comparing shots A and C, we find that the recurrence intervals of SA events in the interseismic stage of LA (e.g., shot C) are longer than those in the preseismic stage (e.g., shot A). This result suggests that the recurrence interval of the events at SA varies even in the interseismic period of LA, as the change in slip velocity in the surrounding region reflects the stage of LA.

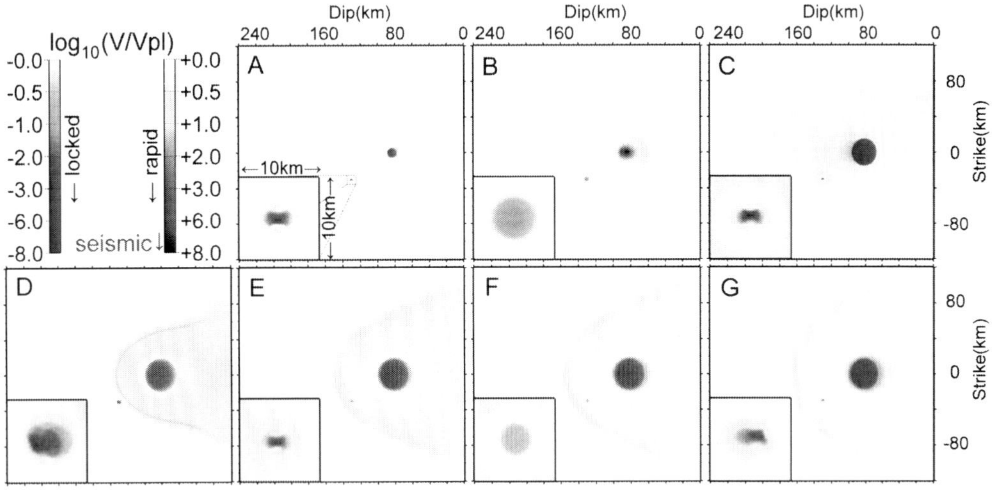

Figure 7. Snapshots of slip velocity for the LA+SA model at the times labeled A to G in Figure 6c after Ariyoshi et al. [2007]. The inset in each snapshot is a close-up of the slip velocity distribution in and around SA.

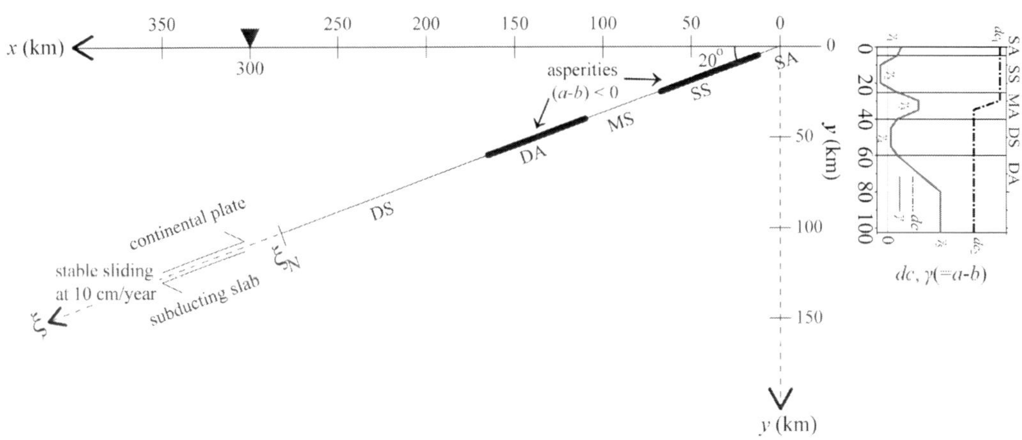

Figure 8. A 2-D model of a subduction zone. (Left Figure) The two thick lines indicate seismogenic segments. The plate interface deeper than 103 km is assumed to slip at a constant velocity relatively between the continental and oceanic plates (10cm/year). (Right Figure) Friction parameters of rate- and state-dependent law dc and γ (= a–b) employed in the simulation. The horizontal axis (y) represents depth from the surface. The values of dc and γ are (dc1, dc2) = (5.0, 3.0) [×10–2 m] and (γ1, γ2, γ3, γ4, γ5) = (+0.2, –0.8, +1.0, –0.3, +2.0) [×10–3]. Segments labeled SA, SS, MA, DS, and DA represent the "shallow aseismic segment", "shallow seismogenic segment", "middle aseismic segment", "deep seismogenic segment", and "deep aseismic segment", respectively. These figures are revised from Ariyoshi et al. [2009].

Since small repeating earthquakes are thought to be used as slip indicators with negligible stress perturbation due to its existence, it is reasonable that RP shown in Figure 5b should be treated as the background total slip rate surrounding SA. Figure 6a shows that the cumulative slip at RP (dashed lines) delineates the cumulative slip at SA (black solid line) for the LA+SA model. The rapid slip amounts for every quarter cycle are attached on the top of Figure 6b, and show that the ratios of rapid/total slip are not constant but significantly low in the postseismic stage of LA. This result suggests that missing events may exist because of worse cross-correlation between slow and usual events in the conventional analyses of repeating earthquakes when the change in the background slip rate is large. Therefore, a comparison of the analyses of repeating earthquakes and GPS data as observed on the 2003 Tokachi earthquake [e.g., Fukuda et al., 2004] is very important in estimating the spatio–temporal distribution of aseismic slip.

A SIMULATION OF CHARACTERISTIC SLIP MODEL

In the last section, we review an example of perturbation on recurrence interval, seismic moment magnitude and rupture initiation point, in addition to temporal slow slip events triggered by the passage of postseismic slip of large interplate earthquake. The reason why slow slip events occur is partly because the postseismic slip propagation front and the distance between the large and small asperities is much longer than the size of the small asperity, which is not applied the case with each size of seismogenic segment is approximately same with spatial interval of same scale. In this section, we show another example of interaction along dip direction as seen off Miyagi prefecture, northeast Japan.

In this simulation, we divide the plate boundary into five segments as shown in Figure 8, according to the sign of γ [e.g., Ariyoshi et al., 2007] : (i) a shallow aseismic segment (SA); (ii) shallow seismogenic segment (SS); (iii) middle aseismic segment (MA); (iv) deep seismogenic segment (DS); and (v) deep aseismic segment (DA). The spatial distributions of the frictional parameters are illustrated in Figure 8, where $(d_{c1}, d_{c2}) = (5.0, 3.0)$ [$\times 10^{-2}$ m] and $(a, \gamma_1, \gamma_2, \gamma_3, \gamma_4, \gamma_5) = (+2.0, +0.2, -0.8, +1.0, -0.3, +2.0)$ [$\times 10^{-3}$] based on laboratory experimental results [e.g., Blanpied et al., 1998].

The geophysical interpretations of the different segments are as follows: (i) low rates of seismicity at the plate boundary beneath shallow accretionary prisms [e.g., Pacheco et al., 1993; Kimura et al., 2007]; (ii and iv) two seismogenic regions as seen off Miyagi Prefecture, Japan [Kawakatsu and Seno, 1983]; and (iii) the seismic gap hypothesized to stem from the existence of frictionally stable ophiolitic basement at intermediate depths as observed in Nicaragua [Walther et al., 2000] and northeastern Japan [Yamamoto et al., 2008]. We will discuss the validity of these interpretations in Section 5.3. The basis for segment (v) is that no inter-plate seismicity deeper than around 60 km has been observed in northeastern Japan [Igarashi et al., 2001]. Additionally, laboratory experiments show that γ changes from negative to positive at temperatures exceeding 600°C and becomes more positive at higher temperatures for olivine [Boettcher et al., 2007], which is consistent with typical thermal structures [e.g., Peacock, 2001; Hacker et al., 2003].

SLIP BEHAVIORS OF CHARACTERISTIC SLIP MODEL

Figure 9 displays the spatiotemporal evolution of slip on the plate boundary from 100 days before the onset of a slip event to 100 days after the event. The figure reveals that post-seismic slip for events occurring after short T_s (s1 and d4) is significantly larger than for other events. To examine the effects of fault interaction, we classify events into three types based on T_S: (A) events after short T_s (s3, d1); (B) events after with T_s (s1, d4); and (C) others (s2, d2, d3). The amount of cumulative post-seismic slip over three years is found to be (A) > (B) > (C), with the amounts for type (A) events several times larger than those for (C). Since M_0' and the pre-seismic slip values (white areas) of the (A) events, s1 and d4, are also larger than for the other events in each seismogenic segment, these results suggest that events occurring after short T_s intervals produce more slip in each of the pre-, co-, and post-seismic stages.

When we treat d1+s1 and s3+d4 as coupled events, the slip amounts of the sub-events are expected to be twice those of single (non-coupled) events (d2, s2, d3) on the basis of the constant stress-drop model [Liu-Zeng, et al., 2005; Kanamori and Anderson, 1975], but the same as those of the coupled event in the characteristic slip model [Kondo et al., 2005]. For co-seismic slip, the difference of the seismic moment between the following and non-coupled events in each seismogenic segment is as much as 13% for SS (s1/s2 = 1.13) and 36% for DS (d4/d2 = 1.36), meaning that co-seismic slip approximately obeys the characteristic slip model. For post-seismic slip, Figure 9 shows that the difference is about 2–4 times for both SS (s3/s1 = 1.89, s3/s2 = 4.08) and DS (d1/d4 = 1.77, d1/d2 = 3.27), implying that the constant stress drop model is more appropriate. This result holds true even when we change the initiation time of the post-seismic stage within a range of 6 hours to one day.

In Figures 9c, 8d and 8e, gray curves represent slow slip events triggered by post-seismic slip of seismic events. The delay in post-seismic slip initiation between SS and DS is caused by the time required to cross the intervening MA portion. The amount of aseismic slip in SS following d2 (gray curve) is about twice that following d3 in spite of their having similar amounts of post-seismic slip in DS. This difference probably reflects differences in T_s (59.2 and 30.8 years, respectively), because the slip deficit accumulated in SS should be proportional to T_s while the co-seismic slip is characteristic.

From white portion of slip distribution in Figure 9, the amplification of post-seismic slip is also applied to pre-seismic slip by the interaction. Therefore, characteristic slip mode makes seismic slip "characteristic" while aseismic slip including pre-seismic and post-seismic slip "amplified" when nearby large earthquake occur on the same plate boundary.

DISCUSSION

In case of slip proportional to fault size, Kato [2008] formulated two adjacent large asperities reproducing the 1968 Tokachi-oki earthquake (M_w 8.2) breaking both the asperities and the 1994 Sanriku-oki earthquake (M_w 7.8) breaking one characteristic asperity [Nagai et al., 2001]. *Ariyoshi et al.* [2011] pointed out that seismic slip of the 1968 earthquake is approximately twice of the 1994 earthquake, considering that the ratio of seismic moment is 8.2/7.8 = 0.4 and the asperity area ratio is almost 2 because both asperity has nearly same area. On the other hand, Kato [2008] showed the simulation result that preseismic slip amount for the 1968 earthquake is nearly the same as the 1994 earthquake. These results suggest that a triggering nearby earthquake in case of "model of slip proportional to fault size" largely affects only on co-seismic slip, not on pre-seismic slip.

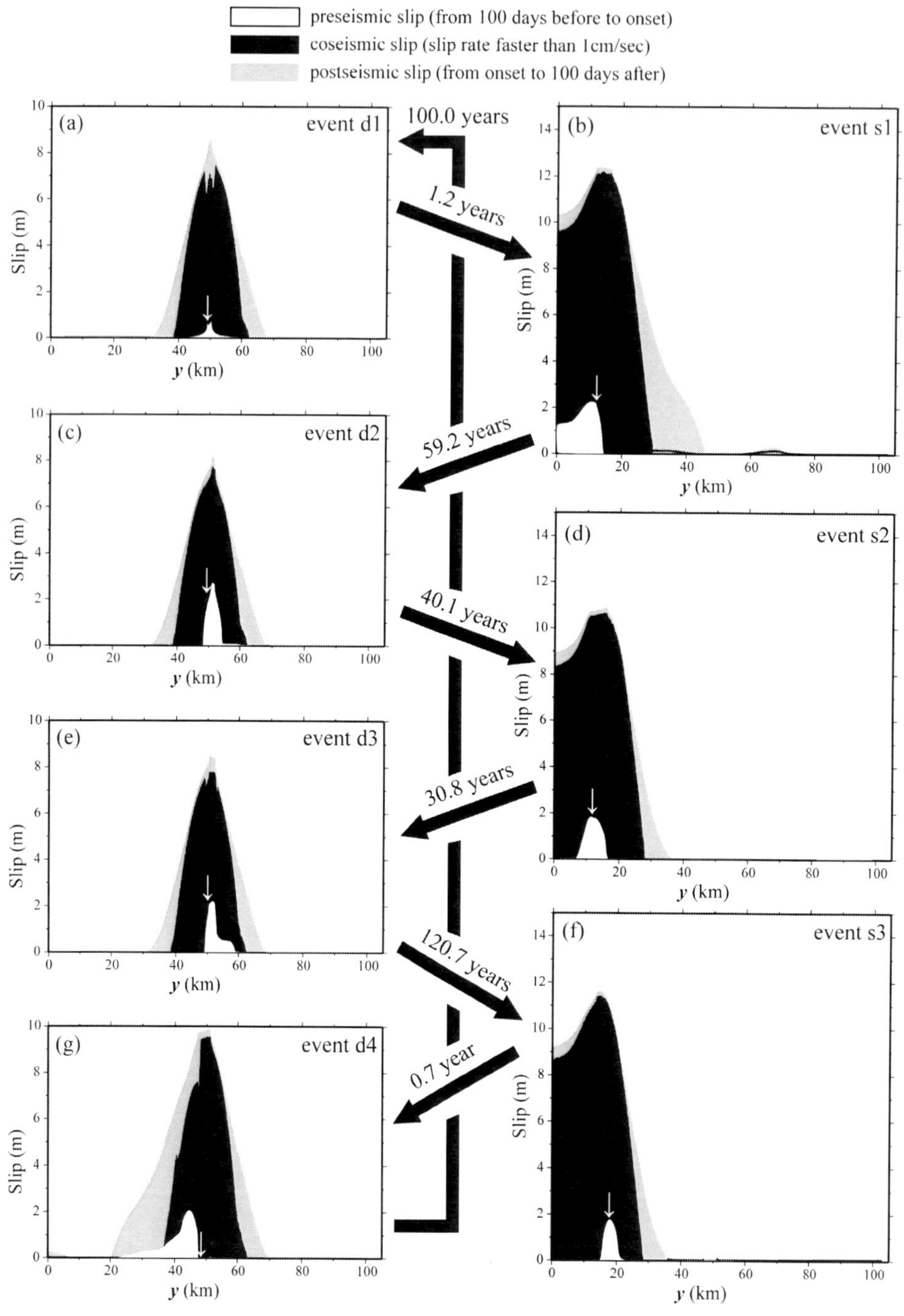

Figure 9. Spatiotemporal slip evolution on the plate boundary from 100 days before to 100 days after each event. Pre-seismic slip is defined as the total slip from 100 days before to the onset of co-seismic slip; co-seismic slip as slip faster than 1 cm/sec; and post-seismic slip as the total slip from onset to 100 days after. This figure is derived from Ariyoshi et al. [2009].

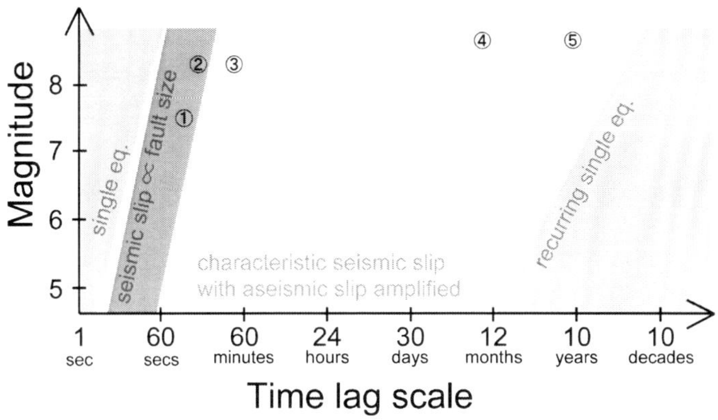

Figure 10. A schematic relation between time lag, magnitude and slip model for the triggered earthquake of coupled earthquakes developed from Ariyoshi et al. [2011]. Encircled numbers represent recently observed examples; (1) the 1968 Tokachi earthquake [Nagai et al., 2001], (2) Nicobar segment of the 2004 Sumatra Andaman earthquake [Lay et al., 2005], (3) Andaman segment of the 2004 Sumatra Andaman earthquake [Lay et al., 2005; Bilham et al., 2005], (4) the 2004 Nias earthquake [Briggs et al., 2006; Wiseman and Bürgmann, 2011], (5) the 2007 Southern Sumatra earthquake [Konca et al., 2008; Wiseman and Bürgmann, 2011].

Table 2. Summary of slip components for the three segments ruptured by the 2004 Sumatra earthquake. Transit time represents time elapsed from the origin time of the 2004 Sumatra earthquake. D_{seis} and D_{slow}, represents slip amounts for co-seismic and aseismic (slow) component, respectively [Lay et al., 2005]. D_{single} represents coseismic slip amount for single event based on previous researches [Bilham et al., 2005] for comparison [Ariyoshi et al., 2011]

Segment	D_{seis} (Transit time)	D_{slow} (Transit time)	D_{single}
Sumatra	7 m (0-50 sec)	not resolved	not known
Nicobar	5 m (230-350 sec)	5 m (230-3500+ sec)	2.7±0.3 m
Andaman	< 2 m (350-600 sec)	5 m (600-3500+ sec)	2-3 m

Figure 10 shows a schematic relation between slip model and time lag of coupled earthquake on the basis of magnitude for the following (triggered) earthquake. As an example, we apply some coupled earthquakes regarding to the 2004 Sumatra Andaman earthquake into the relation in Figure 10. Table 2 summarized slip components for the three sub-faults of the 2004 Sumatra earthquake which ruptured three seismogenic segments along trench — the Sumatra, Nicobar, and Andaman segments — and the average amount of co-seismic slip in each segment was roughly estimated at about 7 m, 5 m and <2 m, respectively [Lay et al., 2005]. The fault size factor of the coupled Nicobar and Sumatra segments relative to the Nicobar segment alone is about 2.35 (where the respective sizes (trench × dip direction) of the Sumatra and Nicobar segments are 420×240 km^2 and 325×170 km^2, respectively, and [{420+325 km} × {240+170 km} / {325 km × 170 km}]$^{0.5}$ ~ 2.35) and the co-seismic slip amount of single-event earthquakes rupturing the Nicobar segment is estimated to be 2.7±0.3 meters based on the 1881 earthquake [Bilham et al., 2005].

This implies that the co-seismic slip amount of the Nicobar segment is approximately proportional to fault size and, therefore, the stress drop model is preferable in describing interaction between the Sumatra and Nicobar segments. On the other hand, the observed co-seismic slip amount for the Andaman portion of the rupture is roughly the same as in the 1941 Andaman earthquake (2–3 m, Bilham et al. [2005]) and post-seismic slip is substantial (~5 m, Lay et al. [2005]), meaning that the characteristic slip model is preferable to account for interaction between the Nicobar and Andaman segments.

The 2005 Nias earthquake adjacent to the Sumatra segment occurred about three months after the 2004 Sumatra Andaman earthquake and was followed by the 2007 Southern Sumatra earthquake [e.g., Wiseman and Bürgmann, 2011; Konca et al., 2008] easterly adjacent to the source region of the Nias earthquake. Both of their post-seismic slips were amplified in western part [Hsu et al., 2006], which were the same as directions of after-slip (post-seismic slip) arrivals [Wiseman and Bürgmann, 2011; Konca et al., 2008]. On the 2005 Nias earthquake (M_w 8.6), its seismic magnitude was largely equal to the 1861 Nias earthquake (M8.5) which occurred as single event, which is largely equal to the 2005. On the 2007 Southern Sumatra earthquake (M_w 8.5), its magnitude was neither as much as previous events occurred in 1833 (M8.9) nor moment magnitude expected from slip deficit [Konca et al., 2008]. These results also imply that the 2005 Nias earthquake and the 2007 Southern Sumatra earthquake were applied to the "characteristic slip model".

Conclusion

Figure 11 shows asperity map of major trench-type megathrust earthquakes around Japan with spatial distribution of co-seismic slip for the 2011 off the Pacific Coast of Tohoku Earthquake. Focusing on the 2011 off the Pacific Coast of Tohoku Earthquake, we find that its source region covers some major asperities (M7~8) which overlap or neighbour each other. Unfortunately, it is well-known that pre-seismic change such as crustal deformation and seismicity has not been observed significantly. These observational results may suggest that the 2011 off the Pacific Coast of Tohoku Earthquake is composed of several asperities which behave as the "model of slip proportional to fault size" as mentioned above. In other words, pre-seismic change of the 2011 off the Pacific Coast of Tohoku Earthquake may be as small as M7 class earthquakes as observed for the 1968 Tokachi-oki earthquake [Kato, 2008]. In off Kanto and Boso, southward of the 2011 off the Pacific Coast of Tohoku Earthquake, there are several major source regions including the 1923 great Kanto earthquake (M7.9) and the 1953 off Boso peninsula earthquake (M7.4) which generated large tsunami. Since both the earthquakes are far away from the source region of the 2011 off the Pacific Coast of Tohoku Earthquake, both the southward major earthquake may behave as the "characteristic slip model". This suggests that their pre-seismic changes may be amplified so as to be detected by oceanfloor observations such as acoustic GPS [Kido et al., 2006] and/or repeating earthquakes [Uchida et al., 2004].

On the megathrust earthquakes along the Nankai Trough, it is thought that Tokai, Tonankai and Nankai earthquakes may occur in the near future and some researchers have pointed out that Hyuga-nada earthquake may be triggered by the M9 class coupled earthquakes composed of the three megathrust earthquakes [e.g., Furumura et al., 2011].

However, size of asperities composing the possible M9 class coupled earthquakes along the Nankai Trough is significantly larger than that of the 2011 off the Pacific Coast of Tohoku Earthquake which may be composed of M7 class as shown in Figure 11. This suggests that pre-seismic change of the possible M9 class coupled megathrust earthquakes along the Nankai Trough may be larger and is expected to be as large as the 1944 Tokai earthquake with detectable pre-seismic change reported by some researchers [e.g., Sato, 1977; Mogi, 2004]. Therefore, real-time monitoring of crustal deformation and seismicity is essential for us to detect the pre-seismic change in advance.

Figure 12 shows an overview of Dense Oceanfloor Network System for Earthquakes and Tsunamis (DONET) toward an anticipated Tonankai Earthquake. DONET has a submarine cabled real-time seafloor observatory network for the precise earthquake and tsunami monitoring. For the purpose of understanding and forecasting the earthquake and related activities underneath the seafloor, the twenty sets of state-of-arts submarine cabled sub-sea measurement instrument will be deployed in seafloor at the interval of 15-20km. All of the twenty sets of preliminary interface have been installed just on July 31, 2011 and are to be prepared in consideration of the improvement of observation capability in the future.

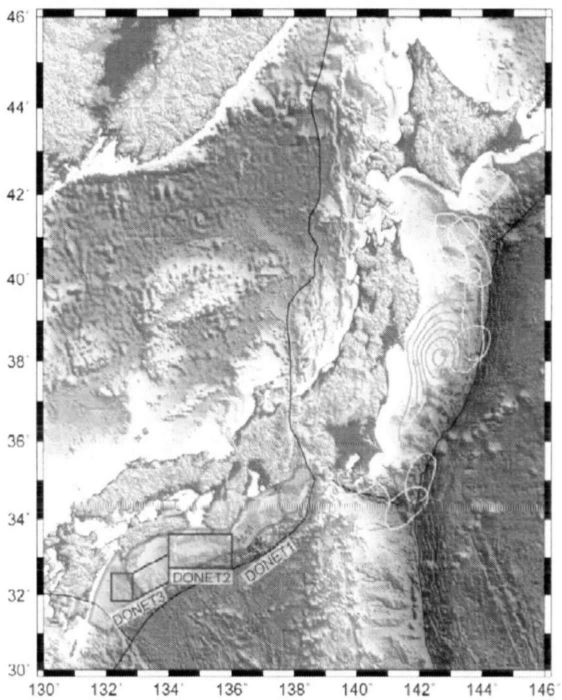

Figure 11. Map of major asperities around Japan. Black curves are plate boundaries [*Peter*, 2003]. Orange contours represent co-seismic slip of the 2011 off the Pacific Coast of Tohoku Earthquake [Geospatial Information Autority of Japan, 2011] as interval of 4 meters. Yellow ellipses are the estimated source regions of past megathrust earthquakes (excluding outer-rise earthquakes) around the 2011 off the Pacific Coast of Tohoku Earthquake [The Headquarters for Earthquake Research Promotion, 2002]. Pink regions from east to west along Nankai trough represent the seismogenic zones of Tokai, Tonankai, Nankai and Hyuganada earthquakes [The Headquarters for Earthquake Research Promotion, 2004]. Purple filled circles and open rectangle regions represent observation points of DONET 1 and regions of DONET 2 and 3, respectively.

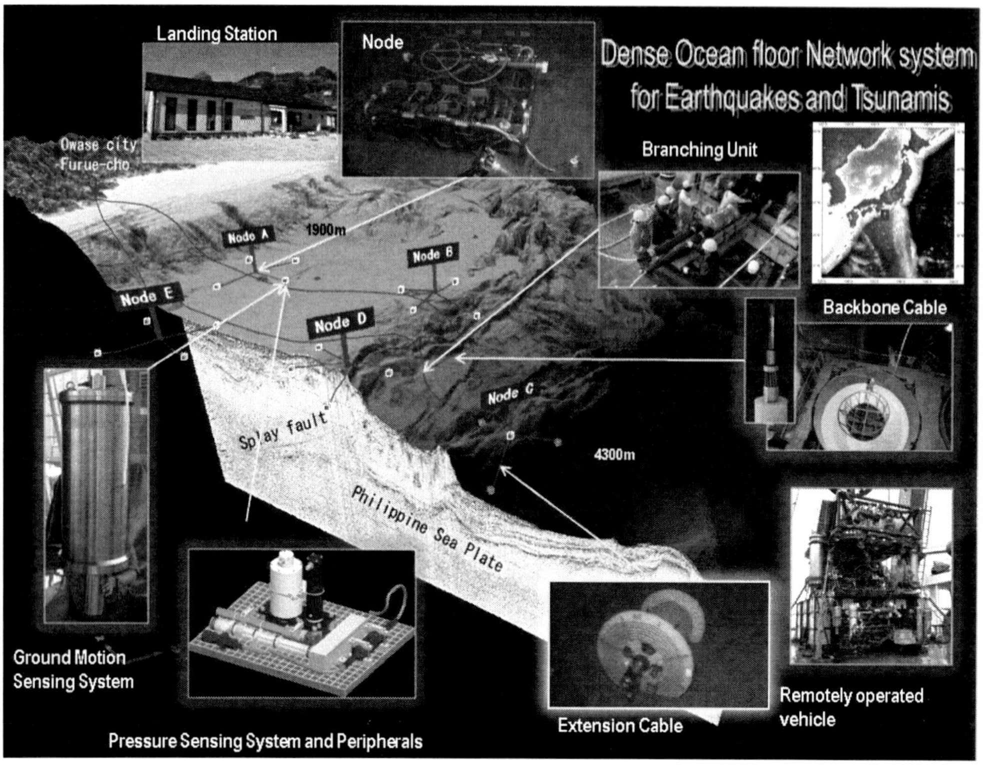

Figure 12. An overview of Dense Oceanfloor Network System for Earthquakes and Tsunamis (DONET) in Tonankai region (DONET 1 in Figure 11).

As described by Kawaguchi et al. [2011], operating large-scale subsea infrastructure over a long period of time (20-30 years) is one of a challenge of underwater technology. The increase of measurement instruments has a big influence on the total system reliability, because of the state-of-arts instrument is a bottleneck to maintain long-term reliability. A novel system design concept is necessary for the observatory network development to make two demands such as 'high reliability system design' and 'state-of-arts measurement' united. The observatory network should be able to replace, maintenance and extend while operating, and should be have a redundancy for the internal or external observatory network component failure. To achieve these requirements, DONET adapt a strategy to combining the following three major components with different system reliability: (i) high reliability backbone cable system, (ii) replaceable science node, and (iii) extendable measurement instruments.

We must develop and expand DONET not only in nationwide (DONET 2, 3 to be installed) but also worldwide of major subduction zones in order to mitigate the catastorophic disasters due to coupled megathrust earthquakes.

ACKNOWLEDGMENTS

The authors would like to thank DONET members and Tohoku University researchers. Naoyuki Kato provided us simulation code of earthquake cycle on 2-D subduction plate boundary model. Akira Hasegawa, Toru Matsuzawa, Ryota Hino and Yasuo Yabe advised us about the strategy of our studies patiently. Publication of this chapter was partly supported by Grant-in-Aid for Young Scientists (KAKENHI) 23710212.

REFERENCES

Ariyoshi, K., T. Matsuzawa, R. Hino, and A. Hasegawa (2007), Triggered non-similar slip events on repeating earthquake asperities: Results from 3D numerical simulations based on a friction law, *Geophys. Res. Lett.*, 34, doi:10.1029/2006GL028323.

Ariyoshi, K., T. Matsuzawa, Y. Yabe, N. Kato, R. Hino, A. Hasegawa, and Y. Kaneda (2009), Character of slip and stress due to interaction between fault segments along the dip direction of a subduction zone, *J. Geodyn.*, 48, 55-67, doi:10.1016/j.jog.2009.06.001.

Ariyoshi, K., T. Matsuzawa, Y. Yabe, N. Kato, R. Hino, and A. Hasegawa (2011), Consideration on the 2011 off the Pacific Coast of Tohoku Earthquake and the 2004 Sumatra Earthquake, *J. A. M. S. T. E. C. Rep. Res. Dev.*, 13, 17-33.

Bilham, R., R. Engdahl, N. Feldl, and S. P. Satyabala (2005), Partial and complete rupture of the Indo-Andaman plate boundary 1847-2004, *Seismo. Res. Lett.*, 76(3), 299-311.

Blanpied, M. L., C. J. Marone, D. A. Lockner, J. D. Byerlee, and D. P. King (1998), Quantitative measure of the variation in fault rheology due to fluid-rock interactions, *J. Geophys. Res.*, 103, 9691-9712.

Boatwright, J. and M. Cocco (1996), Frictional constraints on crustal faulting, *J. Geophys. Res.*, 101, 13895-13909.

Boettcher, M. S., G. Hirth, and B. Evans (2007), Olivine friction at the base of oceanic seismogenic zones, *J. Geophys. Res.* 112, B01205, doi:10.1029/2006JB004301.

Coffin, M. F., L. M. Gahagan, and L. A. Lawver (1998), Present-day Plate Boundary Digital Data Compilation, *University of Texas Institute for Geophysics Technical Report*, 174, 5.

Dieterich, J. H. (1979), Modeling of rock friction: 1. Experimental results and constitutive equations, *J. Geophys. Res.*, 84, 2161–2168.

Fujii, Y. and M. Matsu'ura (2000), Regional Difference in Scaling Laws for Large Earthquakes and its Tectonic Implication, *Pure Appl. Geophys.*, 157, 2283-2302.

Fukuda, J., T. Higuchi, S. Miyazaki, and T. Kato (2004), A new approach to time-dependent inversion of geodetic data using a Monte Carlo mixture Kalman filter, *Geophys. J. Int.* 159, doi: 10.1111/j.1365-246X.2004.0238.x.

Furumura, T., K. Imai, and T. Maeda (2011), A revised tsunami source model for the 1707 Hoei earthquake and simulation of tsunami inundation of Ryujin Lake, Kyushu, Japan, *J. Geophys. Res.*, 116, B02308, doi:10.1029/2010JB007918.

Geospatial Information Authority of Japan (2011), The 2011 off the Pacific coast of Tohoku Earthquake: Postseismic Slip Distribution Model (Preliminary), *http://www.gsi.go.jp/cais /topic110315.2-index-e.html.*

Hacker, B. R., S. M. Peacock, G. A. Abers, and S. D. Holloway, (2003), Subduction factory 2. Are intermediate-depth earthquakes in subducting slabs linked to metamorphic dehydration reactions?, *J. Geophys. Res.* 108, doi:10.1029/2001JB001129.

Hsu, Y., M. Simons, J. Avouac, J. Galetzka, K. Sieh, M. Chlieh, D. Natawidjaja, L. Prawirodirdjo, and Y. Bock (2006), Frictional afterslip following the 2005 Nias–Simeulue earthquake, Sumatra, *Science*, 312, 1921–1926.

Igarashi, T., T. Matsuzawa, N. Umino, and A. Hasegawa (2001), Spatial distribution of focal mechanisms for interplate and intraplate earthquakes associated with the subducting Pacific plate beneath the northeastern Japan arc: A triple-plated deep seismic zone, *J. Geophys. Res.* 106, 2177-2191.

Igarashi T., T. Matsuzawa, and A. Hasegawa, (2003), Repeating earthquakes and interplate aseismic slip in the northeastern Japan subduction zone, *J. Geophys. Res.*, 108, 2249, doi:10.1029/2002JB001920.

Ishii, M., P. M. Shearer, H. Houston, and J. E. Vidale (2005), Extent, duration and speed of the 2004 Sumatra-Andaman earthquake imaged by the Hi-Net array, *Nature*, 435, 933-936, doi:10.1038/nature03675.

Japan Meteorological Agency (2011), The 2011 off the Pacific coast of Tohoku Earthquake ~15 th report~ (in Japanese), *http://www.jma.go.jp/jma/press/1103 /13b/kaisetsu201103131255.pdf.*

Kanamori, H. and D. L. Anderson (1975), Theoretical basis of some empirical relations in seismology, *Bull. Seism. Soc. Am.*, 65, 1073–1095.

Kato, N. (2004), Interaction of slip on asperities: Numerical simulation of seismic cycles on a two-dimensional planar fault with nonuniform frictional property, *J. Geophys. Res.*, 109, B12306, doi:10.1029/2004JB003001.

Kato, N. (2008), Numerical simulation of recurrence of asperity rupture in the Sanriku region, northeastern Japan, *J. Geophys. Res.*, 113, B06302, doi:10.1029/2007JB005515.

Kawaguchi, K., S. Kaneko, T. Nishida, and T. Komine (2011), Construction of Real-time Seafloor Observatory for Earthquakes and Tsunami Monitoring, *Praxis book on "Seafloor Observatories"*, in press.

Kawakatsu, H. and T. Seno (1983), Triple seismic zone and regional variation of seismicity along the northern Honshu arc, *J. Geophys. Res.* 88, 4215-4230.

Kido, M., H. Fujimoto, S. Miura, Y. Osada, K. Tsuka, and T. Tabei (2006), Seafloor displacement at Kumano-nada caused by the 2004 off Kii Peninsula earthquakes, detected through repeated GPS/Acoustic surveys, *Earth Planets Space,* 58, 911-915.

Kimura, G., Y. Kitamura, Y. Hashimoto, A. Yamaguchi, T. Shibata, K. Ujiie, and S. Okamoto (2007), Transition of accretionary wedge structures around the up-dip limit of the seismogenic subduction zone, *Earth Planet. Sci. Lett.* 255, 471-484, doi:10.1016/j.epsl.2007.01.005.

Konca, A. O., J. Avouac, A. Sladen, A. J. Meltzner, K. Sieh, P. Fang, Z. Li, J. Galetzka, J. Genrich, M. Chlieh, D. H. Natawidjaja, Y. Bock, E. J. Fielding, C. Ji, and D. V. Helmberger (2008), Partial rupture of a locked patch of the Sumatra megathrust during the 2007 earthquake sequence, *Nature*, 456, 631–635.

Kondo, H., Y. Awata, Ö. Emre, A. Doğan, S. Özalp, F. Tokay, C. Yildirim, T. Yoshioka, and K. Okumura (2005), Slip Distribution, Fault Geometry, and Fault Segmentation of the 1944 Bolu-Gerede Earthquake Rupture, North Anatolian Fault, Turkey, *Bull. Seism. Soc. Am.*, 95, 1234-1249.

Lay, T., H. Kanamori, C. J. Ammon, M. Nettles, S. N. Ward, R. C. Aster, S. L. Beck, S. L. Bilek, M. R. Brudzinski, R. Butler, H. R. DeShon, G. Ekström, K. Satake, and S. Sipkin (2005), The great Sumatra-Andaman earthquake of 26 December 2004, *Science,* 308, 1127–1133, doi:10.1126/science.1112250.

Leonard, M. (2010), Earthquake Fault Scaling: Self-Consistent Relating of Rupture Length, Width, Average Displacement, and Moment Release, *Bull. Seism. Soc. Am.* 100(5A), 1971–1988, doi:10.1785/0120090189.

Liu-Zeng, J., T. Heaton, and C. DiCaprio, (2005), The effect of slip variability on earthquake slip-length scaling, *Geophys. J. Int.* 162(3), 841-849, doi:10.1111/j.1365-246X.2005.02679.x.

Mogi,K., (2004) Two grave issues concerning the expected Tokai Earthquake, *Earth Planets Space,* 56, li–lxvi.

Nagai, R., M. Kikuchi, and Y. Yamanaka (2001), Comparative study on the source processes of recent large earthquakes in Sanriku-oki region: The 1968 Tokachi-oki earthquake and the 1994 Sanriku-oki earthquake, *J. Seismol. Soc. Jpn.*, 54, 267– 280, in Japanese.

National Geophysical Data Center (2011), Global Significant Earthquake Database, *http://www.ngdc.noaa.gov/hazard/earthqk.shtml.*

Okada, Y. (1992), Internal deformation due to shear and tensile faults in a halfspace, *Bull. Seism. Soc. Am.*, 82, 1018–1040.

Pacheco, J. F., L. R. Sykes, and C. H. Scholz, C. H. (1993), Nature of seismic coupling along simple plate boundaries of the subduction type. *J. Geophys. Res. 98*, 14133-14159.

Peacock, S. M., (2001), Are the lower planes of double seismic zones caused by serpentine dehydration in subducting oceanic mantle?, *Geology* 29(4), 299-302.

Peter, B. (2003), *An updated digital model of plate boundaries: Geochemistry, Geophysics, Geosystems*, 4(3), 1027, doi:10.1029/2001GC000252.

Press, W. H., B. P. Flannery, S. A. Teukolsky, and W. T. Vetterling (1992), *Numerical Recipes,* 2nd. ed., Cambridge Univ. Press, New York.

Rice, J. R. (1993), Spatio-temporal complexity of slip on a fault, *J. Geophys. Res.*, 98, 9885–9907.

Romanowicz, B. (1992), Strike-slip earthquakes on quasi-vertical transcurrent faults: inferences for general scaling relations, *Geophys. Res. Lett.*, 19, 481–484.

Ruina, A. (1983), Slip instability and state variable friction laws, *J. Geophys. Res.*, 88, 10,359–10,370.

Sato, H. (1977), Some precursors prior to recent great earthquakes along the Nankai trough, *J. Phys. Earth*, 25, S115–S121 (Suppl.).

Savage, J. C. (1983), A dislocation model of strain accumulation and release at a subduction zone, *J. Geophys. Res.*, 88, 4984–4996.

Scholz, C. H. (1982), Scaling laws for large earthquakes; consequences for physical models, *Bull. Seism. Soc. Am.* 72, 1–14.

Sieh, K. (1996), The repetition of large-earthquake ruptures, *Proc. Natl. Acad. Sci.*, 93, 3764-3771.

The Headquarters for Earthquake Research Promotion (2002), Long-term Evaluation of seismic activity off Boso to Sanriku (in Japanese), *http://www.jishin.go.jp/main/chousa /kaikou_pdf/sanriku_boso.pdf.*

The Headquarters for Earthquake Research Promotion (2004), Long-term Evaluation of occurrence potentials or subduction-zone earthquakes around Hyuga-nada and Nansei island (in Japanese), *http://www.jishin.go.jp/main/chousa/04feb_hyuganada/index.html.*

The Headquarters for Earthquake Research Promotion (2011), The 2011 off the Pacific Coast of Tohoku Earthquake, *http://www.jishin.go.jp/main/chousa/11apr_sanriku-oki3/index-e.htm.*

Uchida, N., A. Hasegawa, T. Matsuzawa, and T. Igarashi (2004), Pre- and post-seismic slow slip on the plate boundary off Sanriku, NE Japan associated with three interplate earthquakes as estimated from small repeating earthquake data, *Tectonophysics,* 385, 1-15.

United States Geological Survey (2004), Magnitude 9.1 - OFF THE WEST COAST OF NORTHERN SUMATRA, *http://earthquake.usgs.gov/earthquakes/eqinthenews/2004/us2004slav.*

Walther, C. H. E., E. R. Flueh, C. R. Ranero, R. von Huene, and W. Strauch (2000), Crustal structure across the Pacific margin of Nicaragua: evidence for ophiolitic basement and a shallow mantle sliver, *Geophys. J. Int.*, 141, 759-777.

Wiseman, K. and R. Bürgmann (2011), Stress and Seismicity Changes on the Sunda Megathrust Preceding the 2007 M_w 8.4 Earthquake, *Bull. Seism. Soc. Am.,* 101(1), 313-326.

Yamamoto, Y., R. Hino, K. Suzuki, Y. Ito, T. Yamada, M. Shinohara, T. Kanazawa, G. Aoki, M. Tanaka, K. Uehira, G. Fujie, Y. Kaneda, T. Takanami, T., and T. Sato (2008), Spatial heterogeneity of the mantle wedge structure and interplate coupling in the N. E. Japan forearc region, *Geophys. Res. Lett.* 35, L23304, doi:10.1029/2008GL036100.

In: Earthquakes
Editor: Kostas Konstantinou
ISBN: 978-1-62081-883-1

Chapter 5

STATISTICAL SEISMICITY IN VIEW OF COMPLEX SYSTEMS

***Tomohiro Hasumi*[1,*]*, Chien-chih Chen*[2]*, Takuma Akimoto*[3] *and Yoji Aizawa*[4]
[1] Division of Environment, Natural resources and Energy, Mizuho Information and Research Institute, Inc., Tokyo Japan
[2] Department of Earth Sciences and Graduate Institute of Geophysics, National Central University, Taoyuan, Taiwan
[3] Department of Mechanical Engineering, Keio University, Yokohama, Japan
[4] Department of Applied Physics, Advance School of Science and Engineering, Waseda University, Tokyo, Japan

Abstract

In this chapter, we investigate the statistical seismicity in view of complex systems, namely, interoccurrence time statistics and intermittency of earthquakes.

To show statistical properties of earthquakes, we analyze three different kinds of data; i) natural earthquake data, ii) synthetic data, and iii) time sequence analysis based on a one-dimensional map. For i), Japan Meteorological Agency (JMA), Southern California Earthquake Data Center (SCEDC), and Taiwan Central Weather Bureau (TCWB) are analyzed. For ii), we investigate synthetic data produced by the two-dimensional (2D) spring-block model. For iii), intermittency and correlation of earthquakes are studied using JMA data and the constructed map.

First, we focus on the interoccurrence time statistics, statistical properties of time intervals between successive earthquakes by analyzing both natural data and the synthetic data. We show that the probability distribution of interoccurrence time depends on the threshold of magnitude, m_c and is described by the superposition of the Weibull distribution and log-Weibull distribution for all earthquake data. As m_c increases, the ratio of the Weibull distribution increases. In other words, the distribution changes from the Weibull to the log-Weibull distribution according to an increase of m_c, i.e., the Weibull - log-Weibull transition. This result reinforces the view that the interoccurrence time statistics contain the Weibull statistics and the log-Weibull statistics. In synthetic data, the distribution changes from the pure log-Weibull regime to the Weibull regime via the superposition regime, i.e., a sharp Weibull - log-Weibull transition. For

*e-mail address: tomohiro.hasumi@mizuho-ir.co.jp

natural data, the distribution presents a smooth Weibull - log-Weibull transition which means that the distribution varies from the superposition regime to the pure Weibull regime. For large earthquakes, the Weibull distribution with the exponent $\alpha_l < 1$ is preferred, which indicates that the sequence of earthquakes is not a Possion process. We give a geophysical interpretation of the Weibull - log-Weibull transition.

Second, to characterize an intermittency and the correlation of interoccurrence times in earthquakes, we analyze the JMA catalog. We demonstrate that interoccurrence times are not independent and identically distributed random variables. However, we find that the tails of all conditional probability distribution functions of interoccurrence times are universal. The tail characterizes intermittency of earthquakes. It is shown that forecasting of the occurrence of large earthquakes is more difficult than that of small earthquakes.

PACS 05.45-a, 05.65.+b, 91.30.Px, 05.10.-a, 02.50.Ey, 05.45.Tp, 05.45.Ac, 91.30.Dk
Keywords: Earthquakes, Seismicity, Complex systems, Interoccurrence time, Spring-block model, Weibull distribution, log-Weibull distribution, and, Weibull - log-Weibull transition, Intermittency, renewal process

1. Introduction

1.1. Statistical Properties of Earthquakes in View of Seismology

1.1.1. Relation Between the Earthquake Intensity and Frequency

One of the well known statistical properties of earthquakes is the relation between the earthquake intensity and frequency. The relation has been proposed since the 1930s. Here we review some examples.

In 1930s, Wadati (1932) proposed that the energy distribution of earthquakes can be described by a power law distribution (Wadati, 1932),

$$P(E) \propto E^{-w} \tag{1}$$

This relation is extracted from the assumption that the amplitude of seismic wave at the distance from the epicenter can be written as the following;$A \propto \sqrt{E}e^{-kr}/r$. Wadati estimated w at 2.1 by analyzing the earthquakes which occurred around Tokyo and by assuming the k set to be 0.

Ishimoto and Iida (1939) proposed that the frequency distribution $n(a)$ of a maximum seismic wave a is described by a power law,

$$n(a) = ka^{-m}, \tag{2}$$

where k is a constant and m stands for the power law exponent which is estimated to be 1.74 by analyzing the 1636 earthquakes from 1935 to 1938. This relation can be rewritten as

$$N(a) = Ka^{-(m-1)}, \tag{3}$$

where $N(a)$ means the cumulative frequency of earthquakes in the range $[a, a+da]$.

Richter defined the magnitude, so-called the "Richter magnitude". Gutenberg and Richter (1956) suggested that the cumulative number of earthquake $n(m)$ and its magnitude m shows the exponential distribution, namely,

$$\log_{10} n(m) = a + bm, \tag{4}$$

where a and b are constants. a stands for the seismic rate. b is almost unity and does not change by varying an area and temporal terms. For example, they analyzed the Southern California and the world-wide earthquake data and calculated the b is $b = 0.88$ for Southern California and $b = 1.12$ for worldwide (Gutenberg, 1956). This relation is so-called the Gutenberg-Richter law. In addition the relation holds when the moment magnitude is introduced so that the Gutenberg-Richter (GR) law is one of the most well-known relations for the statistical seismicity. The GR law indicates that the cumulative frequency decreases $1/10$ as the magnitude m increases from m to $m+1$.

Kagan (1999) stressed that seismic moment is a key factor rather than the magnitude, because the seismic moment is measureable and its definition is unique. He proposed the seismic moment M_0 and frequency n relation, namely,

$$\log_{10} n = \alpha - \beta M_0, \tag{5}$$

where α and β are constants. β is estimated to be $\beta = 0.60 \pm 0.02$ by analyzing the shallow earthquakes all over the world (Kagan, 1999). It should be noted that relation between the b-value of the GR law and β in equation (5) can be written as $b = 1.5\beta$, because the relation between seismic moment M_0 and moment magnitude M_w can be described by $\log M_w = M_0/1.5 + 9.1$. Hence, the $\beta = 0.60$ is corresponding to $b = 0.90$.

1.1.2. Statistical Properties of Aftershocks

Generally, after a mainshock, many earthquakes occur around epicenters, which are so-called aftershocks. In this subsection, we review the statistical properties of aftershocks.

Omori (1894) first proposed the relation between a rate of aftershocks dN/dt and time from a mainshock occurrence time t, namely,

$$\frac{dN}{dt} \sim t^{-p}, \tag{6}$$

where p is so-called p-value. Omori estimated p at 1.2 by analyzing the two different types of the sequence of aftershocks. This relation is often called as Omori law.

Enya (1901) also proposed the statistical properties of aftershocks, which states that the number of aftershock n_1 from t to $t+1$ can be described by the following relation;

$$n_1 = K \ln\{1 + 1/(t+c)\}, \tag{7}$$

where K and c are constants. This equation is similar to the Omori law.

After the work of Omori, Utsu (1961) was proposed the modified version of the Omori law, namely,

$$\frac{dN}{dt} = K(c+t)^{-p}, \tag{8}$$

where K is a constant and c means a characteristic time. This relation is so-called modified Omori law and still use for prediction of aftershocks. Utsu (1961) also investigated the time interval distribution between the mainshock and maximum aftershock showed the power law relation.

$$P(\tau) \sim t^{-1}, \tag{9}$$

Båth (1965) investigated the difference between the magnitude of mainshock and of the maximum aftershock, denoted here Δm, and proposed the following relation,

$$\Delta m \approx 1.2. \tag{10}$$

In seismology, this relation is used to detect the maximum aftershock.

It should be noted that according to the Omori law, aftershocks never disappear even when time after a mainshock is sufficiently long. It seems to be difficult to distinguish an earthquake from an aftershock. Thus, in order to study the aftershocks, it is necessary to mention how to extract aftershocks. In seismology, the way to extract aftershocks is called declustering, and its detail information is shown for example (Gardner and Knopoff, 1974; Reasenberg, 1985; Zhuang et. al 2004).

1.2. Statistical properties of earthquakes in view of statistical physics

1.2.1. Simulation Model Study

In the 1980s, Bak, Tang, and Wiesenfeld proposed the concept of self-organized criticality (SOC) (Bak et al. 1987; Bak et al. 1988), according to which a non-equilibrium open system gradually evolves into a critical state, where a distribution of physical quantity shows a power law. A crust is a non-equilibrium open system, because energy is supplied by plate motion and released by an earthquake. Introducing a sand-pile model involving SOC and using the Cellular Automaton (CA) simulation method, they demonstrated an empirical power law which is related to the GR law (Bak and Tang, 1989). On the basis of their result, they suggested that earthquakes can be categorized into SOC phenomena. Inspired by their study, a number of SOC-based earthquake models have been proposed to reproduce statistical properties of earthquakes (Rundle et al. 2003; Rundle et al. 2000). A long-range connective sand-pile (LRCS) model has been very recently proposed (Chen et al. 2008; Chen et al. 2008; Lee et al. 2008; Lee et al. 2009) and revealed some underlying mechanisms of possible precursory phenomena as observed in real earthquake fault systems. The LRCS model differs from the original sand-pile model in the aspect of releasing toppling grains to remote, instead of nearest neighboring, sites.

A spring-block model was firstly proposed by Burridge and Knopoff (1967), hereafter referred to as the BK model. Burridge and Knopoff investigated both a laboratory scale experiment and a numerical simulation and then they demonstrated that the released energy distribution shows a power law by making the model which consists 10 blocks connected by the spring. The background idea of the model is that earthquakes are caused by a stick-slip motion of the fault.

After the concept of SOC, Carlson and Langer showed a power law distribution like GR law, in a restricted magnitude region (Carlson and Langer 1989; Carlson and Langer

1989), based on the one-dimensional (1D) BK model with non-linear friction force, hereafter denoted by BK (original) model. It has been shown that the power law distribution of magnitude only concerns small creeping events (Carlson et al. 1991; Carlson et al. 1994). For large magnitude region, the magnitude distribution shows a sharp peak indicating that a characteristic earthquake (all blocks slip at the same time) occurs. However, the peak of the magnitude distribution disappear by varying model parameters and by assuming inhomogeneous parameters. Carlson investigated the time interval between characteristic earthquakes in the 1D BK (original) model, and found that the characteristic earthquakes occur quasi-periodically (Carlson and Langer 1991). Based on the 2D BK (original) model extended by Carlson, the magnitude distribution (Carlson 1991) and statistical properties of the shear stress drops (Kumagai et al 1999) were reported. The most important finding of the work by Carlson and Langer is that they can extract the power law magnitude distribution (in a limited magnitude region) even if they assume the non-homogeneous fault systems. The earthquakes produced by the 1D and 2D BK (original) model exhibit both the characteristic and SOC.

This model calculation has also been performed by the CA simulation method, from here on referred to as the BK (CA) model, and the statistical properties of earthquakes, such as the magnitude distribution (Ito and Matsuzaki 1990; Nakanishi 1990; Nakanishi 1991; Otsuka 1972) and the time interval statistics (Brown et al. 1991; Sánchez et al. 2002) were discussed. Note that the 1D BK (CA) model is not a good model for reproducing the Gutenberg-Richter law, see for instance (Schmittbuhl et al. 1996). The studies of the interoccurrence time statistics based on its modified versions were reported (Preston et al. 2000, Christensen and Olami 1992; Hedges and Takacs 2005; Hainzl et al. 1999; Hainzl et al. 2000). The fundamental difference between the BK (original) model and the BK (CA) model is how to dissipate a friction energy; the friction force is governed by the constitutive law for the BK (original) model, and by the equipartition law for the BK (CA) model. The BK (original) model still has a great advantage of involving a constitutive friction law that well accounts for the laboratory experiment as well as the inertia.

The BK model in itself traces back to the late 1980s. We realize that the model simplifies the complex fault dynamics, so that some features of the fault, such as no radiation damping and long-range interactions, are inherently neglected. However, since the model can efficiently extract the statistical properties of earthquakes, such as the GR law and the Omori law (Nakanishi 1992), it has been attracted much attention. Recent reports based on the BK (original) model have focused on the long range interaction (stress transfer) (Xia et al. 2005; Xia et al. 2008; Mori and Kawamura 2008), a fractal structure of faults (Hasumi et al. 2006), the statistical properties of the magnitude distribution and the interoccurrence time based on the rate and state dependent constitutive law (Ohmura and Kawamura 2006), the correlation of seismicity (Mori and Kawamura 2005; Mori and Kawamura 2006; Mori and Kawamura 2008; Kawamura 2006). However, the comprehensive application of the 2D BK (original) model to the statistical properties of the interoccurrence time still remains to be done. According to recent papers (Saichev and Sornette 2006; Saichev and Sornette 2007), the interoccurrence time statistics derived from the modified versions of the BK model (Nakanishi 1992; Hedges and Takacs 2005; Hainzl et al. 1999; Hainzl et al. 2000) reproduce the GR law and the Omori law in nature. On the other hand, in the case of the 2D BK (original) model, whether the natural interoccurrence time statistics can be extracted or not is still contoroversial, because this model can not produce aftershocks.

1.2.2. Unified Scaling Law for Earthquakes and its Implication

Since 1990s, statistical properties of time interval between successive earthquakes, hereinafter interoccurrence time have been focused from the statistical physics viewpoint. Ito (1995) showed that the interoccurrence time distribution exhibits a power law $\sim t^{-1.4}$ by analyzing the Southern California earthquake data covering from 20° to 45° N for latitude, and from 100° to 125° W for longitude from 1971 to 1985. In his work, the fitting accuracy of the power law distribution is not displayed so that it is difficult to conclude that the power law distribution is the best [please see the Figure.2 in ref (Ito 1995)]. Bak, Christensen, Danon, and Scanlon (2002) proposed the unified scaling law for earthquakes interoccurrence time by analyzing the the Southern California earthquake data covering from 20° to 45° N for latitude, and from 100° to 125° W for longitude from 1984 to 2000. The scaling law of the interoccurrence time can be related to the b-value of the GR law, the p-value of the Omori law, and the fractal dimension of the epicenter, denoted here d_f. Inspire of this work, Corral (2003 and 2004) demonstrated the interoccurrence time distribution, $P(\tau)$, can be written as

$$P(\tau) = Rf(R\tau), \tag{11}$$

$$f(R\tau) = C\frac{1}{(R\tau)^{1-\gamma}}\exp(-(R\tau)^{\delta}/B), \tag{12}$$

where R is the seismicity rate. He used the world wide earthquake data for instance, NEIC, SCSN (North California), JUNTEC (Japan), and BGS. It should be noted that the interoccurrence times were analyzed for the events with the magnitude m above a certain threshold m_c under the following two assumptions: (a) earthquakes can be considered as a point process in space and time; (b) there is no distinction between foreshocks, mainshocks, and aftershocks. It was found that $f(R\tau)$ follows the generalized gamma distribution. In equation (3), C is the normalized constant and is $C = 0.50 \pm 0.05$. γ, δ, and B are parameters estimated to be $\gamma = 0.67 \pm 0.05$, $\delta = 0.98 \pm 0.05$, and $B = 1.58 \pm 0.15$. The theoretical framework of this equation can be extracted by assuming the GR law and Omori law (Saichev and Sornette 2006; Saichev and Sornette 2007), and by assuming the stochastic process where the next earthquake depends only on the previous one and renormalization group method (Corral 2005).

As for aftershocks, the interoccurrence time distribution shows the generalized gamma distribution (Shcherbakov et al. 2005). They found that the sequence of aftershocks is similar to the non-homogeneous passion process by governing the Omori law. On the other hand, the sequence of earthquakes with aftershock removed follows a Possion process which is equivalent to the exponential distribution of the interoccurrence time (Gardner and Knopoff 1974; Enescu et al. 2008).

Abe and Suzuki studied the statistical properties of earthquakes on the basis of the non extensive statistical mechanics proposed by the Tsallis (1988) by analyzing the Japan and Sourthen California earthquake data from the 2001 to 2004 (Abe and Suzuki 2005). They showed that the interoccurrence time distribution $Pr(t > \tau)$ can be described by the

q-exponential distribution with $(q > 1)$, corresponding to the power law,

$$\Pr(t > \tau) = \frac{1}{(1+\varepsilon\tau)^{\gamma}} = e_q(-\tau/\tau_0) = [(1+(1-q_t)(-\tau/\tau_0))^{\frac{1}{1-q_t}}]_+, \quad (13)$$

where q_t, τ_0, γ, and ε are positive constants and $([a]_+ \equiv \max[0,a])$. On the contrary, the epicenter distribution between successive earthquakes also can be written as the q-exponential distribution with $(q < 1)$ (Abe and Suzuki 2003), namely,

$$\Pr(d > r) = e_q(-r/r_0) = [(1+(1-q_r)(-r/r_0))^{\frac{1}{1-q_r}}]_+, \quad (14)$$

where q_r and r_0 are positive constants. After their study, Darooneh and Dadashinia demonstrated that $Pr(t > \tau)$ and $\Pr(d > r)$ are well fitted by the q-exponential distribution with $(q > 1)$ and $(q < 1)$, respectively by analyzing the Iranian earthquakes, which is the same in Japan (Darooneh and Danashinia 2008). As we mentioned in section 2, this statistical properties can be obtained by analyzing the 2D BK (original) model (Hasumi 2007; Hasumi et al. 2008; Hasumi 2009a).

Recent works show that interoccurrence times are not independent random variables, but have long-term memory (Bunde et al. 2005; Livina et al. 2005; Lennartz et al. 2008; Akimoto et al. 2010). In general, interevent times such as return times become correlated random variables when original random variables are correlated. Since an interoccurrence time depends on the past, it is difficult to determine the distribution of interoccurrence times theoretically. Therefore, the determination of the distribution of interoccurrence times is still an open problem. One of the progress analyses is the conditional distribution of interoccurrence times. However, the condition needs the knowledge of all the past when the underlying stochastic process has a long-term memory. Moreover, an effect of a threshold of magnitude on the interoccurrence time statistics is unknown.

1.3. Statistical Properties of Earthquakes in View of Deterministic Chaotic Systems

Chaotic behaviors appearing in deterministic systems often reveal very clear statistical laws, and the origin of their probabilistic aspects is successfully understood in ergodic-theoretical framework. Earthquakes and their time evolution are also considered to be a kind of statistical phenomena which demonstrates the probabilistic nature, but their statistical laws still remain empirical ones at the present time though they are useful for future predictions and correspondence. We have to admit that it is not so easy to construct the dynamical model which precisely describes the details of the physical changes in earth crusts, however, it is an important subject for us to elucidate the universal aspects in seismological statistics and their dynamical origins from the viewpoint of deterministic chaos theory.

Here and in what follows in this section, we concentrate our discussions to the empirical laws in seismological statistics, that are clearly shown by a sort of scaling relations; Wadati [see equation (1)], Omori [see equation (6)], Enya [see equation (7)], Ishimoto-Iida [see equation (2)], and Utsu [see equation (9)]. It will be shown that all these scaling relations are reproduced by a simple dynamical model based on the so called Modified Bernoulli Map at least qualitatively, and that these statistical laws are universal in the critical regime which is induced by a big earthquake. The model proposed here is a toy-model, but it is emphasized that the model displays not only the ergodic-theoretical universal aspects observed in a wide class of dynamical systems, but also in the aftershock statistics of the critical regime (Aizawa 1995).

1.3.1. Chaotic Dynamics which Generates Random Coin-Tossing

One of the simplest random processes is coin-tossing, which generates a series of symbols, $\{\sigma_k\}$, where k stands for the discrete time at k-th trial, and $\sigma_k = \pm 1$ (for head or tail). The probability for the continuous occurrence of n-heads (or tails) is the exponential distribution, and the mean occurrence rate of head (or tail) approaches to $1/2$ when the number of trials goes to large enough. Statistical properties of the above coin-tossing is reproduced by a chaotic dynamical system, so called Bernoulli map, which is defined by

$$x_{k+1} = \begin{cases} 2x_k & (0 \leq x < 1/2), \\ 2x_k - 1 & (1/2 \leq x \leq 1), \end{cases} \tag{15}$$

where x_k is the real number and k (integer) the discrete time. The Bernoulli map is ergodic and strong mixing under the invariant distribution $\rho(x) = 1$ $(0 \leq x \leq 1)$, and that when the symbolic dynamics is defined by $\sigma_k = +1$ for $(0 \leq x < 1/2)$, or -1 $(1/2 \leq x \leq 1)$, the sequence $\{\sigma_k\}$ has the same statistical properties as the above mentioned coin-tossing.

1.3.2. The Modified Bernoulli Map

Let us consider the generalization of the Bernoulli map (Aizawa and Kohyama 1984a; Aizawa 1984b; Aizawa 1989; Aizawa 2000; Shinkai and Aizawa 2006) to include long-term memories in the coin-tossing process. The modified Bernoulli map (MBM) is given by,

$$x_{k+1} = \begin{cases} x_k + 2^{B-1}x_k^B & (0 \leq x < 1/2), \\ x_k - 2^{B-1}(1-x_k)^B & (1/2 \leq x \leq 1), \end{cases} \tag{16}$$

and the symbolic dynamics $\{\sigma_k\}$ in the same way as the Bernoulli map. When the parameter $B = 1$, the MBM is the original Bernoulli map, but in what follows, we consider the case for $B > 1$. The statistical properties of the MBM are well understood by ergodic-theoretical analysis. For instance, the invariant distribution $\rho(x)$ is well approximated by,

$$\rho(x) \propto x^{-(B-1)} + (1-x)^{-(B-1)}, \ (0 \leq x \leq 1), \tag{17}$$

and the waiting time probability $P(n)$, which describes the probability for the continuous n-occurrence of the state $\sigma = +1$ (or $\sigma = -1$), obeys,

$$P(n) \propto n^{-\beta}, \ \left(\beta = \frac{B}{B-1}\right) \tag{18}$$

This explains that the strong correlations or the long-term memories are generated in the MBM. Figure 1 shows typical time series for various cases.

A striking point is that the sequence $\{\sigma_k\}$ is non-stationary and the mean waiting time $\langle n \rangle$ is divergent for $B \geq 2$, but is stationary with $\langle n \rangle =$ finite for $B < 2$. The critical regime appears near $B \cong 2$.

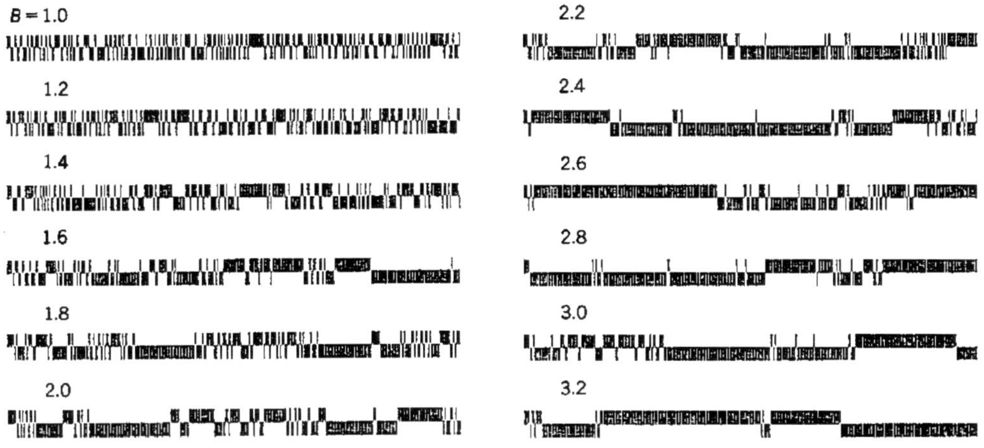

Figure 1. Symbolic time series of the MBM for various B values Aizawa 1995).

1.3.3. A Metaphor Model which Simulates Seismological Statistics

Here we propose a metaphorical model based on the chaotic behaviors in the MBM, which reproduces some scaling aspects appeared in many seismological studies. First, we assume that the symbolic variable $|\sigma_k|$ is proportional to the rate of energy accumulation in the earth crust, and the accumulated energy during the period of $\sigma > 0$ (or $\sigma < 0$) is release at once by an earthquake when the symbolic variable σ_i changes its sign, i.e., $\sigma_{i+1} = -\sigma_i$. This assumption means that the released energy by the earthquake E is related to the waiting time n defined in equation (18). Though the mechanism is still unclear, the following form is plausible from the consideration of nonlinear visco-elastic spring energy,

$$E \propto n^{1/\kappa'}, \tag{19}$$

where the exponent κ' is a parameter. Therefore, the energy distribution of the earthquake $P(E)$ becomes,

$$P(E) \propto E^{-\kappa}, \quad (\kappa = \kappa'(\beta - 1) + 1). \tag{20}$$

The parameter B is considered to express the strength of earth-crusts or the tolerance to the strain in crust. Actually, it is shown in Figure 1 that the number of bigger earthquakes increases when the value B becomes large. The power law distribution in the earthquake energy is first pointed out by Wadati (1932) for several cases, and suggested to be $\kappa \cong 2.1$.

The second assumption of our metaphor model is that the strength parameter B is sharply reduced by the occurrence of a big earthquake (mainshock), namely, the value just before the big earthquake, say B_{pre}, is larger than that after the big earthquake B_{post}, i.e., $B_{pre} > B_{post}$, and that the value B goes back slowly to the original value B_{pre}. If we assume the exponential recovery (relaxation time is τ) of the parameter B, the value B at time k, say $B(k) = B_k$, is given by

$$B_k = B_{pre} - (B_{pre} - B_{post})e^{-k/\tau}, \quad (k = 0, 1, 2, \ldots) \tag{21}$$

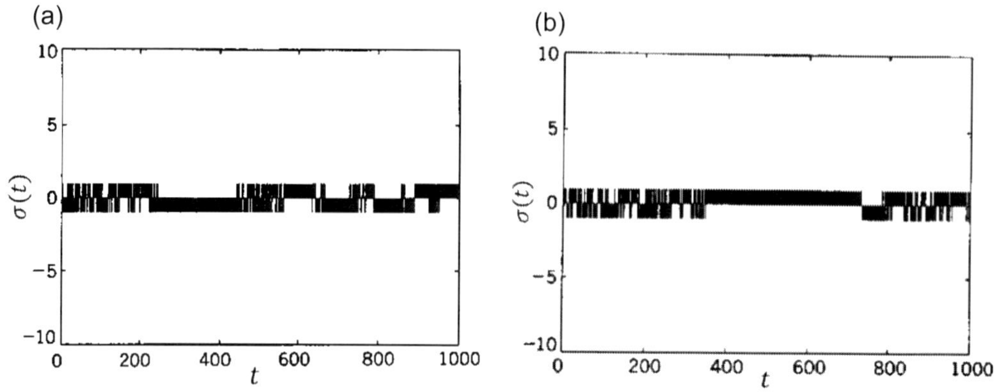

Figure 2. Symbolic time series after the main shock at $k = 0$ for two different values of x_0 (Aizawa 1995).

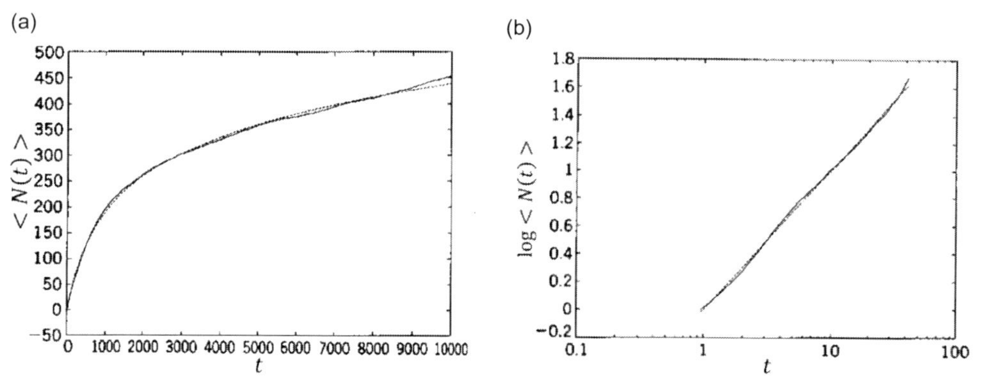

Figure 3. Cumulative numbers of aftershocks for $1 \leq k \leq t$. (a) normal plot (long time) and (b) logarithmic plot (short time); Numerical plots are almost completely adjusted by a logarithmic curve (doted line) consistent with the Omori law (Aizawa 1995).

when k stands for the time measured from the moment of the main shock ($k = 0$). From these assumptions, the MBM dynamics after the occurrence of the big earthquake is expressed by,

$$x_{k+1} = \begin{cases} x_k + 2^{B_k - 1} x_k^{B_k} & (0 \leq x < 1/2), \\ x_k - 2^{B_k - 1}(1 - x_k)^{B_k} & (1/2 \leq x \leq 1), \quad (k = 0, 1, 2, \ldots). \end{cases} \tag{22}$$

Equations (21) and (22) are the metaphorical chaotic model which is used in what follows.

1.3.4. Empirical Statistical Laws in Aftershocks and Critical Regime

Numerical results obtained by equations (21) and (22) are shown in Figures 2, 3, 4, and 5 under the following conditions; $B_{pre} = 5/2$, $B_{post} = 3/2$, and $\tau = 10^3$. Figure 2 shows two sample paths of $\{\sigma_k\}$ for different initial data of x_0, where we can see that the inter occur-

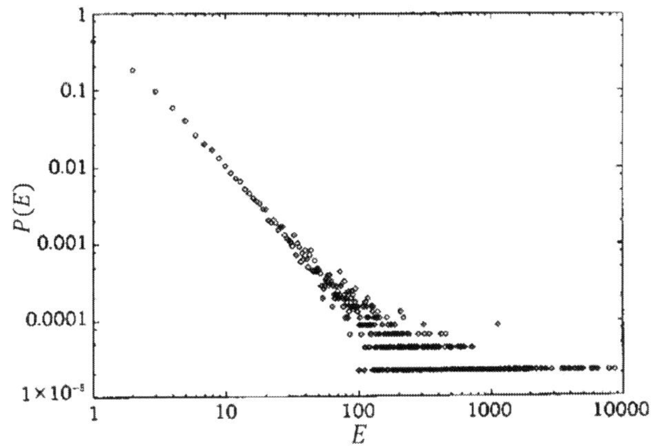

Figure 4. Energy distribution of aftershocks. ($P(E) \simeq E^{-\kappa}, \kappa \cong 2.0$.) (Aizawa 1995)

rence time between two flip-flop jumps in $\{\sigma_k\}$ is very short when k is small enough. This corresponds to the fact that a series of small aftershocks are induced by a big earthquake. The accumulated number of the aftershocks in the interval $1 \leq k \leq t$, say $N(t)$, is a random variable depending on the initial data. The average number $\langle N(t) \rangle$ is shown in Figure 3, where the numerical data is very well adjusted by a logarithmic function,

$$\langle N(t) \rangle \propto \log(at+b), \tag{23}$$

where a and b are fitting parameters. It is important to point that this is quite consistent with the Omori formula (Omori 1894), and that the formula by Enya is naively adopted in our metapher model. By changing the values of B_{pre} and B_{post}, we can obtain the generalized Omori formula; $\langle N(t) \rangle \propto (at+b)^{\gamma}$, but we do not touch this problem here.

Figure 4 shows the energy distribution of aftershocks $P(E)$ at $\kappa' = 1$. The result is well fitted by,

$$P(E) \propto E^{-\kappa}, \;\; (\kappa \cong 2.0 \, at \;\; \kappa' = 1) \tag{24}$$

If we take the effect by κ' into account, the exponent obeys $\kappa = \kappa' + 1$. In these numerical analysis, we used the simulation data for $1 \leq k \leq T \; (= 10 \times \tau)$, and the aftershocks occurred in $k > T$ are omitted because they are not the aftershocks induced by the main shock occurred at $k = 0$.

The reason why the exponent κ takes $\kappa \cong 2.0$ (at $\kappa' = 1$) can be analyzed by using the intrinsic properties of the MBM, where the bifurcation parameter B plays an essential roles in the statistical behaviors of the symbolic dynamics $\{\sigma_k\}$. As mentioned in Equation (18), there occurs a phase transition between stationary regime ($B < 2$) and nonstationary regime ($B > 2$), where the critical fluctuations dominate the statistical behaviors in the recovering process $B(k)$ in equation (21), namely, $B(k)$ passes always through the critical value $B = 2$ at a certain time under the condition of $B_{pre} > 2.0$, and $B_{post} < 2.0$. As the result, the waiting time distribution in equation (18) becomes $P(n) \propto n^{-2}$, and the energy distribution in a wide critical regime becomes,

$$P(E) \propto E^{-(\kappa'+1)}. \tag{25}$$

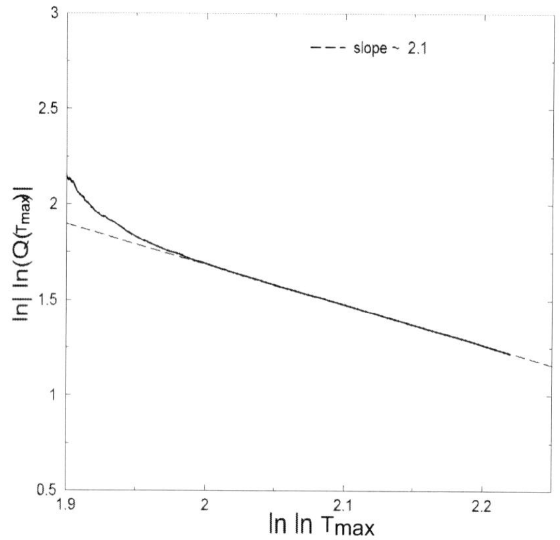

Figure 5. Waiting time distribution of the maximum aftershock $P(T_{max})$ which obeys a log-Weibull distribution.

Furthermore, by using equation (25), the energy distribution of the maximum aftershock E_{max} can be analyzed theoretically; $E_{max} = \text{Max}[E_1, E_2, \cdots E_N]$ where E_j is the j-th aftershock energy and we take N large enough, the distribution $P(E_{max})$ becomes,

$$P(E_{max}) \propto E_{max}^{-(\kappa'+1)} L(E_{max}), \tag{26}$$

where $L(x)$ is a slowly varying function $L(x) \propto e^{Cx^{-\kappa'}}$ and C is a constant (Feller 1966). Equation (25) is the same as the Ishimoto-Iida formula if we assume that the maximum amplitude a_{max} is related to E by $E \propto a_{max}^{\kappa''}$ where $\kappa'' = 2$ in the case of elastic matters, but in general case with nonlinear elasticity κ'' should be a parameter.

An important point in the above consideration is that the earth crust is in a critical state just after a big earthquake happened, and as the result there appear some characteristic scaling laws in statistical properties.

The last problem in our interests is to estimate the onset time of the maximum aftershock T_{max} by use of our metaphor model based on Equations (21) and (22). As we have not yet succeeded to derive any theoretical results regarding the onset time T_{max}, we only show the numerical results briefly. Figure 5 is the cumulative distribution function $Q(x) = \exp\left[\{\log(x/a) - b\}^{-(c-1)}\right]$ of the onset time $P(T_{max})$, which is well adjusted by a log-Weibull distribution,

$$P(T_{max}) \propto \frac{1}{T_{max}} \frac{1}{(\log T_{max})^c} L(\log T_{max}), \tag{27}$$

where c is a constant parameter $c > 1$ and $L(x)$ is a slowly varying function and T_{max} is large enough. The dominant scaling term in the r.h.s. in Equation (27) is T_{max}^{-1}, and this is consistent with the formula suggested by Utsu in aftershock statistics. The onset time of the maximum aftershock is one of the most crucial problems in seismology, and must

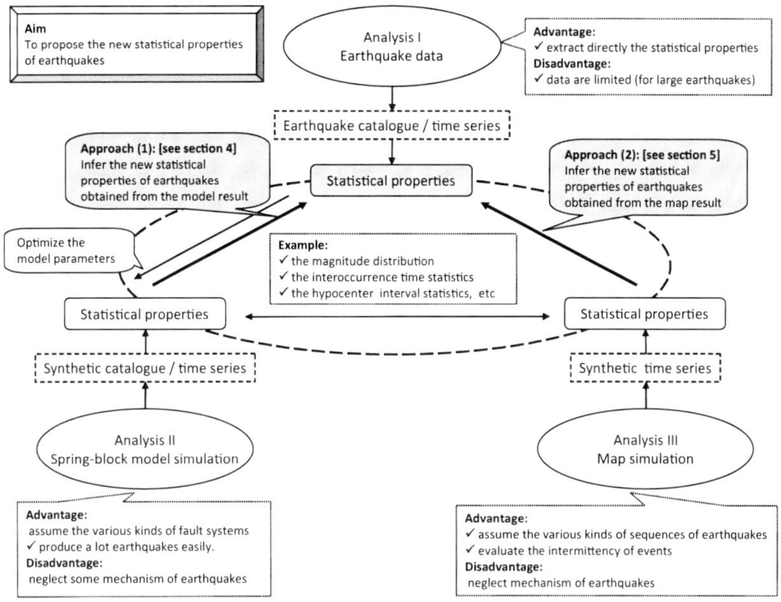

Figure 6. A schematic diagram of our research.

be carefully studied in relation to the Omori formula, especially the fluctuations from the mean behavior $N(t) - < N(t) >$ are strongly correlated to the onset of large aftershocks. Furthermore, the onset time T_{max} depend on the magnitude of the initial mainshock. There are so many difficult problems remained in determining the distribution $P(T_{max})$, but we can expect to obtain the enough information about statistical properties of aftershocks if we find out any useful dynamical models. The metaphor model proposed in the present paper is a challenge toward this goal from the recent development in chaotic dynamical system theory (Akimoto and Aizawa 2003, Akimoto and Aizawa 2005). Further details and results approached from the viewpoint of chaos theory are discussed in section 5.

2. Concept of Our Research

As shown in Section 1, the statistical seismicity has been studied from not only the seismological viewpoint but also the statistical physics (the concept of self-organized criticality, fractal, non-additive statistical mechanics, and a proposition of the unified scaling law) and chaotic systems viewpoint. Recently, a complex system approach has attracted much interest from several kinds of field, for example, earthquakes, economics, and neuro-sciences. The aim of this approach is to figure out or to clarify the statistical features of the system by taking into account the interactions between the components or sub-systems. In this chapter, we focus on the statistical seismicity from complex systems on the basis of previous works.

A schematic diagram of our research is shown in Figure 6. As can be seen from the Figure 6, our concept consists of three components corresponding to the analysis; I. earthquake data, II. synthetic data generated by the 2D spring-block model, and III. time series created by the map we proposed. The advantage and disadvantage of each analysis are mentioned bellow.

I. Earthquake data

Advantage: we can extract directly the statistical properties of earthquakes.

Disadvantage: the number of data is limited, especially for large earthquakes.

II. The spring-block model simulation

Advantage: we can assume the various kinds of fault by changing the model parameters, and produce a lot of earthquake data easily when the simulation time takes longer.

Disadvantage: some physical and geochemical features of fault systems neglected.

III. A map simulation

Advantage: we can treat various kinds of the sequence of earthquake like event by varying the parameter of the map, and evaluate the intermittency of the event.

Disadvantage: the source mechanism of earthquakes neglected.

One of our motivations of this study is to reveal the statistical properties of earthquakes, which are not reported before. We take two approaches;

Approach (1) we infer the statistical properties of earthquakes from those obtained by the synthetic data [see Section 4].

Approach (2) we infer the statistical properties of the sequence of earthquakes from those extracted by the map we proposed [see Section 5].

3. Spring-Block Model and Earthquake Catalogues

3.1. Review of the spring-block model

In Figure 7 (a), we schematically illustrate the 2D spring-block model which we simulated in this study. The model is composed of blocks, two plates, two different kinds of coil springs, k_c^x and k_c^y, and a leaf spring, k_p, corresponding respectively to segments of a fault, geological plates, a compression stress, and a shear stress. The fundamental idea of the model is that earthquakes are caused by the stick-slip motion of the fault. It is noted that this model is virtually the same as Otsuka's model (Otsuka 1972) but is different from Carlson's ($k_c^x = k_c^y$) (Carlson 1991). We assume that the slip direction of blocks is restricted only to the y-direction. The equation of motion at site (i, j) can be written as

$$\begin{aligned} m\frac{d^2y_{i,j}}{dt^2} &= k_c^x(y_{i+1,j}+y_{i-1,j}-2y_{i,j})+k_c^y(y_{i,j-1}+y_{i,j+1}-2y_{i,j})-k_py_{i,j} \\ &- F\left(v+\frac{dy_{i,j}}{dt}\right), \end{aligned} \tag{28}$$

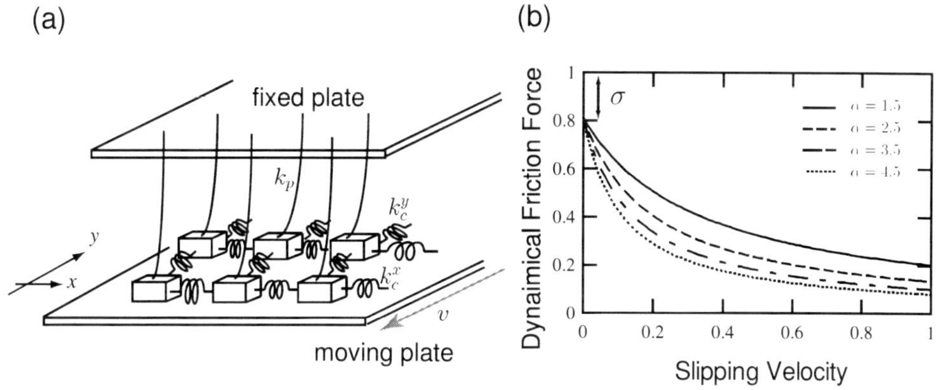

Figure 7. (a) 2D spring-block model. k_c^x, k_c^y, and k_p are spring constant. The friction force acts on the surface between the block and the bottom plate. (b) α-dependence of the non-linear dynamical friction function. $\sigma = 0.01$ throughout the simulation.

where m and $y_{i,j}$ are, respectively, mass and displacement of the block and F is a dynamical force between the block and the bottom plate.

Now we rewrite equation (28) into a dimensionless form. The dimensionless dynamical friction force ϕ is obtained as $\phi(\dot{y}/v_1) = F(\dot{y})/F_0$, where F_0 and v_1 are the maximum friction force and the characteristic velocity, respectively. A dimensionless time t' and displacement $U_{i,j}$ are defined by

$$t' = \omega_p t = \sqrt{k_p/m}\, t, \quad U_{i,j} = \frac{y_{i,j}}{D_0} = \frac{y_{i,j}}{F_0/k_p}.$$

Then the dimensionless form is

$$\begin{aligned}\frac{d^2U_{i,j}}{dt'^2} &= l_x^2(U_{i+1,j}+U_{i-1,j}-2U_{i,j}) + l_y^2(U_{i,j-1}+U_{i,j+1}-2U_{i,j}) - U_{i,j} \\ &- \phi\left(2\alpha\left(\nu + \frac{dU_{i,j}}{dt'}\right)\right),\end{aligned} \qquad (29)$$

and,

$$l_x^2 = \frac{k_c^x}{k_p}, \quad l_y^2 = \frac{k_c^y}{k_p}, \quad \nu = \frac{v}{D_0\omega_p} = \frac{v}{\hat{v}}, \quad 2\alpha = \frac{D_0\omega_p}{v_1} = \frac{\hat{v}}{v_1},$$

where $\hat{v}$ is the slipping velocity, being on the order of 1m/s in a real crust. In this work, we choose the friction force as the velocity-weakening constitutive law which states that as the slipping velocity increases, the dynamical friction force decreases. This constitutive law was observed in the accurate rock fracture experiment (Scholz 2002; Yoshida 1997). Then, we use the dimensionless dynamical friction force ϕ given by

$$\phi(\dot{U}) = \begin{cases} (-\infty, 1] & \dot{U} = 0, \\ \dfrac{(1-\sigma)}{\{1+2\alpha[\dot{U}/(1-\sigma)]\}} & \dot{U} > 0. \end{cases} \qquad (30)$$

The formulation given in equation. (30) was introduced by Carlson et al. (1991) (see Figure 7 (b)). This can be regarded as an ideal friction function involving the stick-slip motion, so that it has often been adopted to 1D and 2D BK (original) models (Nakanishi 1992; Mori and Kawamura 2005; Mori and Kawamura 2006; Xia et al. 2005; Hasumi et al. 2006; Carlson 1991; Kumagai et al. 1999; Hasumi 2007, Hasumi: 2009a; Kawamura:2006; Mori and Kawamura 2008; Xia et al. 2008; Mori and Kawamura 2008; Hasumi et al. 2009b; Hasumi et al. 2010, Hasumi and Aizawa 2008). To exclude a back slip, ($\dot{U}_{i,j} < 0$), we treat ϕ as a tunable parameter when $\dot{U} = 0$.

The model is governed by five parameters, l_x, l_y, α, σ, and ν. l_x and l_y are dimensionless stiffness parameters in the x- and y- direction, respectively. α means a decrement of the dynamical friction force with increasing slipping velocity. σ is a stress gap between the normalized maximum friction force ($= 1$) and dynamical friction force $\phi(0)$. ν is a dimensionless loading velocity. l_x and l_y are related to Lame's constants, λ and μ, as (Yamashita 1976)

$$l_x^2 = \left(\frac{\Delta z}{\Delta x}\right)^2, l_y^2 = \frac{5\lambda + 6\mu}{\lambda + 2\mu}\left(\frac{\Delta z}{\Delta y}\right)^2, \tag{31}$$

where $\Delta x, \Delta y$, and Δz are infinitesimal lengths in the x-,y-, and z- directions, respectively. According to the experiments (Scholz 2002; Yoshida et al. 1997), α is positive and on the order of 10^0 (=1). We can treat ν as being equal to zero, when a slip event occurs because ν, in nature $\sim 10^{-9}$, is far smaller than a slipping velocity of ~ 1. This approximation ensures that no other event takes place during an ongoing event. Obviously, ν is related to loading time and interoccurrence time. We fixed $\nu = 0.01$ because we will rescale it later.

In this report, $\Delta x = \Delta y = \Delta z$ is used with assumptions of $\lambda = \mu$. We solved equations (29) and (30) by the 4th-order Runge-Kutta method under a free boundary condition. A small irregularity of block displacements is considered at an initial stage of the calculation. We use a 10^5 order of earthquake-like events after some period when the initial randomness does not influence the statistical properties. Then the interoccurrence time statistics are systematically studied by changing l_x and l_y, and by α increasing the threshold of magnitude m_c. In this model, a slip of blocks is considered as an earthquake. An event starts when a block begins to slip in the direction toward $+y$ and ends when all blocks stop slipping. Event magnitude m in the model is defined as

$$m = m_0 + \log_{10}\left(\sum_{i,j}^{n} \delta u_{i,j}\right)/1.5, \tag{32}$$

where $\delta u_{i,j}$ and n are the total slip at (i, j) and the total number of slipping blocks, respectively (Ben-Menahem 1981). m_0 is set at 0.7 for shifting m to a positive value.

3.2. Earthquake Catalogue Database

We analyzed three natural earthquake catalogues, Japan Metrological Agency (JMA), Southrn California Earthquake Data Center (SCEDC), Taiwan Central Weather Breau (TCWB) and one synthetic catalogue created by the 2D spring-block model. Information on each catalogue is displayed in Table 1. In this table, m_{min} stands for the minimum magnitude in the catalogue and m_c^0 is the magnitude of completeness, that is the lowest magnitude

Table 1. Information on the used earthquake catalogues

Name	Coverage	Term	Number of earthquakes	m_{min}	m_c^0	m_{max}
JMA	25° –50° N and 125° –150° E	01/01/2001–31/10/2007	130244	2.0	2.0	8.0
SCEDC	32° –37° N and 114° –122° W	01/01/2001–31/12/2007	83936	0.0	1.4	5.7
TCWB	21° –26° N and 119° –123° E	01/01/2001–31/12/2007	148155	0.0	1.9	7.1
Synthetic	system size (e.g. 50 × 50)	–	297040	0.0	0.3	2.8

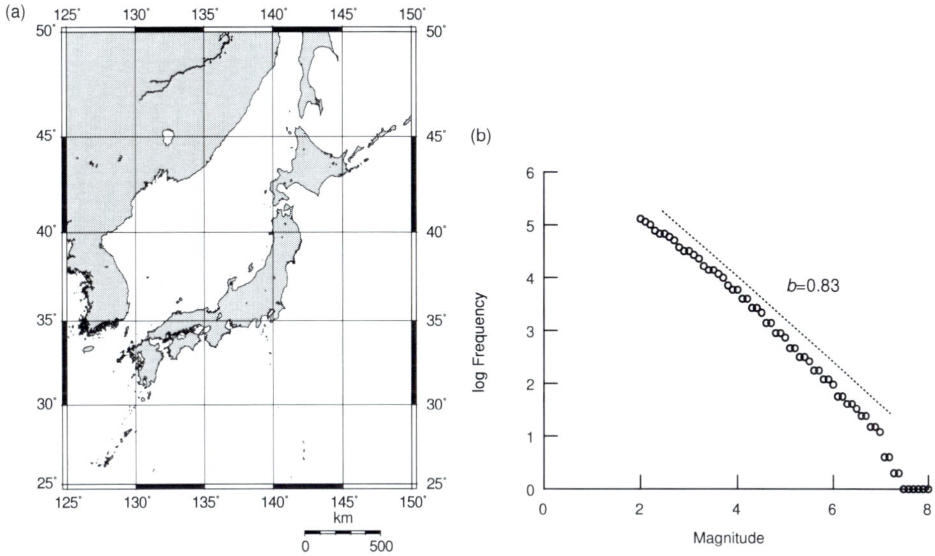

Figure 8. Information on the Japan Metrological Agency (JMA) earthquake catalogues. (a) covering region. The number of each cell means the number of earthquakes. (b) the magnitude distribution. $b = 0.83$ is calculated from the slope of the distribution.

at which the GR law holds. We basically consider events with magnitude greater than and equal to m_c^0 because events whose magnitudes are smaller than m_c^0 are supposedly incomplete for recording. m_{max} means the maximum magnitude of an earthquake in a region, where we analyzed. Detail information of each catalogue is described below.

3.2.1. Japan Metrological Agency (JMA) Catalogue

JMA catalogue is maintained by the Japan Metrological Agency, covering from 25° to 50° N for latitude, and from 25° to 150° E for longitude [see Figure 8 (a)] during from 1923 to latest. This catalogue consists of an occurrence of times, a hypocenter, a depth, and a magnitude. In this chapter, we use the data from 1st January 2001 to 31st October 2007. As can be seen from Figure 8 (b), the distribution of magnitude obeys the Gutenberg - Richter law and m_c^0 is estimated to be 2.0.

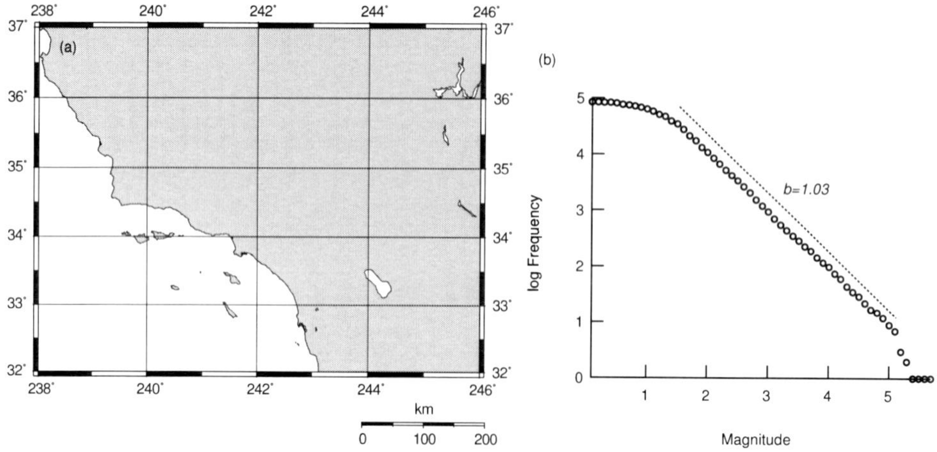

Figure 9. Southern California Earthquake Data Center (SCEDC) earthquake catalogue information. (a) covering region. (b) the magnitude distribution. $b = 1.03$ is calculated from the slope of the distribution

3.2.2. Southern California Earthquake Data Center (SCEDC) Catalogue

SCEDC catalogue is maintained by the Southern California Earthquake Data Center, covering from 32° to 37° N for latitude, and from 114° to 122° W for longitude [see Figure 9 (a)] during from 1932 to latest. The information of an earthquake, such as an occurrence of times, a hypocenter, a depth, and a magnitude, is listed. Here, we analyze the earthquake data from 1st January 2001 to 31st December 2007. In Figure 9 (b), we demonstrate the magnitude distribution, and we obtain $b = 1.03$.

3.2.3. Taiwan Center Weather Bureau (TCWB) Catalogue

TCWB catalogue is maintained by the Central Weather Bureau, covering from 21° to 26° N for latitude, and from 119° to 123° E for longitude [see Figure 10 (a)]. This catalogue consists of an occurrence of times, a hypocenter, a depth, and a magnitude. We use the data from 1st January 2001 to 31st December 2007. As shown in Figure 10 (b), the Gutenberg - Richter law is valid in a magnitude range, $1.9 \leq m \leq 6.7$. b-value is calculated from the slope of the distribution, and is estimated to be 0.90.

3.2.4. Synthetic Catalogue (2D Spring-Block Model)

The synthetic catalogue is generated by the 2D spring-block model with the velocity-weakening friction law (Carlson 1991). As we described in 3.1, the 2D spring-block model is characterized by five parameters: the stiffness l_x^2 and l_y^2, the decrement of the friction force α, the plate velocity ν, and the difference between the maximum friction force and dynamical friction force σ. As shown in figure, we demonstrate the magnitude distribution for different stiffness parameters, l_x^2 and l_y^2, and friction parameter α. In the case of $l_x^2 = 1, l_y^2 = 3$, and $\alpha = 3.5$, the distribution shows the power law in the magnitude region,

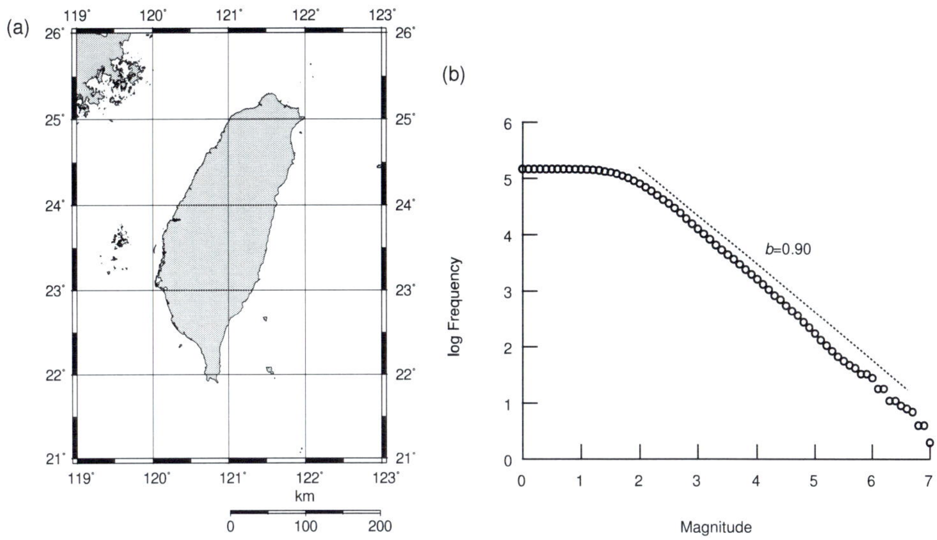

Figure 10. Information on the Taiwan Central Weather Bureau (TCWB) earthquake catalogue. (a) covering region. (b) the magnitude distribution. $b = 0.90$ is calculated from the slope of the distribution.

$0.5 \lessapprox m \lessapprox 1.0$, which is the broadest. In addition, the b-value of the GR law is estimate to be $b = 1.10$, which reproduces the realistic aspect of the GR law. Previously, it was shown that for $l_x^2 = 1, l_y^2 = 3$, and $\alpha = 3.5$, the earthquakes produced by the 2D spring-block model can exhibit the the constant stress drop (Kumagai et al. 1999). At the same time, we have revealed that the interoccurrence time distribution and the hypocenter interval distribution are well fitted by the q-exponential distribution with $(q > 1)$ (Hasumi 2007) and $q < 1$ (Hasumi 2009), respectively, which reproduces semi-quantitatively the previous works (Abe and Suzuki 2005; Abe and Suzuki; 2003). Thus, we have set those parameters as $l_x^2 = 1, l_y^2 = 3, \alpha = 3.5, \nu = 0.01$ and $\sigma = 0.01$.

4. The Weibull - Log-Weibull Transition of Interoccurrence Times

In this section, we investigate the new statistical properties of earthquakes by using the Approach (1), which states that we infer the statistical properties of earthquakes from those obtained by the synthetic data in Section 2. Here, we focus on the time intervals between successive earthquakes, interoccurrence times, by analyzing the synthetic data and three natural data.

4.1. Background idea

The study of the time interval between successive earthquakes has been widely investigated in order to infer when the next earthquake happens. Previously, time intervals can be categorized by two types; interoccurrence time and the recurrence time (Abaimov et al.

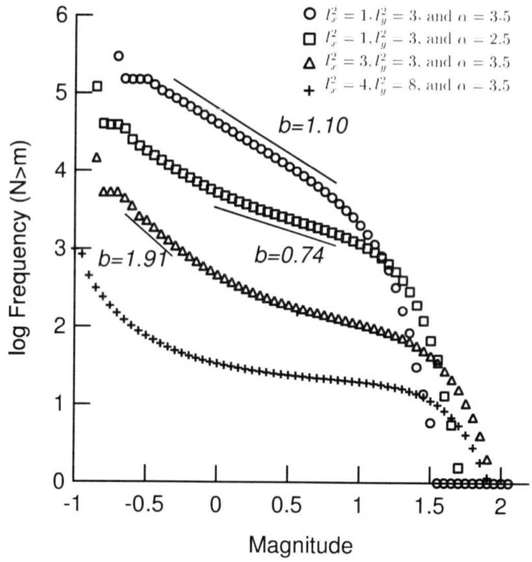

Figure 11. The cumulative number of earthquakes as a function of magnitude obtained from the 2D BK model. Circles (○), squares (□), triangles (△), and plus signs (+), correspond respectively to the case of $l_x^2 = 1, l_y^2 = 3$, and $\alpha = 3.5$, $l_x^2 = 1, l_y^2 = 3$, and $\alpha = 2.5$, $l_x^2 = 3, l_y^2 = 3$, and $\alpha = 3.5$, and $l_x^2 = 4, l_y^2 = 8$, and $\alpha = 3.5$. All the plots except for the case of $l_x^2 = 1, l_y^2 = 3$, and $\alpha = 3.5$ are shifted vertically for clarity.

2007). Interoccurrence times are the interval times between earthquakes on all faults in a region, and recurrence times are the time intervals between earthquakes in a single fault or fault segment. For seismology, recurrence times mean the interval times of characteristic earthquakes that occur quasi-periodically in a single fault.

Previous works of the interoccurrence time statistics were shown in Section 1.2.. As can be seen in Section 1.2, the studies of interoccurrence times can be classified into two types; the detection the distribution functions, the discussion on the "long-term memory". Since an interoccurrence time depends on the past, it is difficult to determine the distribution of interoccurrence times theoretically. Therefore, the determination of the distribution of interoccurrence times is still an open problem. Moreover, an effect of a threshold of magnitude on the interoccurrence time statistics is unknown. We study the distribution of interoccurrence times by changing the threshold of magnitude both the synthetic data and natural data.

4.2. Methodology and Applicable Distributions

4.2.1. Methodology

Our method is similar to the that of previous works (Hasumi et al. 2009a, Hasumi et al. 2009b, Hasumi et al. 2010) (see Figure 12).

1. We divided the spatial areas into a window of L degrees in longitude and L degrees

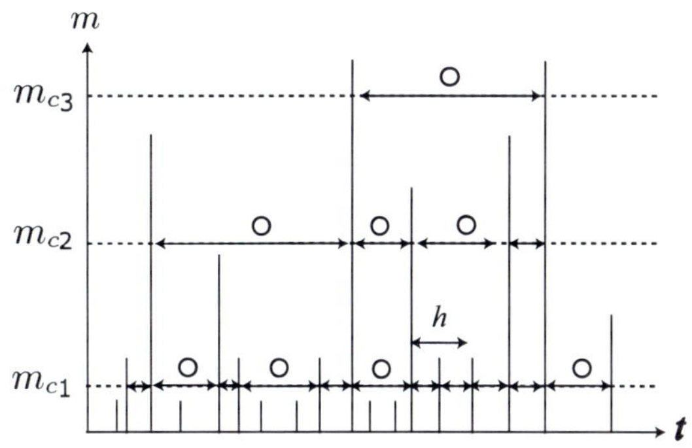

Figure 12. Schematic diagram of the interoccurrence time of our analysis for different threshold of magnitude m_c. Circles (◯) satisfy the condition. We analyze interoccurrence times greater than or equals to h.

in latitude.

2. For each bin, earthquakes with magnitude m above a certain cutoff magnitude m_c were considered.

3. We analyzed interoccurrence times greater than and equals to h day.

For each bin, we analyzed interoccurrence times using at least 100 events to avoid statistical errors. h is fixed to be 0.5 and 0 for analysis of natural data and the synthetic data, respectively. We confirmed that the results do not change for the case of $h = 0.25$ and $h = 0.75$. L is fixed firstly 5 by analyzing the JMA data. We also discuss the system size dependence later (please see in 4.4.1.) It is noted that for SCEDC and TCWB, we analyze earthquake covering the whole region. As shown in Figure 4, we investigated the interoccurrence time statistics for different 17 regions (14 regions in Japan, Southern California, Taiwan, and the systhetic). In this works, we do not distinguish earthquakes, mainshocks, foreshocks, aftershocks. Based on the previous studies (Corral 2004; Enescu et al. 2008), regularly, the interoccurrence times for aftershocks is shorter than 0.5 day, so that interoccurrence times which we have analyzed might be excluded afterschock effect.

4.2.2. Applicable Distributions

One of our main goals in this chapter is to determine the distribution function of the interoccurrence time. Here, we will focus our attention on the applicability of the Weibull distribution P_w, the log-Weibull distribution P_{lw}, the power law distribution P_{pow} (Abe and Suzuki 2005), the gamma distribution P_{gam} (in the case of $\delta = 1$ in the paper (Corral 2004), and the log normal distribution P_{ln} (Matthews et al. 2002). In this subsection, we mention the physical and mathematical meaning of these distributions.

Weibull distribution In 1939, Waloddi Weibull proposed the distribution to quantify the strength of materials (Weibull 1939). This distribution, hereinafter the Weibull distribution, can extract the one-dimensional chain model. The concept of the Weibull distribution is that the failure point decided the most weakness point of the chain, although the distribution of force acting the chain obeys the Gauss distribution. He found that the Weibull distribution appears the distribution of failure and strength in the several materials (Weibull 1951).

The probability distribution $p_w(\tau)$ and the cumulative distribution $P_w(\tau)$ of the Weibull distribution can be described by

$$p_w(t) = \left(\frac{t}{\beta_1}\right)^{\alpha_1 - 1} \frac{\alpha_1}{\beta_1} \exp\left[-\left(\frac{t}{\beta_1}\right)^{\alpha_1}\right], \tag{33}$$

$$P_w(t) = \int_0^t p_w(\tau)d\tau = 1 - \exp\left[-\left(\frac{t}{\beta_1}\right)^{\alpha_1}\right], \tag{34}$$

where α_1 and β_1 are parameters. For $\alpha_1 = 1$ and $\alpha_1 < 1$, the tail of the Weibull distribution is equivalent to the exponential distribution and the stretched exponential distribution, respectively. The hazard rate at t_0, denoted here $h(t_0)$ defined by

$$\begin{aligned} h(t_0) &= \frac{\text{probability that failure occurs at } t_0}{\text{probability that failure does not occur at } t_0} \\ &= \frac{p_w(t_0)}{1 - P_w(t_0)} = \frac{\alpha_1}{\beta_1}\left(\frac{t_0}{\beta_1}\right)^{\alpha_1 - 1}. \end{aligned} \tag{35}$$

It should be noted that $h(t_0)$ is categorized by three types depends on the α_1,

$$\begin{aligned} h(t_0) &= \frac{\alpha_1}{\beta_1}\left(\frac{t_0}{\beta_1}\right)^{\alpha_1 - 1} \\ &= \begin{cases} > 0, \ (\alpha_1 > 1) \text{ increases following the power law.} \\ = \text{constant}, \ (\alpha_1 = 1) \text{ constant for all time.} \\ < 0, \ (0 < \alpha_1 < 1) \text{ decreases following the power law.} \end{cases} \end{aligned} \tag{36}$$

It seems to be natural that the recurrence time distribution shows the Weibull distribution, because the probability of next big earthquakes increases as time goes by, corresponding to the view that $h(t)$ increases as t increases. In seismology, the distribution of ultimate strain (Hagiwara 1974), the recurrence time distribution (Bakun et al. 2005; Abaimov et al. 2007; Abaimov et al. 2008), and the damage mechanics of rocks (Nanjo et al. 2005) show the Weibull distribution. In numerical studies, the recurrence time distribution in the 1D (Abaimov et al. 2007) and 2D (Hasumi et al. 2009b) spring-block model, and in the "Virtual California model" (Yakovlev et al. 2006) also exhibits the Weibull distribution.

Log-Weibull Distribution In this chapter, log-Weibull distribution is introduced by the logarithmic modification of the Weibull distribution. Previous work showed that the log-Weibull distribution can be extracted the chain-reaction model (Huillet and Raynaud 1999) The probability distribution $p_{lw}(t)$ can be described by

$$p_{lw}(t) = \frac{(\log(t/k))^{\alpha_2 - 1}}{(\log \beta_2)^{\alpha_2}} \frac{\alpha_2}{t} \exp\left[-\left(\frac{\log(t/k)}{\log \beta_2}\right)^{\alpha_2}\right], \tag{37}$$

where α_2, β_2 and k are parameters. For $\alpha_2 = 1$, $p_{lw}(t)$ is equivalent to the power law distribution. k is introduced not to diverge the distribution. The cumulative distribution, $P_{lw}(t)$, is written as

$$P_{lw}(t) = \int_0^t p_{lw}(\tau)d\tau = 1 - \exp\left[-\left(\frac{\log(t/k)}{\beta_2}\right)^{\alpha_2}\right]. \tag{38}$$

It should be noted that the stochastic process governed by the log-Weibull distribution depends on the parameter α_2. For $\alpha_2 > 1$ and for $\alpha_2 \leq 1$, the stochastic process is stationary and non-stationary, respectively.

Previously, the log-Weibull distribution for the first time in seismology was adapted by the magnitude distribution by analyzing the French and Japanese earthquake data (Huillet and Raynaud 1999). However, the physical meaning of the distribution remains open.

Power Law Distribution Generally, a physical quantity possesses a no characteristic scale when the probability distribution of the physical quantity shows the power law distribution. In this work, the power law distribution is defined as

$$p_{pow}(t) = \frac{1}{(1+\beta_3 t)^{\alpha_3}}, \tag{39}$$

where α_3 and β_3 are parameters.

Previously, some works reported that the interoccurrence time shows the power law distribution (Ito 1995; Abe and Suzuki 2005).

Gamma Distribution The gamma distribution is often used for waiting time for events. This distribution is obtained by superposition of the exponential distribution. The probability distribution p_{gam} and the cumulative distribution P_{gam} are the following;

$$p_{gam}(t) = t^{\alpha_2 - 1}\frac{\exp(-t/\beta_2)}{\Gamma(\alpha_2){\beta_2}^{\alpha_2}}, \tag{40}$$

$$P_{gam}(t) = \int_0^t p_{gam}(\tau)d\tau = \frac{\gamma(\alpha_2, t/\beta_2)}{\Gamma(\alpha_2)}, \tag{41}$$

where α_3 and β_3 are parameters, and $\Gamma(x)$ and $\gamma(x)$ are the gamma function, the incomplete gamma function, respectively. The decay of the gamma distribution corresponds approximately to the power law distribution for a short time domain, and the exponential distribution for a long time domain.

For seismology, the gamma distribution is used as the fitting distribution of the sequence of large earthquakes (Utsu 1984). Recently, the interoccurrence time statistics obeys the generalized gamma distribution which is obtained by adding the one parameter of the gamma distribution (Corral 2004).

Log-Normal Distribution The log-normal distribution is a probability distribution of a random variable whose logarithm is normally distributed. The tail of the log-normal distribution is longer than that of the normal (Gauss) distribution. A variable might be modeled as log-normal if it can be thought of as the multiplicative product of many independent random variables each of which is positive. The probability distribution, $p_{ln}(t)$ and the cumulative distribution, $P_{ln}(t)$ can be described by

$$p_{ln}(t) = \frac{1}{t\beta_5\sqrt{2\pi}} \exp\left[-\frac{(\ln(t)-\alpha_5)^2}{2\beta_5^2}\right], \tag{42}$$

$$P_{ln}(t) = \int_0^t p_{ln}(\tau)d\tau = \frac{1}{2} + \frac{1}{2}\mathrm{erf}\left[-\frac{(\ln(t)-\alpha_5)}{\beta_5\sqrt{2}}\right], \tag{43}$$

where α_5 and β_5 are constants. erf(x) is so-called the error function defined by

$$\mathrm{erf}(x) = \frac{2}{\pi}\int_0^x e^{-t^2}dt. \tag{44}$$

Previous study showed that the recurrence time distribution is well fitted by the log-normal distribution (Nishenko and Buland 1987).

It should be noted that in this work we applied these distributions for the interoccurrence times, so that the variable is not t but τ, so that these distribution can be written as

$$P_w(\tau) = \left(\frac{\tau}{\beta_1}\right)^{\alpha_1-1}\frac{\alpha_1}{\beta_1}\exp\left[-\left(\frac{\tau}{\beta_1}\right)^{\alpha_1}\right], \tag{45}$$

$$P_{lw}(\tau) = \frac{(\log(\tau/h))^{\alpha_2-1}}{(\log\beta_2)^{\alpha_2}}\frac{\alpha_2}{\tau}\exp\left[-\left(\frac{\log(\tau/h)}{\log\beta_2}\right)^{\alpha_2}\right], \tag{46}$$

$$P_{pow}(\tau) = \frac{1}{(1+\beta_3\tau)^{\alpha_3}}, \tag{47}$$

$$P_{gam}(\tau) = \tau^{\alpha_4-1}\frac{\exp(-\tau/\beta_4)}{\Gamma(\alpha_4)\beta_4{}^{\alpha_4}}, \tag{48}$$

$$P_{ln}(\tau) = \frac{1}{\tau\beta_5\sqrt{2\pi}}\exp\left[-\frac{(\ln(\tau)-\alpha_5)^2}{2\beta_5^2}\right], \tag{49}$$

where α_i, β_i, and h are constants and characterize the distribution.

4.2.3. Goodness-of-Fit Test

In order to determine the best fitting for the distribution of the interoccurrence time data, we used three goodness-fit-test; the root mean square (rms) test, the Kolomogorov-Smirnov (KS) test, and the Anderson-Darling (AD) test. Here, we introduce three goodness-fit-tests that we applied in this study.

Root Mean Square (RMS) Test The definition of the rms value is

$$\mathrm{rms} = \sqrt{\frac{\sum_{i=1}^n (x_i - x_i')^2}{n-k}}, \tag{50}$$

where x_i is actual data and x_i' is estimated data obtained from $P(\tau)$. n and k indicate the numbers of the data points and of the fitting parameters, respectively. In this study, the rms value is calculated using the cumulative distribution for decreasing the fluctuation of the data. As readily known, the most appropriate distribution is, by definition, the smallest rms value.

Kolomogorov-Smirnov (KS) Test The test statistic for Kolomogorov-Smirnov test, hereinafter DKS, which means the maximum deviation of statics, defined as,

$$\mathrm{DKS} = \max_i |y_i - y'_i|, \tag{51}$$

where y_i and y'_i mean the actual data of the cumulative distribution and the data estimated from the fitting distribution, respectively. Then, the significance level of probability of the goodness-of-fit, Q, is defined as

$$Q = 2\sum_{i=1}^{\infty}(-1)^{i-1}e^{2i^2\lambda^2}, \tag{52}$$

where λ is written as

$$\lambda = \mathrm{DKS}\left(\sqrt{n'} + 0.12 + \frac{0.11}{\sqrt{n'}}\right), \tag{53}$$

where n' stands for the number of data points. It is known that the preferred distribution shows the smallest value of DKS and the largest value of Q (Press et al. 1995).

Anderson-Darling (AD) Test The AD test gives more weight to the tails of the distribution than the KS test. The Anderson Darling statistic, A^2 is defined as

$$A^2 = -N - \frac{1}{N}\sum_{i=1}^{N}(2i-1)[\ln(F(z_i)) + \ln(1 - F(z_{n-i+1}))], \tag{54}$$

where N is the sample number, $F(x)$ is the cumulative distribution function, and z_i is the ith interoccurrence time. Note that z is put in order. The most suitable distribution exhibits the smallest A^2.

4.3. Results

4.3.1. Synthetic Data

We attempt to trace a change in the interoccurrence time statistics by changing m_c. Our previous work (Hasumi 2007) was mainly focused on the no-threshold case, $m_c = -\infty$. The cumulative distributions of the interoccurrence times for different m_c are displayed in Fig. 13 for $l_x^2 = 1, l_y^2 = 3, \sigma = 0.01, \nu = 0.01$, and $\alpha = 3.5$. For (a), (b), and (c), m_c is set at $m_c = 0.3$, 0.6, and 1.1, respectively. The results of the rms value for a different distribution function for $m_c = 1.1$ and $m_c = 0.3$ are listed in Table 2. As can be seen from the table, for small m_c, (e.g., $m_c = 0.3$) the interoccurrence time distribution obeys the log Weibull distribution, whereas for a large m_c (e.g., $m_c = 1.1$) the Weibull distribution is preferred. However, the fitting accuracy of the Weibull distribution becomes worse when m_c is decreased. At the same time, the fitting accuracy of the log Weibull distribution becomes worse as m_c is increased. Hence, we think that for the intermediate case (b), the distribution can be described by the superposition of the Weibull distribution and the log Weibull distribution because the Weibull components and the log-Weibull components of $P(\tau)$ do not disappear suddenly. Actually, $P(\tau)$ is well fitted by the superposition of the Weibull distribution and the log-Weibull distribution.

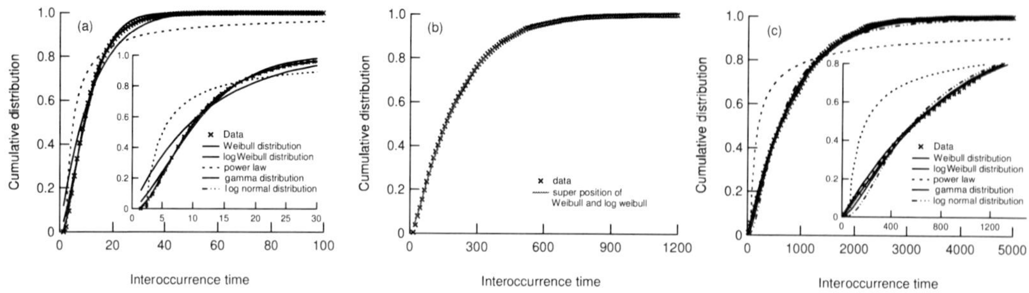

Figure 13. The cumulative distribution of interoccurrence times in the 2D BK model for different m_c. (a), (b), and (c) correspond respectively to the log Weibull regime $m_c = -0.3$, superposition regime $m_c = 0.6$, and the Weibull regime $m_c = 1.1$. The original figure is in (Hasumi et al. 2009a).

Table 2. The results of the rms value and fitting parameters for different distribution functions (Hasumi et al. 2009a)

	distribution	α_i	β_i	rms	ln rms	D_{KS}	ln D_{KS}
	$P_w\ (i=1)$	1.08± 0.01	8.29×10^2± 6.68	6.4×10^{-3}	−5.05	0.019	−3.96
$m_c = 1.1$	$P_{lw}\ (i=2)$	8.05± 0.20	1.62×10^3± 26.6	1.3×10^{-2}	−4.34	0.040	−3.21
1138 events	$P_{pow}\ (i=3)$	1.56 ± 0.08	6.36×10^1± 16.8	1.3×10^{-1}	−2.07	0.35	−1.05
70 data points	$P_{gam}\ (i=4)$	1.02± 0.004	8.21×10^2± 9.8	9.5×10^{-3}	−4.65	0.026	−3.65
	$P_{ln}\ (i=5)$	6.33± 0.01	0.92± 0.08	2.4×10^{-2}	−3.71	0.077	−2.56
	$P_w\ (i=1)$	1.31± 0.02	5.02×10^1± 0.54	7.4×10^{-3}	−4.90	0.036	−3.32
$m_c = 0.3$	$P_{lw}\ (i=2)$	5.92± 0.01	9.74×10^1± 1.70	8.5×10^{-4}	−7.08	0.0027	−5.91
19545 events	$P_{pow}\ (i=3)$	1.70 ± 0.08	5.57± 1.10	9.5×10^{-2}	−2.35	0.035	−1.05
80 data points	$P_{gam}\ (i=4)$	1.01± 0.01	4.86×10^1± 1.80	2.3×10^{-2}	−3.78	0.11	−2.23
	$P_{ln}\ (i=5)$	3.59± 0.01	0.79± 0.02	6.5×10^{-3}	−5.03	0.025	−3.69

Taken together, the probability distribution of the interoccurrence time can be expressed explicitly by the following;

$$P(\tau) = p \times \text{Weibull distribution} + (1-p) \times \text{log-Weibull distribution}, \\ = p \times P_w + (1-p) \times P_{lw}, \qquad (55)$$

where p means the rate of the Weibull distribution in the range, $0 \le p \le 1$ and depends on m_c. As for $p = 1$, $P(\tau)$ obeys the Weibull distribution, whereas for $p = 0$, $P(\tau)$ follows the log-Weibull distribution. $P(\tau)$ is characterized by five parameters, $\alpha_1, \alpha_2, \beta_1, \beta_2$, and p, while for the pure Weibull and log-Weibull regimes, the number of fitting parameters is two; α_1 and β_1 for P_w, and α_2 and β_2 for P_{lw}.

4.3.2. Earthquake Data

JMA data Firstly. we analyze the JMA data. Here, we show the result of Okinawa region (125°–130°E and 25°–30°N). It is noted that the result is obtained by analyzing the

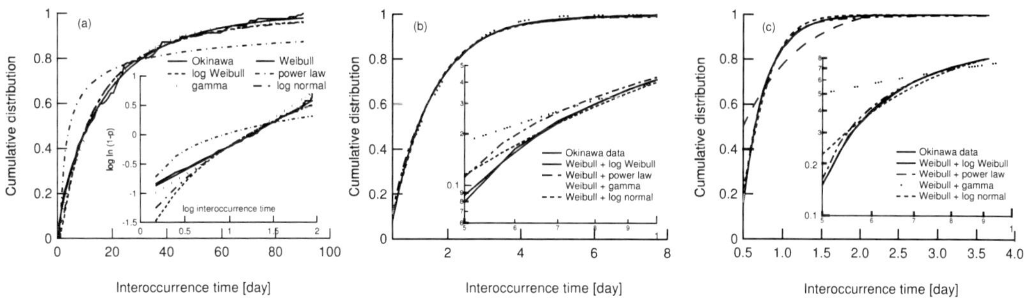

Figure 14. Distribution of interoccurrence time at Okinawa region for different m_c and distribution functions. (a) $m_c = 4.5$, (b) $m_c = 3.0$, and (c) $m_c = 2.0$. Inset figure of (a) shows the Weibull plot and (b) and (c) represent the log-log scale of the figures. The original figure is in (Hasumi et al. 2009b).

Table 3. The results of the rms value, DKS , A^2, and optimal parameters for different distribution functions in Figure 14 (a). The error bars are 95% confidence limits (Hasumi et al. 2009b)

	distribution	α_i	β_i	rms (cdf)	D_{KS}	A^2	rms (Weibull-plot)
	P_w $(i=1)$	0.82± 0.01	17.5± 0.28	0.014	0.039	0.89	0.065
$m_c = 4.5$	P_{lw} $(i=2)$	2.95± 0.12	32.4 ± 1.08	0.028	0.107	1.90	0.099
127 events	P_{pow} $(i=3)$	1.46 ± 0.04	0.99± 0.24	0.112	0.384	14.8	0.19
91 data points	P_{gam} $(i=4)$	0.94± 0.01	17.6 ± 0.44	0.022	0.066	1.51	0.098
	P_{ln} $(i=5)$	2.37± 0.04	1.22 ± 0.05	0.027	0.088	1.06	0.093

different areas. The total number of earthquakes in Okinawa is 12 024.

The cumulative distributions of the interoccurrence times for different m_c in Okinawa region is displayed in Figures 14. We carried out three statistical tests, the rms, the KS test, and AD test, so as to determine the distribution function. The results for large magnitude $(m_c = 4.5)$ in Okinawa are shown in Table 3. For Okinawa, we found that the most suitable distribution is the Weibull distribution in all tests. In general, there is a possibility that the preferred distribution is not unique but depends on the test we use. However, the results obtained in Table 3 provide evidence that the Weibull distribution is the most appropriate distribution. It follows that the Weibull distribution is preferred.

However, the fitting accuracy of the Weibull distribution becomes worse with a gradual decrease in m_c. We now propose a possible explanation which states that "the interoccurrence time distribution can be described by the superposition of the Weibull distribution and another distribution, hereafter referred to as the distribution $P_X(\tau)$,

$$\begin{aligned} P(\tau) &= p \times \text{Weibull distribution} + (1-p) \times \text{distribution X} \\ &= p \times P_w(\tau) + (1-p) \times P_X(\tau) \end{aligned} \quad (56)$$

where p is a parameter in the range, $0 \leq p \leq 1$ and stands for the ratio of P_w divided by $P(\tau)$. The interoccurrence time distribution obeys the Weibull distribution for $p = 1$. On the

Table 4. The interoccurrence time statistics of earthquakes in Okinawa region. The error bars are 95% confidence limits (Hasumi et al. 2009b)

m_c Region	distribution X	Weibull distribution α_1	β_1	Distribution X α_i	β_i	p	rms-value $[\times 10^{-3}]$	ln rms
4.5 Okinawa	P_{lw} $(i=2)$	0.82 ± 0.01	17.5 ± 0.28	–	–	1	14	−4.27
	P_{pow} $(i=3)$	0.82 ± 0.01	17.5 ± 0.28	–	–	1	14	−4.27
	P_{gam} $(i=4)$	0.82 ± 0.01	17.5 ± 0.28	–	–	1	14	−4.27
	P_{ln} $(i=5)$	0.82 ± 0.01	17.5 ± 0.28	–	–	1	14	−4.27
4.0 Okinawa	P_{lw} $(i=2)$	0.91 ± 0.01	8.28 ± 0.12	–	–	1	11	−4.51
	P_{pow} $(i=3)$	0.91 ± 0.01	8.28 ± 0.12	–	–	1	11	−4.51
	P_{gam} $(i=4)$	0.91 ± 0.01	8.28 ± 0.12	–	–	1	11	−4.51
	P_{ln} $(i=5)$	0.91 ± 0.01	8.28 ± 0.12	–	–	1	11	−4.51
3.5 Okinawa	P_{lw} $(i=2)$ [1]	1.09 ± 0.03	3.74 ± 0.13	1.87 ± 0.25	4.73 ± 0.70	0.78 ± 0.03	5.1	−5.28
	P_{pow} $(i=3)$ [2]	1.09 ± 0.03	3.74 ± 0.14	1.81 ± 0.07	0.64 ± 0.07	0.95 ± 0.02	7.3	−4.92
	P_{gam} $(i=4)$ [2]	1.09 ± 0.03	3.74 ± 0.14	–	–	1	12	−4.42
	P_{ln} $(i=5)$ [2]	1.09 ± 0.03	3.74 ± 0.14	0.86 ± 0.01	0.93 ± 0.01	0.66 ± 0.06	5.7	−5.17
3.0 Okinawa	P_{lw} $(i=2)$ [1]	1.43 ± 0.04	1.81 ± 0.04	1.34 ± 0.06	2.46 ± 0.08	0.60 ± 0.02	3.8	−5.57
	P_{pow} $(i=3)$ [2]	1.38 ± 0.04	1.64 ± 0.02	2.20 ± 0.08	0.55 ± 0.02	0.79 ± 0.02	8.8	−4.73
	P_{gam} $(i=4)$ [2]	1.38 ± 0.04	1.64 ± 0.02	–	–	1	17	−4.07
	P_{ln} $(i=5)$ [2]	1.38 ± 0.04	1.64 ± 0.02	0.19 ± 0.006	1.46 ± 0.01	0.08 ± 0.08	6.4	−5.05
2.5 Okinawa	P_{lw} $(i=2)$ [1]	1.76 ± 0.04	1.16 ± 0.02	1.23 ± 0.03	1.67 ± 0.02	0.47 ± 0.01	3.0	−5.81
	P_{pow} $(i=3)$ [2]	1.90 ± 0.09	1.02 ± 0.01	2.83 ± 0.08	0.51 ± 0.01	0.59 ± 0.04	9.8	−4.63
	P_{gam} $(i=4)$ [2]	1.90 ± 0.05	1.02 ± 0.008	1.09 ± 0.03	0.94 ± 0.04	0.99 ± 0.09	23	−3.77
	P_{ln} $(i=5)$	–	–	-0.19 ± 0.008	0.53 ± 0.01	0	12	−4.42
2.0 Okinawa	P_{lw} $(i=2)$ [1]	1.75 ± 0.06	0.78 ± 0.02	1.18 ± 0.03	1.43 ± 0.02	0.42 ± 0.02	2.3	−6.07
	P_{pow} $(i=3)$ [2]	2.56 ± 0.18	0.77 ± 0.01	3.60 ± 0.09	0.48 ± 0.004	0.39 ± 0.04	7.3	−4.92
	P_{gam} $(i=4)$ [2]	2.56 ± 0.18	0.77 ± 0.01	1.09 ± 0.06	0.68 ± 0.05	0.96 ± 0.08	25	−3.09
	P_{ln} $(i=5)$	–	–	-0.41 ± 0.01	0.39 ± 0.03	0	15	−4.20

1 We used the parameter estimation procedure A.

2 We used the parameter estimation procedure B.

other hand, it follows the distribution $P_X(\tau)$ for $p = 0$. Here, the log-Weibull distribution, the power law distribution, the gamma distribution, and the log normal distribution are candidates for the distribution $P_X(\tau)$.

Next we shall explain the parameter estimation procedures;

Procedure A the optimal parameters are estimated so as to minimize the differences between the data and the test function by varying five parameters, $\alpha_1, \beta_1, \alpha_i, \beta_i$ and p.

Procedure B At first, the Weibull parameters, α_1 and β_1 are estimated. Secondly, the parameters of $P_X(\tau)$, α_i and β_i, are optimized. Finally, p is estimated.

At first, we take the procedure A. If the ratio of the standard deviation divided by the mean for a parameter exceeds 0.1, another estimation procedure, (B), is performed.

According to those procedures (A) and (B), we obtain the fitting of results of $P(\tau)$. The results for Okinawa are listed in Table 4. We assume that the Weibull distribution is a fundamental distribution, because p becomes unity for large m_c, which means that the effect of the distribution $P_X(\tau)$ is negligible. As observed in Table 4, the log-Weibull distribution is the most suitable distribution for the distribution $P_X(\tau)$ according to the two goodness-of-fit tests. Thus, we find that the interoccurrence times distribution can be described by the superposition of the Weibull distribution and the log-Weibull distribution, namely,

$$\begin{aligned} P(\tau) &= p \times \text{Weibull distribution} + (1-p) \times \text{log-Weibull distribution}, \\ &= p \times P_w + (1-p) \times P_{lw}, \end{aligned} \tag{57}$$

$P(\tau)$ is controlled by five parameters, $\alpha_1, \alpha_2, \beta_1, \beta_2$, and p.

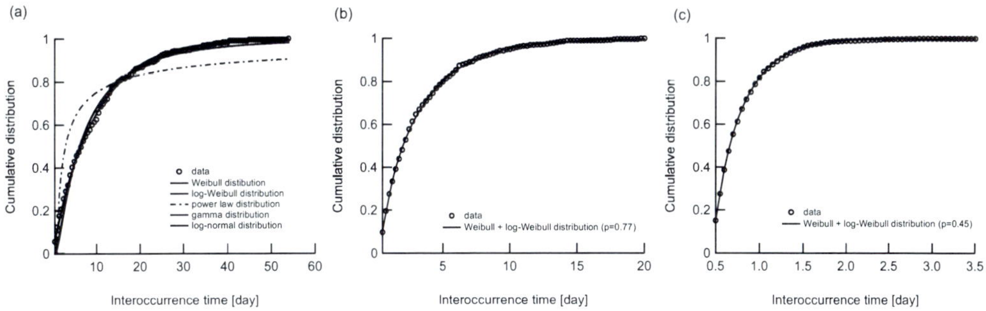

Figure 15. Distribution of interoccurrence time at Southern California for different m_c and distribution functions. (a) $m_c = 3.5$, (b) $m_c = 3.0$, and (c) $m_c = 2.0$. In (a), the pure Weibull regime. For (b) and (c) the superposition regime.

Table 5. Results of rms value, DKS, and Q for different distribution functions in the case of $m_c = 3.5$ for Southern California earthquakes. The error bars mean the 95% confidence level of fit (Hasumi et al. 2010)

m_c Region	distribution X	Distribution X α_i	β_i [day]	RMS test rms [$\times 10^{-3}$]	KS test DKS	Q
3.5 SCEDC	P_w $(i=1)$	0.95 ± 0.01	9.46 ± 0.08	7.6	0.006	0.999
	P_{lw} $(i=2)$	2.84 ± 0.10	17.6 ± 0.50	27	0.112	0.125
	P_{pow} $(i=3)$	1.57 ± 0.05	0.87 ± 0.03	11	0.229	1.89×10^{-5}
	P_{gam} $(i=4)$	0.99 ± 0.005	9.53 ± 0.09	9.8	0.030	0.999
	P_{ln} $(i=5)$	1.81 ± 0.03	1.05 ± 0.03	24	0.084	0.403

SCDEC data Secondly, we analyze the interoccurrence time statistics using the SCEDC data (114°–122°W and 32°–37°N). The total number of earthquakes is 83 936.

The cumulative distributions of interoccurrence time for $m_c = 3.5$, $m_c = 3.0$ and $m_c = 2.0$ are shown in Figure 15 (a), (b) and (c), respectively. By the rms test and KS test, we confirmed that the Weibull distribution is preferred for large m_c ($m_c = 4.0$) [see Table 5], which is the same result as that from JMA data. Unfortunately, the fitting accuracy of the Weibull distribution gets worse by decreasing a threshold m_c. We propose the following hypothesis, "the interoccurrence time distribution can be described by the superposition of the Weibull distribution and the distribution $P_X(\tau)$". As shown in Table 6, the log-Weibull distribution is the most suitable for the distribution $P_X(\tau)$, because the smallest rms-value, the smallest DKS value and the largest Q value can be obtained.

Thus, we find that the interoccurrence times distribution in SCEDC data also can be written as the superposition of the Weibull distribution and the log-Weibull distribution (see equation (57)). $P(\tau)$ is controlled by five parameters, $\alpha_1, \alpha_2, \beta_1, \beta_2$, and p.

TCWB data Finally, the TCWB data (119°–123°E and 21°–26°N). was analyzed to investigate the interoccurrence time statistics in Taiwan. The total number of earthquakes is 148 155.

Figure 16 shows the cumulative distribution of interoccurrence time for $m_c = 4.5$,

Table 6. Interoccurrence time statistics by analyzing the SCEDC data. ± means the 95% confidence level of fit (Hasumi et al. 2010)

m_c Region	distribution X	Weibull distribution α_1	β_1	distribution X α_i	β_i	p	rms $[\times 10^{-3}]$
4.0 SCEDC	P_{lw} $(i=2)$	0.98 ± 0.02	26.7 ± 0.52	–	–	1	23
	P_{pow} $(i=3)$	0.98 ± 0.02	26.7 ± 0.52	–	–	1	23
	P_{gam} $(i=4)$ [2]	0.98 ± 0.02	26.7 ± 0.52	0.99 ± 0.01	26.7 ± 0.50	0.34 ± 2.58	23
	P_{ln} $(i=5)$	0.98 ± 0.02	26.7 ± 0.52	–	–	1	23
3.5 SCEDC	P_{lw} $(i=2)$	0.95 ± 0.01	9.46 ± 0.08	–	–	1	7.6
	P_{pow} $(i=3)$	0.95 ± 0.01	9.46 ± 0.08	–	–	1	7.6
	P_{gam} $(i=4)$	0.95 ± 0.01	9.46 ± 0.08	–	–	1	7.6
	P_{ln} $(i=5)$	0.95 ± 0.01	9.46 ± 0.08	–	–	1	7.6
3.0 SCEDC	P_{lw} $(i=2)$ [1]	1.00 ± 0.02	3.41 ± 0.10	1.52 ± 0.14	3.75 ± 0.38	0.77 ± 0.02	4.1
	P_{pow} $(i=3)$ [2]	0.97 ± 0.02	3.08 ± 0.04	1.82 ± 0.06	0.59 ± 0.06	0.91 ± 0.02	6.8
	P_{gam} $(i=4)$ [2]	0.97 ± 0.02	3.08 ± 0.04	0.99 ± 0.004	3.07 ± 0.02	0.09 ± 0.60	9.1
	P_{ln} $(i=5)$ [2]	0.97 ± 0.02	3.08 ± 0.04	0.71 ± 0.02	1.01 ± 0.04	0.57 ± 0.06	5.4
2.5 SCEDC	P_{lw} $(i=2)$ [2]	1.44 ± 0.02	1.45 ± 0.02	1.22 ± 0.04	2.16 ± 0.04	0.58 ± 0.01	2.3
	P_{pow} $(i=3)$ [2]	1.44 ± 0.06	1.35 ± 0.02	2.36 ± 0.08	0.52 ± 0.02	0.72 ± 0.02	7.2
	P_{gam} $(i=4)$ [2]	1.44 ± 0.06	1.35 ± 0.02	–	–	1	17
	P_{ln} $(i=5)$ [2]	–	–	0.02 ± 0.06	0.70 ± 0.02	0	6.1
2.0 SCEDC	P_{lw} $(i=2)$ [1]	1.83 ± 0.04	0.81 ± 0.01	1.13 ± 0.02	1.47 ± 0.01	0.45 ± 0.01	1.7
	P_{pow} $(i=3)$ [2]	2.36 ± 0.16	0.81 ± 0.01	3.40 ± 0.10	0.48 ± 0.008	0.44 ± 0.04	7.1
	P_{gam} $(i=4)$ [2]	2.36 ± 0.16	0.81 ± 0.01	1.11 ± 0.06	0.72 ± 0.06	0.97 ± 0.10	25
	P_{ln} $(i=5)$	–	–	-0.38 ± 0.01	0.42 ± 0.04	0	15

1 We used the parameter estimation procedure A.

2 We used the parameter estimation procedure B.

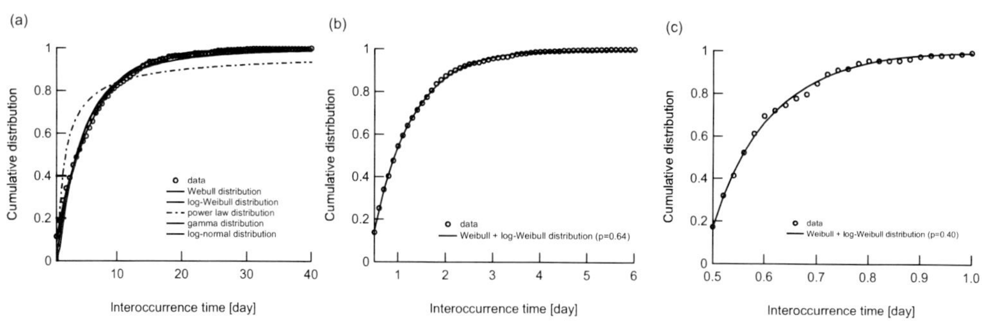

Figure 16. Distribution of interoccurrence time at Taiwan for different m_c and distribution functions. (a) $m_c = 4.5$, (b) $m_c = 3.5$, and (c) $m_c = 2.5$. In (a), the pure Weibull regime. For (b) and (c) the superposition regime.

$m_c = 3.5$, and $m_c = 2.5$, respectively. For large m_c, the Weibull distribution is preferred on the basis of the rms and KS test [see Table 7]. As the threshold of magnitude m_c decreases, the fitting accuracy of the Weibull distribution is getting worse, as is common in JMA and SCEDC. According to a hypothesis that the interoccurrence time distribution can be described by the superposition of the Weibull distribution and another distribution, we investigate the distribution $P_X(\tau)$. As shown in Table 8, the log-Weibull distribution is preferred as the distribution on the basis of the two goodness-of-fit tests.

Table 7. Results of rms value, DKS, and Q for different distribution functions in the case of $m_c = 4.5$ for Taiwan earthquakes. The error bars mean the 95% confidence level of fit (Hasumi et al. 2010)

m_c		Distribution X		RMS test	KS test	
Region	distribution X	α_i	β_i [day]	rms [$\times 10^{-3}$]	DKS	Q
	P_l $(i = 1)$	0.92 ± 0.01	5.44 ± 0.04	5.8	0.019	1
	P_{lw} $(i = 2)$	2.28 ± 0.12	9.93 ± 0.21	31	0.129	0.13
5.0	P_{pow} $(i = 3)$	1.67 ± 0.07	0.66 ± 0.12	11	0.20	0.003
TCWB	P_{gam} $(i = 4)$	0.99 ± 0.005	5.52 ± 0.07	10	0.040	0.999
	P_{ln} $(i = 5)$	1.25 ± 0.04	1.07 ± 0.04	22	0.079	0.690

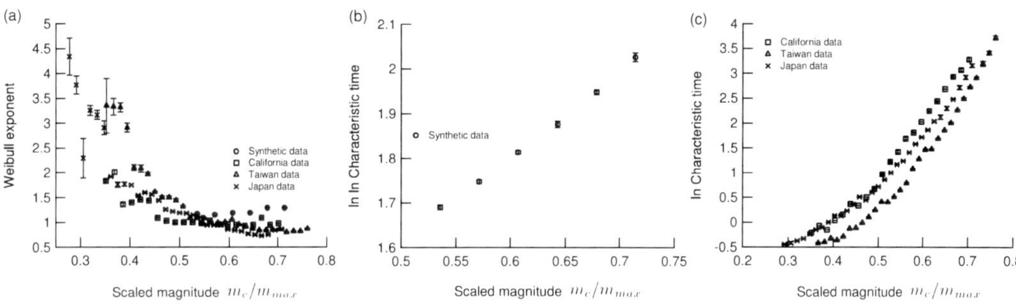

Figure 17. Change of the Weibull component as the function of m_c/m_{max}. The result of α_1 for all four data, β_1 for the synthetic and β_1 for natural data are shown in (a), (b), and (c), respectively (Hasumi et al. 2010).

Hence, it is found that the interoccurrence times distribution in TCWB data also can be described by the superposition of the Weibull distribution and the log-Weibull distribution (see equation (57)). That is to say, $P(\tau)$ is controlled by five parameters, $\alpha_1, \alpha_2, \beta_1, \beta_2$, and p.

4.3.3. The Weibull and Log-Weibull Transition of Interoccurrence Times

As we demonstrated before, the interoccurrence time distribution depends on the threshold of magnitude m_c and can be described by the superposition of the Weibull distribution and the log-Weibull distribution by analyzing the four earthquake catalogues, namely, synthetic, JMA, SCEDC, TCWB. For large m_c, the interoccurrence time distribution obey the Weibull distribution for $\alpha_1 < 1$ (natural data) and $\alpha_1 > 1$ (synthetic data). Here, we further investigate the change of the Weibull and the log-Weibull components as a function of m_c. Note that those results of Figure 17, 18, and 19 of the JMA data are obtained from a divided region spanning 140°–145°E and 35°–40°.

The changes in the fitting paramcters of the Weibull and log-Weibull distributions, as a function of m_c/m_{max}, are shown in Figure 17 and 18. We have introduced the scaled magnitude, defined as m_c/m_{max}, for unifying the data presentation of individual catalogues. We demonstrate the relation between the Weibull parameters (α_1, β_1) as a function of m_c/m_max

Table 8. Interoccurrence time statistics by analyzing the TCWB data. ± means the 95% confidence level of fit (Hasumi et al. 2010)

m_c	distribution X	Weibull distribution α_1	β_1	distribution X α_i	β_i	p	rms $[\times 10^{-3}]$
5.0 TCWB	P_{lw} $(i=2)$	0.86 ± 0.02	15.3 ± 0.22	–	–	1	12
	P_{pow} $(i=3)$	0.86 ± 0.02	15.3 ± 0.22	–	–	1	12
	P_{gam} $(i=4)$	0.86 ± 0.02	15.3 ± 0.22	–	–	1	12
	P_{ln} $(i=5)$	0.86 ± 0.02	15.3 ± 0.22	–	–	1	12
4.5 TCWB	P_{lw} $(i=2)$	0.92 ± 0.01	5.44 ± 0.04	–	–	1	5.8
	P_{pow} $(i=3)$	0.92 ± 0.01	5.44 ± 0.04	–	–	1	5.8
	P_{gam} $(i=4)$	0.92 ± 0.01	5.44 ± 0.04	–	–	1	5.8
	P_{ln} $(i=5)$	0.92 ± 0.01	5.44 ± 0.04	–	–	1	5.8
4.0 TCWB	P_{lw} $(i=2)$ [1]	1.00 ± 0.04	2.20 ± 0.10	1.84 ± 0.14	4.14 ± 0.34	0.68 ± 0.04	3.5
	P_{pow} $(i=3)$ [2]	1.09 ± 0.02	2.25 ± 0.04	1.94 ± 0.08	0.53 ± 0.06	0.88 ± 0.02	7.4
	P_{gam} $(i=4)$ [2]	1.09 ± 0.02	2.25 ± 0.04	2.04 ± 0.02	0.98 ± 0.04	0.83 ± 0.06	9.4
	P_{ln} $(i=5)$ [2]	1.09 ± 0.02	2.25 ± 0.04	0.44 ± 0.01	0.91 ± 0.01	0.40 ± 0.06	3.4
3.5 TCWB	P_{lw} $(i=2)$ [1]	1.44 ± 0.06	1.28 ± 0.02	1.24 ± 0.10	2.08 ± 0.12	0.64 ± 0.04	5.0
	P_{pow} $(i=3)$ [2]	1.50 ± 0.06	1.24 ± 0.02	2.42 ± 0.10	0.50 ± 0.02	0.73 ± 0.04	8.4
	P_{gam} $(i=4)$ [2]	–	–	2.01 ± 0.01	0.56 ± 0.02	0	14
	P_{ln} $(i=5)$ [2]	–	–	−0.05 ± 0.006	0.66 ± 0.008	0	6.1
3.0 TCWB	P_{lw} $(i=2)$ [1]	2.08 ± 0.14	0.75 ± 0.02	1.19 ± 0.06	1.49 ± 0.02	0.49 ± 0.04	3.4
	P_{pow} $(i=3)$ [2]	2.63 ± 0.20	0.78 ± 0.01	3.56 ± 0.14	0.48 ± 0.008	0.53 ± 0.04	7.5
	P_{gam} $(i=4)$ [2]	2.63 ± 0.20	0.78 ± 0.01	1.92 ± 0.10	0.37 ± 0.04	0.94 ± 0.12	24
	P_{ln} $(i=5)$ [2]	–	–	−0.40 ± 0.008	0.38 ± 0.01	0	12
2.5 TCWB	P_{lw} $(i=2)$ [1]	3.35 ± 1.08	0.59 ± 0.06	1.01 ± 0.18	1.18 ± 0.04	0.40 ± 0.12	14
	P_{pow} $(i=3)$ [2]	5.09 ± 0.72	0.61 ± 0.01	6.21 ± 0.22	0.48 ± 0.004	0.39 ± 0.04	14
	P_{gam} $(i=4)$ [2]	5.09 ± 0.72	0.61 ± 0.01	1.67 ± 4.84	0.68 ± 0.73	0.96 ± 0.14	43
	P_{ln} $(i=5)$	–	–	−0.57 ± 0.01	0.19 ± 0.02	0	32

1 We used the parameter estimation procedure A.

2 We used the parameter estimation procedure B.

in Figure 17. As shown in Figure 17 (a), α_1 gradually decreases as m_c/m_{max} increases for the case of natural data, namely, JMA, SCEDC and TCWB. Additionally, for large m_c/m_{max}, α_1 obtained from the natural data is less than unity, indicating that the occurrence of natural earthquakes is not a Poisson process, whereas α_1 for the synthetic data gradually increases and is greater than unity. However, it should be also noted that α_1 derived from the synthetic data could be less than 1 by tuning the parameters of the 2D spring-block model (Hasumi et al. 2009b). On the other hand, as m_c/m_{max} increases, β_1 for the synthetic data (Figure 17 (b) increases double exponentially, while β_1 for natural data (Figure 17 (c)) increases exponentially.

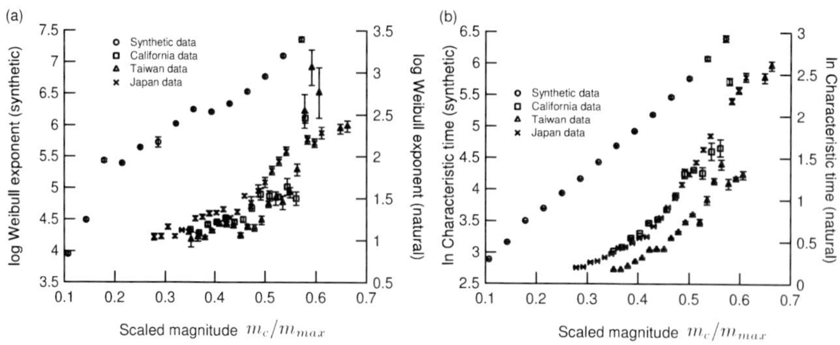

Figure 18. Change of the log-Weibull component as the function of m_c/m_{max}. The result of α_2 and β_2 are presented in (a) and (b), respectively (Hasumi et al. 2010).

We also showin Figure 18 the log Weibull parameters (α_2, β_2) as a function of m_c/m_{max}. For all catalogues, as m_c/m_{max} increases, α_2 (Figure 18 (a) and β_2 (Figure 18 (b)) increase linearly and exponentially, respectively.

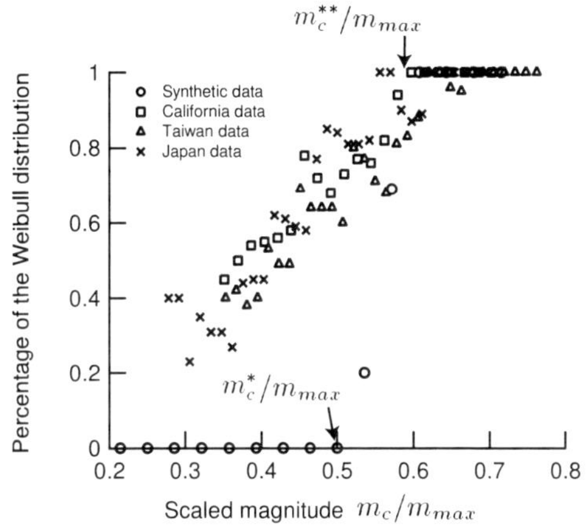

Figure 19. Change of the percentage of the Weibull distribution as the function of m_c/m_{max}. (Hasumi et al. 2010).

Most importantly, as clearly shown in Figure 19, the percentage p of the Weibull distribution in $P(\tau)$ gradually increases as m_c/m_{max} increases, which demonstrates that as the threshold of magnitude is varied the probability distribution for $P(\tau)$ changes and the dominant distribution is transferred from the log-Weibull distribution to the Weibull distribution. We thus define two crossover (transition) magnitudes m_c^*/m_{max} and m_c^{**}/m_{max} representing the scaled magnitudes of the distribution changed from the pure log-Weibull regime to the superposition regime and from the superposition regime to the pure Weibull regime, respectively (Figure 19). As shown in Figure 19, the transition from the log-Weibull regime to the Weibull regime for the synthetic data (circles) appears clearly, so that both the first (m_c^*/m_{max}) and the second (m_c^{**}/m_{max}) crossover magnitudes can be well estimated. As for the cases of natural catalogues, due to incomplete events smaller than the magnitude of completeness, the pure log Weibull regime cannot be observed (Figure 19) and we can thus determine the second crossover point m_c^{**}/m_{max} only. It looks like that the values of m_c^{**}/m_{max} are coincidently around 0.60 for all of four catalogues, indicating its region-independence.

4.4. Discussions

4.4.1. Finite Size Effect

To investigate the region-size or system size dependency of the Weibull - log-Weibull transition, we further change the region-size or system size.

Synthetic data The system size dependence of the interoccurrence times is studied to discuss the finite size effect (Hasumi et al. 2009a). In this time, the number of blocks N

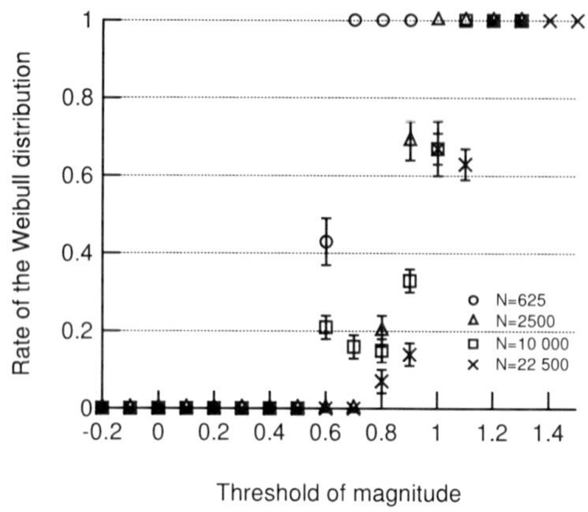

Figure 20. The ratio of Weibull distribution p as a function of m_c for different numbers of blocks N ($\circ$: 625, $\triangle$: 2500, $\square$: 10 000, and $\times$: 22 500). Please see the original figure in (Hasumi et al 2009a).

is changed from 625 (25×25) to 22 500 (150×150), while other parameters are fixed at $l_x^2 = 1, l_y^2 = 3$, and $\alpha = 3.5$; N is taken to be $N = 625$ (25×25), $N = 2500$ (50×50), $N = 10\ 000$ (100×100), and $N = 22\ 500$ (150×150). As clearly seen from Figure 20, p changes from 0 to 1 as m_c is increased, suggesting the fact that the Weibull - log Weibull transition appears in all cases. Crossover magnitudes, m_c^* and m_c^{**} are then evaluated; $m_c^* = 0.5$ and $m_c^{**} = 0.7$ for $N = 625$ ($\circ$), $m_c^* = 0.7$ and $m_c^{**} = 1.0$ for $N = 2500$ ($\triangle$), $m_c^* = 0.5$ and $m_c^{**} = 1.1$ for $N = 10\ 000$ ($\square$), and $m_c^* = 0.7$ and $m_c^{**} = 1.2$ for $N = 22\ 500$ ($\times$). Thus, we conclude that the interoccurrence time statistics, especially the Weibull - log-Weibull transition, are retained for a large system size.

JMA data As for JMA, we change the window size L is varied from 3° to 25° (Hasumi et al. 2009b). Hasumi et al. (2009c), used JMA data we used is from 1st January 2001 to 31st October 2007. We use the data covering the region 140°-143° E and 35°-38° N for $L = 3$, 140°-145° E and 35°-40° N for $L = 5$, 140°-150° E and 35°-45° N for $L = 10$, and 125°-150° E and 25°-50° N for $L = 25$. As for $L = 25$, the data covers the whole region of the JMA catalogue. The result of fitting parameters of $P(\tau)$, the crossover magnitude m_c^{**}, and the rms value are listed in Table 9. It is demonstrated that in all the cases, the Weibull exponent α_1 is less than unity and the Weibull - log-Weibull transition appears. m_c^{**} depends on L, namely $m_c^{**} = 3.9$ for $L = 3$, $m_c^{**} = 4.0$ for $L = 5$, $m_c^{**} = 4.2$ for $L = 10$, and $m_c^{**} = 5.0$ for $L = 25$. Therefore we can conclude that the interoccurrence time statistics, namely the Weibull - log-Weibull transition, presented here hold from $L = 3$ to $L = 25$.

4.4.2. Scaling Relation Between m_c^{**} and m_{max}

To study the feature of the Weibull - log-Weibull transition, we summarize our results obtained from 17 different regions (14 regions in Japan, Southern California, Taiwan, and the synthetic.) Interestingly, m_c^{**} is proportional to the maximum magnitude of an earthquake in a region, where we analyzed, denoted here m_{max} [see Figure 21]. We then obtain a

Table 9. Interoccurrence time statistics for different system size L by analyzing the JMA data from 1st January 2001 to 31st October 2007 (Hasumi et al. 2009b)

L	Region	m_c	α_1	β_1 [day]	rms	m_c^{**}
$L = 3$	140°-143° E and 35°-38° N	3.9	0.88 ± 0.01	19.4 ± 0.18	0.011	4.6
$L = 5$	140°-145° E and 35°-40° N	4.0	0.75 ± 0.02	10 ± 0.19	0.014	4.7
$L = 10$	140°-150° E and 35°-45° N	4.2	0.94 ± 0.005	8.36 ± 0.04	0.0077	4.9
$L = 25$	125°-150° E and 25°-50° N	5.0	0.93 ± 0.01	17.8 ± 0.10	0.041	5.7

Table 10. List of the crossover magnitude, m_c^{**} and the plate velocity (Fowler 1990; Seno et al. 1993). The notation of PH, EU, PA, and NA represent PHilippine Sea plate, EUrasian plate, PAcific plate, and North American plate, respectively (Hasumi et al. 2010)

Region	relative plate motion	velocity [mm/yr]	m_c^{**}
Taiwan	PH-EU	71	4.40
East Japan	PA-PH	49	3.80 [1]
West Japan	PH-EU	47	3.80 [2]
California	PA-NA	47	3.40

1 We take an average using three regions; 25°–30°N, 140°–145°E ($m_c^{**} = 3.7$), 30°–35°N, 140°–145°E ($m_c^{**} = 3.7$), 35°–40°N, 140°–145°E ($m_c^{**} = 4.0$).
2 We take an average using five regions; 25°–30°N, 125°–130°E ($m_c^{**} = 3.7$), 25°–30°N, 130°–135°E ($m_c^{**} = 3.3$), 30°–35°N, 130°–135°E ($m_c^{**} = 4.3$), 30°–35°N, 135°–140°E ($m_c^{**} = 4.1$), 35°–40°N, 135°–140°E ($m_c^{**} = 3.6$).

region-independent relation between m_c^{**} and m_{max},

$$m_c^{**}/m_{max} = 0.54 \pm 0.04 \tag{58}$$

This relation can be useful to interpret the Weibull - log-Weibull transition of geophysical meaning.

4.4.3. Interpretation of the Weibull - Log-Weibull Transition

Although the scaled crossover magnitudes m_c^{**}/m_{max} is region-independent, the crossover magnitude m_c^{**} from the superposition regime to the pure Weibull regime probably depends on the tectonic region. To investigate the Weibull - log-Weibull transition further, we consider the plate velocity with m_c^{**}, which can shed light on the geophysical implication of the region-dependent m_c^{**}. As shown in Table 10, we propose the conjecture that m_c^{**} is on the average proportional to the plate velocity. Since the number of observations is only 4, more observations will be needed in order to reach to a conclusion. That means that the maximum magnitude m_{max} for a tectonic region is more or less proportional to the plate velocity since $m_c^{**}/0.54 = m_{max}$. Such an interesting consequence is reminiscent of the early study by Ruff and Kanamori (1980). They showed a relation that the magnitude of

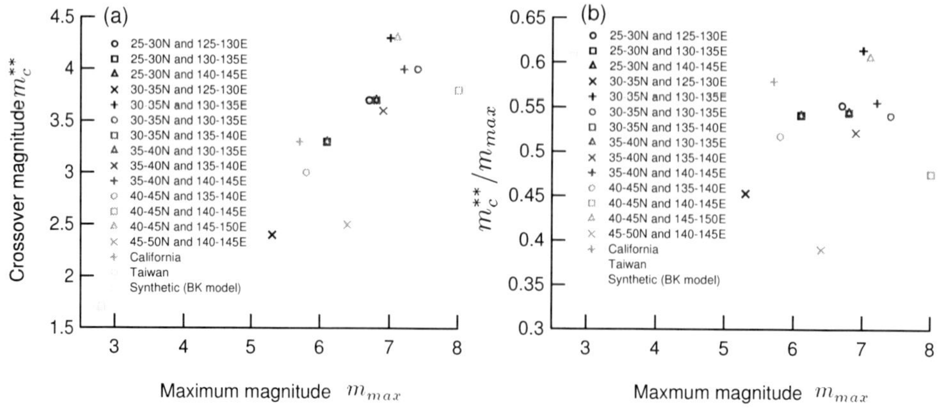

Figure 21. Relation between m_c^{**} and m_{max}. In (a), m_c^{**} is proportional to the m_{max}. In (b), we obtain a region-independent constant for the scaled crossover magnitude m_c^{**}/m_{max}, which is 0.54 ± 0.06. The original figure is in (Hasumi et al. 2010)

characteristic earthquake occurred in the subduction-zone, M_w is directly proportional to the plate-velocity, V, and is directly inversely proportional to plate-age T, namely,

$$M_w = -0.000953T + 1.43V + 8.01. \tag{59}$$

The relation $m_c^{**}/0.54 = m_{max}$ can thus be explained on the basis of their early observation about the velocity-dependence of the characteristic earthquake magnitude. The physical interpretation of the Weibull - log-Weibull transition remains open. However, it might suggest that the occurrence mechanism of earthquake could probably depend on its magnitude then, inevitably, the distribution of the interoccurrence time statistics changes as the threshold of magnitude m_c is varied. It is well known that the Weibull distribution for life-time of materials can be derived in the framework of damage mechanics (Weibull 1951; Ghosh 1999; Turcotte 2003; Wong et al. 2006; Abaimov et al. 2007). Our present results thus suggest that larger earthquakes might be caused by the damage mechanism driven by the plate motion, whereas the effect of the plate-driven damaging process might become minor for smaller earthquakes. Hence, the transition from the Weibull regime to the log-Weibull regime could be interpreted from the geophysical sense as the decrement of the plate-driven damaging mechanics.

4.4.4. A Universal Relation and Intrinsic Meanings of the Gutenberg-Richter Parameter

Here we consider the interrelation between the Gutenberg-Richter law (Gutenberg and Richter, 1941) denoted in this subsection $P(m) \propto e^{-bm}$ and the Weibull distribution for the interoccurrence time ($P(\tau) \propto t^{\alpha-1} \cdot e^{(\tau/\beta)^{\alpha}}$). We assume that these two statistics are correct over wide ranges, and the parameters (α, β) are depending on the magnitude, i.e., $\alpha(m)$ and $\beta(m)$, then the following relation is easily obtained from the calculation of the mean

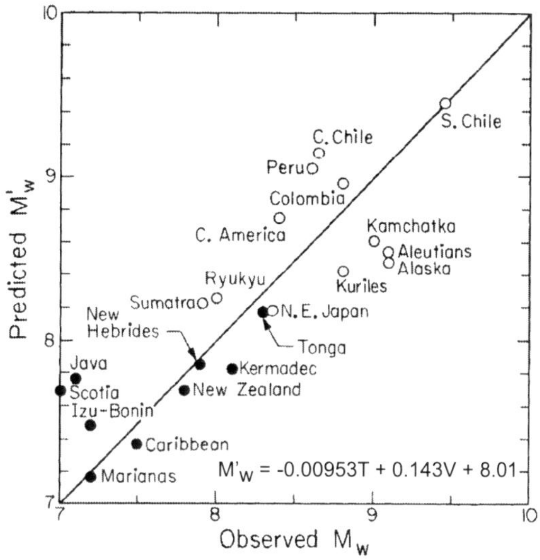

Figure 22. Relation between the predicted magnitude M'_w and observed magnitude M_w (Ruff and Kanamori 1980). This result indicates that the M_ws directly proportional to the plate-velocity, V, and is directly inversely proportional to plate-age T.

interoccurrence time between two earthquakes whose magnitude is larger than m ,

$$\beta(m_1)e^{-bm_1}\Gamma\left(1+\frac{1}{\alpha_1}\right)=\beta(m_2)e^{-bm_2}\Gamma\left(1+\frac{1}{\alpha_2}\right), \tag{60}$$

where m_1 and m_2 are arbitrary values of m. This implies that the quantity defined by $\beta(m_1)e^{-bm_1}\Gamma\left(1+\frac{1}{\alpha_1}\right)$ is a universal constant when we consider the local earthquakes in a relatively small area.

One of the most important results derived from equation (60) is that the GR parameter b is determined by two parameters, in other words, the parameters (α,β) depend on the magnitude m as well as on the GR parameter b,

$$\begin{aligned}\alpha &= f_\alpha(m,b)\\ \beta &= f_\beta(m,b),\end{aligned} \tag{61}$$

where the functional forms of f_α and f_β characterize the time series of earthquakes under consideration.

It is difficult to determine those forms completely from any seismological relations known so far, but it is possible for us to obtain the universal aspects of f_α and f_β by a perturbational approach. Here we consider a particular solution of equation (61) which satisfies the following conditions; $f_\beta(m,b)=\exp\left[b(m-m_c)+c\right]$ and $f_\alpha(m_c,b)=1$, namely, the characteristic time β is a exponentilally increasing function of m, and the interoccurrence time distribution is an exponential one ($\alpha=1$) at $m=m_c$, where b' and c are constant parameters. By use of this simplification, equation (60) is rewritten by putting $m_1=m_c$ and

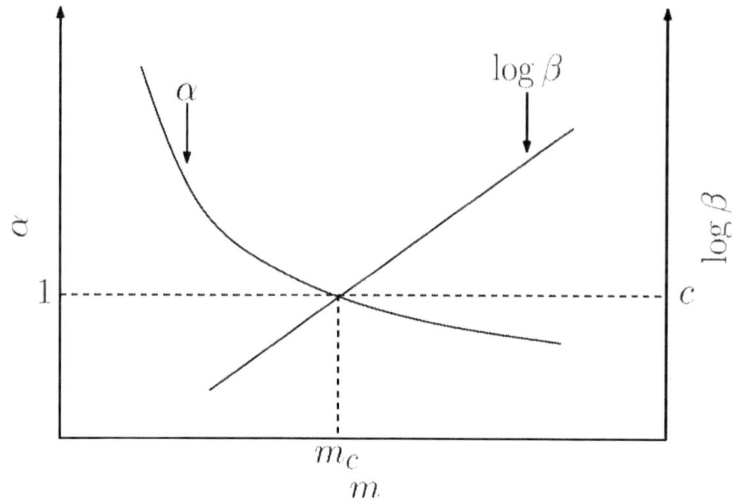

Figure 23. Schematic picture of universal behavior of $f_\alpha(m,b)$ and $f_\beta(m,b)$ near $m \simeq m_c$.

$m_2 = m$,

$$(b'-b)(m-m_c) = -\log\Gamma\left(1+\frac{1}{\alpha(m)}\right) \cong \frac{1}{2}\Delta - \frac{3}{4}(\Delta)^2 + \cdots, \quad (\Delta = \alpha(m) - 1). \tag{62}$$

Here we used the Taylor expansion near $m \cong m_c$ (i.e., $\alpha(m) \cong \alpha(m_c)$). Figure 23 shows the schematic result of equation (62). One can see that the universal relation is recognized in many cases treated in this chapter (section 4.3.3.), though the exponential growth of β, $\log\beta(m) \simeq b'(m-m_c)+c$ is a little bit accelerated.

We have to remind that the solution mentioned above is not unique, but many other solutions for equation (61) are possible under the universal relation of equation (60). Further details will be studied in our forthcoming paper (Aizawa and Hasumi 2011).

4.4.5. Comparison With Previous Works

Finally, we compared our results with those of previous studies. The unified scaling law shows a generalized gamma distribution [see in equations (11), and (12)] which is approximately the gamma distribution, because δ in Corral's paper (2004) is close to unity ($\delta = 0.98 \pm 0.05$). For a long time domain, this distribution decays exponentially, supporting the view that an earthquake is a Poisson process. However, we have demonstrated that the Weibull distribution is more appropriate than the gamma distribution on the basis of two goodness-fit-tests. In addition, for large m_c, the distribution in a long time domain is similar to the stretched exponential distribution because α_1 is less than unity, suggesting that an occurrence of earthquake is not a Poisson process but has a memory. We provide the first evidence that the distribution changes from the Weibull to log-Weibull distribution by varying m_c, i.e., the Weibull - log-Weibull transition. Recently, Abaimov et al. (2007)

showed that the recurrence time distribution is also well-fitted by the Weibull distribution rather than the Brownian passage time (BPT) distribution (Matthews et al. 2002) and the log normal distribution. Taken together, we infer that both the recurrence time statistics and the interoccurrence time statistics show the Weibull distribution.

4.5. Summary of the Section

Here we summarize this section. We investigate the new statistical properties of earthquakes by analyzing the four earthquake catalogues, namely, the synthetic data produced by the 2D spring-block model, JMA (Japan Meteorological Agency), SCEDC (Southern California Earthquake Data Center), and TCWB (Taiwan Central Weather Bureau). We focus on the interoccurrence time, the time intervals between successive earthquakes, by changing the threshold of magnitude, m_c, which has not been reported before.

At first, we analyze the synthetic data created by the spring-block model with parameters, $l_x^2 = 1, l_y^2 = 3, \alpha = 3.5, \sigma = 0.01$, and $\nu = 0.01$. It is found that the interoccurrence time distribution depends on m_c, and can be described by the superposition of the Weibull distribution and log-Weibull distribution. For large m_c the distribution obeys the Weibull distribution with $\alpha_1 > 1$. We infer that this result can also obtained by analyzing the natural data by using the Approach (1). Thus, we analyze the three natural data, JMA, SCEDC, and TCWB so as to confirm whether the interoccurrence time statistics hold or not. As a result, the interoccurrence time distribution also depends on m_c and can be written as the superposition of the Weibull distribution and the log-Weibull distribution. For large m_c the distribution obeys the Weibull distribution with $\alpha_1 < 1$. In the case of synthetic data, as m_c increases, the percentage of the Weibull distribution, p gradually increases from 0 (the pure log-Weibull regime) to 1 (the pure Weibull regime), which indicates that the interoccurrence time distribution exhibits the sharp Weibull - log-Weibull transition. On the contrary, by analyzing the natural data, the pure log-Weibull regime does not appear. The interoccurrence time distribution shows the smooth transition, which means that interoccurrence time distribution changes from the superposition regime to the pure Weibull regime when m_c is increased.

This Weibull - log-Weibull transition holds by varying the system size and area size for synthetic and natural data, respectively. We propose the scaling relation between the crossover magnitude, m_c^{**} and the maximum magnitude, m_{max}, namely, $m_c^{**}/m_{max} = 0.54 \pm 0.04$ on the basis of 17-regions, where we have analyzed (14-regions; JMA, SCEDC, TCWB, and the synthetic). This relation is useful to discuss the interpretation of the Weibull - log-Weibull transition. Taken together the previous works, we might suggest that the transition of the Weibull regime to the log Weibull regime could be interpreted from the geophysical sense as the decrement of the plate-driven damaging mechanics. Finally, we first propose a universal relation between the interoccurrence time statistics, namely the Weibull components, (α_1, β_1) and the b-value of the Gutenberg-Richter law. The details are presented in our forthcoming paper (Aizawa and Hasumi 2011).

5. Intermittent Phenomena of Earthquakes

In this section, we investigate intermittency of earthquakes by using the Approach (2), which states that we infer the statistical properties of the sequence of earthquakes from those extracted by the map we proposed. In this time, we construct a map which we proposed on the basis of the deterministic chaotic dynamics in order to characterize the intermittency of earthquakes in Japan.

5.1. Background Idea

In intermittent phenomena, regular motions called *laminar states* are disturbed by irregular *bursts*. The residence time distribution of laminar states are characterized by a power law, which generates a $1/f$ spectrum. Intermittency is a ubiquitous phenomenon in non-equilibrium statistical physics, biological and atomic physics, seismology. In particular, the fluorescence of quantum dots (Haase et al. 2004) and nanocrystals (Brokmann et al. 2003), and ion channel gating (Lowen et al. 1999) are intermittent phenomena characterized by two states, i.e., on and off states. Earthquakes are also considered as an intermittent phenomenon. Actually, a $1/f^{\delta}$ spectrum is observed in the P-wave and S-wave velocity in function of depth (Shiomi et al. 1997). Also, the residence time distribution of the laminar state, in which the number of earthquakes per unit time is lower than a threshold, obeys a power law (Bottiglieri and Godano 2007), and the intermittency for the occurrence of earthquakes in Irpinia, Italy, has been quantified using the correlation codimension (Telesca et al. 2001). Moreover, intermittency appears in the stick-slip model of earthquakes (Ryabov and Ito 2001).

Renewal processes are point processes where time intervals between renewals are independently and identically distributed (i.i.d.) random variables. Renewal theory (Cox: 1962) is useful to study intermittent phenomena characterized two states (Godréche and Luck 2001) and the continuous time random walk (Miyaguchi and Akimoto 2011). If the interoccurrence times for earthquakes are i.i.d. random variables, the occurrence of earthquakes is considered as a renewal process, where renewals mean occurrences of earthquakes (see Figure 24). This assumption will be used without verifications, as in the previous analysis. However, it is doubtful because interoccurrence times for the occurrence of earthquakes are correlated (Livina et al. 2005). Therefore, it is an important problem to clarify whether the occurrence of earthquakes is a renewal process and how interoccurrence times are correlated.

The conditional mean and probability distribution when the previous interoccurrence time is known are useful to investigate the correlation of interoccurrence times. Here, we study the conditional mean and probability distribution for interoccurrence time of earthquakes and characterize intermittency to estimate the difficulty of forecasting the occurrence of earthquakes. Constructing one-dimensional maps from renewal processes, we characterize intermittency of earthquakes on the basis of the constructed maps (Akimoto et al. 2010). Renewal processes are classified into six different regimes according to difficulty to forecast the next event: (i) non-stationary essential singular intermittency, (ii) non-stationary very strong intermittency, (iii) stationary strong intermittency, (iv) stationary weak intermittency, (v) stationary very weak intermittency, and (vi) stationary non-

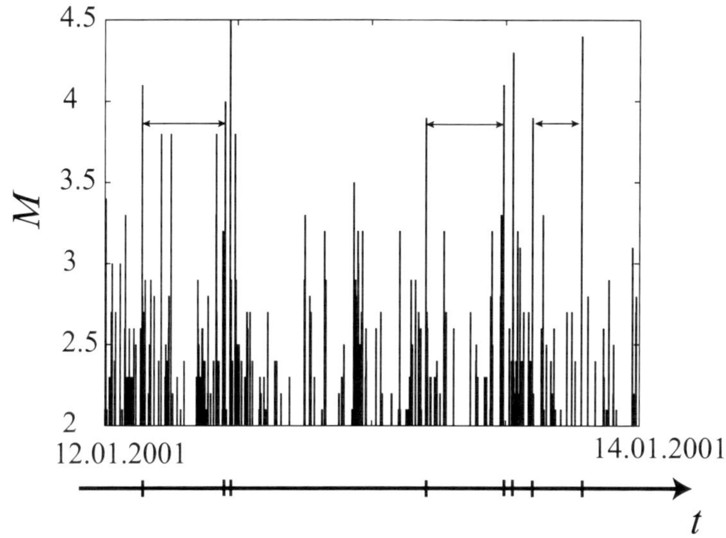

Figure 24. Sequence of earthquakes with magnitude $M \geq 2$ for JMA catalog from January 12, 2001 to January 14, 2001. Three time intervals between renewals are indicated by arrows, where renewals are defined as the occurrence of an earthquakes with magnitude $M \geq 3.9$.

intermittent chaos. Moreover, the degree of activity of earthquakes is studied from the viewpoint of a dynamical instability of the map.

5.2. Intermittency and the Lyapunov Exponent in Renewal Processes

To study the difficulty to forecast the occurrence of earthquakes, we construct one-dimensional maps from discrete time renewal processes. Interoccurrence times between renewals correspond interoccurrence times of earthquakes if they are i.i.d. random variables. Let $f(m)$ be the probability distribution function of a interoccurrence time m ($m = 1, 2, \cdots$), and $F(m) = \sum_{k=1}^{m} f(k)$ and $\mathscr{F}(m) = 1 - F(m)$, where $F(0) = 0$ and $\mathscr{F}(0) = 1$. Then, we have

$$f(m) = \mathscr{F}(m-1) - \mathscr{F}(m). \tag{63}$$

Using this relation, we can construct a one-dimensional map on $[0, c]$ $(c > 1)$ where the residence times in $[0, 1]$ are i.i.d. random variables with probability $f(m)$ (see Fig. 25). In particular, the map is given by a piecewise linear map $T : [0, c] \to [0, c]$ defined by

$$\begin{aligned} x_{n+1} &= T(x_n) \\ &= \begin{cases} \dfrac{a_{k-1} - a_k}{a_k - a_{k+1}}(x_n - a_k) + a_{k-1}, & x_n \in [a_{k+1}, a_k), \\ \\ \dfrac{x_n - 1}{c - 1}, & x_n \in [1, c], \end{cases} \end{aligned} \tag{64}$$

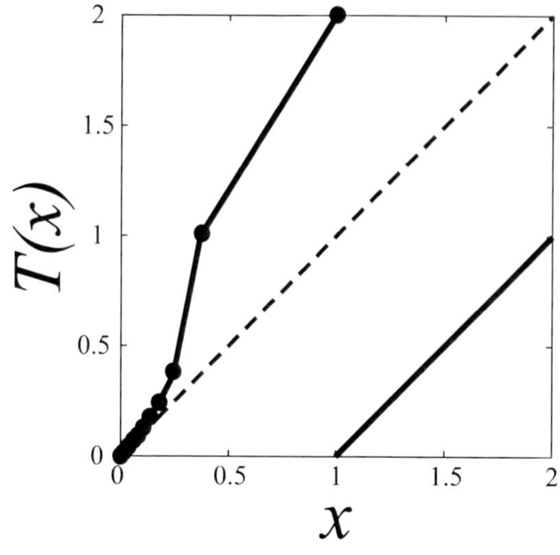

Figure 25. Piecewise linear map $T(x)$ for the Weibull distribution ($a = 0.5, \tau = 5, c = 2.$). Circles indicate the end points of the straight line segments. (see Figure 1 in (Akimoto et al. 2010))

where a sequence a_k is given by $a_k = \mathscr{F}(k)$ and $\mathscr{F}(-1) = c$. In fact, a point in the interval $[a_m, a_{m-1}]$ is mapped into the interval $[1, c]$ by $m+1$th iterations, and then the probability of the residence time m in the interval $[0, 1]$ is given by $a_m - a_{m-1}$.

Next, we focus on the constructed map near the fixed point $T(x) = x$ $(x = 0)$. The derivative of the map is given by

$$T'(x)|_{x\in[a_n,a_{n-1})} = \frac{a_{n-1} - a_n}{a_n - a_{n+1}} = \frac{f(n)}{f(n+1)}. \tag{65}$$

Since $x \cong a_n = \mathscr{F}(n)$, $a_{n-1} - a_n \cong 0$ near the fixed point, we obtain the asymptotic form,

$$T'(x) \sim \frac{f(\mathscr{F}^{-1}(x))}{f(\mathscr{F}^{-1}(x)+1)} \quad \text{as } x \to 0. \tag{66}$$

In particular, the asymptotic behavior for a power law distribution ($\mathscr{F}(m) \sim m^{-\beta}$) is given by

$$T'(x) - 1 \propto x^{1/\beta} \quad \text{as } x \to 0. \tag{67}$$

This map is the piecewise linear version of the Pomeau-Manneville map (Manneville and Pomeau 1979), which is a typical example of intermittent maps (Tasaki and Gaspard 2002; Miyaguchi and Aizawa 2007). The asymptotic behavior is given by

$$T'(x) \sim e^{1/\tau} \quad \text{as } x \to 0 \text{ for } \mathscr{F}(m) \sim e^{-m/\tau}, \tag{68}$$

and

$$T'(x) - 1 \propto (-\log x)^{(a-1)/a} \quad \text{as } x \to 0 \tag{69}$$

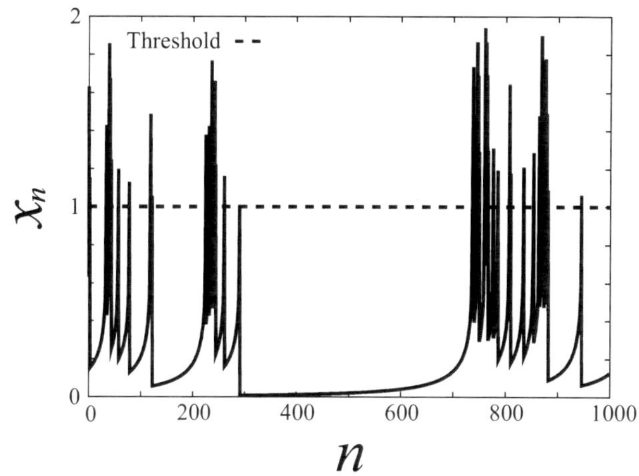

Figure 26. Time series of piecewise linear intermittent maps, where the map is constructed using the Weibull distribution ($\mathscr{F}(m) = \exp(-(m/5)^{0.35})$.

for the Weibull distribution ($\mathscr{F}(m) \sim e^{-(m/\tau)^a}$) ($a \neq 1$). In the case of the log-Weibull distribution ($\mathscr{F}(m) \sim e^{-(\log m/\tau)^b}$), the asymptotic behavior is given by

$$T'(x) - 1 \propto \exp(-\tau(-\log x)^{1/b}) \quad \text{as } x \to 0. \tag{70}$$

We refer to equations (69) and (70) as the Weibull map and the log-Weibull map, respectively. We note that the Weibull map with exponent $a < 1$ and the log-Weibull map have the indifferent fixed point, i.e., $T'(0) = 1$.

In intermittent chaos, orbits are trapped near indifferent fixed points for an extremely long time (*laminar state*), and then irregular chaotic motions (*bursts*) occur (see Fig. 26). The residence time distribution of the laminar state is determined by the structure of a map near the indifferent fixed points. In general, the Lyapunov exponent, which characterizes dynamical instability, of intermittent chaos is small. In other words, strong intermittent chaos has a small Lyapunov exponent. However, residence times near the indifferent fixed point in strong intermittent chaos depend sensitively on the reinjection point. Therefore, the instability of residence times is large in intermittent maps. Here, we characterize intermittency from the asymptotic behavior of the derivative at the indifferent fixed point ($x = 0$):

$$T'(x) - 1 \propto A(x)x^{\alpha} \quad \text{as } x \to 0, \tag{71}$$

where $T(x)$ is the one-dimensional map constructed by a renewal process and $A(x) = o(1)$ and $x^{\alpha} = o(A(x))$. The degree of intermittency is classified into six types:

(i) $\alpha = \infty$, i.e., $T'(x) - 1 \propto e^{-x^{-\alpha'}}$ as $x \to 0$ ($\alpha' > 0$): non-stationary essential singular intermittency,

(ii) $\alpha \geq 1$: non-stationary very strong intermittency,

(iii) $0 < \alpha < 1$: stationary strong intermittency,

(iv) $\alpha = 0$ and $A(x) = o(L(x))$ as $x \to 0$: stationary weak intermittency,

(v) $\alpha = 0$ and $A(x) \sim L(x)$ as $x \to 0$: stationary very weak intermittency,

(vi) $T'(x) - 1 > 0$ as $x \to 0$: stationary non-intermittent chaos,

where the function $L(x)$ is a positive function satisfying $L(tx)/L(t) \to 1$ as $t \to 0$. The intensity of intermittency can be quantified by the exponent α. It is difficult to forecast the next event when the exponent α is large, because slightly different reinjections near the fixed point make the residence time in $[0,1]$ completely different in the case of large α. With regard to type (iv), to be more precise,

$$T'(x) - 1 = O(\exp(-(-\log x)^{\gamma})) \quad \text{as } x \to 0, \tag{72}$$

we can quantify the intensity of intermittency by the exponent γ. For type (v),

$$T'(x) - 1 = O\left(\frac{1}{(-\log x)^{\eta}}\right) \quad \text{as } x \to 0, \tag{73}$$

so that the intensity of intermittency is determined by the exponent η. Exponents α, γ and η represent the degree of the difficulty to forecast events in renewal processes because the sensitive dependence of the interoccurrence time on reinjection points is determined by these exponents.

Note that the renewal function does not increase linearly with time; that is, the occurrence of renewals is not stationary, when the mean of the interoccurrence times is not finite (Cox 1962). Accordingly, the occurrence of renewals becomes non-stationary in the case of $\gamma \geq 1$. For $\gamma > 1$, intermittency is classified into the non-stationary essential singular intermittent regime.

With the aid of the construction of one-dimensional maps from renewal processes, we can estimate a dynamical instability, i.e., the Lyapunov exponent, in renewal processes. The Lyapunov exponent is the exponential rate of separation of nearby orbits. In one-dimensional maps, the Lyapunov exponent λ is defined by

$$\lambda = \lim_{n \to \infty} \frac{1}{n} \sum_{k=1}^{n} \ln |T'(x_k)|, \tag{74}$$

where $T(x)$ is a one-dimensional map constructed from a renewal process. The slope of the piecewise linear map on $[a_1, a_0)$ and $[1, c]$ is given by $(c-1)/(1-a_1)$ and $1/(c-1)$, respectively. Therefore, the Lyapunov exponent does not depend on c because $\ln(c-1)/(1-a_1) + \ln 1/(c-1) = -\ln(1-a_1)$. However, the Lyapunov exponent strongly depends upon the unit of time in renewal processes. Physical meaning of the Lyapunov exponent is the degree of activity of events. In other words, a large Lyapunov exponent implies a high activity of an underlying dynamical system.

5.3. Intermittency and Correlation of the Occurrence of Earthquakes

We analyze the occurrence of earthquakes using the JMA catalog in the area enclosed within 25°-50°N latitude and 125°-150°E longitude with magnitude $M \geq 2$. We are interested in the interoccurrence time distribution for a tail part to construct a one-dimensional piecewise linear map. Similar to previous studies (Corral 2004; Hasumi et al. 2009c), we consider earthquakes with magnitude above a certain threshold M_c. In other words, we study the interoccurrence time statistics in which magnitude is greater than M_c. Here, we use the earthquake data from January 1, 2001 to November 30, 2008.

To verify whether the occurrence of earthquakes is a renewal process, we analyze the return map of interoccurrence times (t_{n+1}, t_n), where t_n is the nth interoccurrence time. Then, to analyze the dependence of the distribution of interoccurrence times $F(t_{n+1})$ on the previous one t_n in detail, we sorted data in order

$$\{t_1,\ldots,t_M\} \to \{t'_1,\ldots,t'_M\}, \tag{75}$$

where $t'_i \le t'_{i+1}$ for $i = 1,\ldots,M-1$. We defined ordered data sets, $\mathcal{T}_{1,N},\ldots,\mathcal{T}_{L,N}$, from $t'_1,\ldots,t'_M$:

$$\mathcal{T}_{1,N} = \{t'_1,\ldots,t'_N\},\ldots,\mathcal{T}_{L,N} = \{t'_{NL+1},\ldots,t'_{(L+1)N}\}, \tag{76}$$

where $NN(L+1) < M$ and $N(L+1) > M$. To avoid statistical error, the number of data N in each ordered data is set to be 10^4. As shown in Figure 27, the return map does not generate a smooth curve. This is unlike the experiment of the dripping water faucet (Shaw 1984), which did generate a smooth curve. Furthermore, interoccurrence times appear to be random. However, the conditional average of an interoccurrence time will be small if the previous interoccurrence time is small. In Figure 27, the conditional averages are taken in each ordered data set: the conditional average on ordered data set $\mathcal{T}_{i,N}$ is given by

$$\mu_{i,N} = \sum_{t_k \in \mathcal{T}_{i,N}} \frac{t_k}{N}. \tag{77}$$

Figure 27 shows that an interoccurrence time depends clearly on the one preceding it. However, the average of an interoccurrence time is constant when the previous one is large, which suggests that large interoccurrence times do not affect the next interoccurrence times.

We consider the interoccurrence times conditioned on that the previous interoccurrence time is in the subset $\mathcal{T}_{i,N}$, i.e., $\mathcal{T}^*_{i,N} = \{t_{k+1} : t_k \in \mathcal{T}_{i,N}, k = 1,\ldots,M\}$. Analyzing the conditional probability distribution functions, $F_1(t),\ldots,F_L(t)$, of interoccurrence times for $\mathcal{T}^*_{1,N},\ldots,\mathcal{T}^*_{L,N}$, we find that they change systematically. Moreover, we find that the conditional probability distribution function $F_L(t)$ obeys the Weibull distribution in almost the entire region, where the Weibull distribution $F(t)$ is defined by

$$F(t) = 1 - \exp(-(t/\tau)^a), \tag{78}$$

(see Figure 28 and 29). Surprisingly, all conditional probability distribution functions, $F_1(t),\ldots,F_L(t)$, obey the *Weibull distribution* in the tail region and $F_i(t)$ fits a Weibull distribution asymptotically as i increases. Therefore, the intermittency of the occurrence of earthquakes is *stationary very weak intermittency* because the conditional probability distribution functions of interoccurrence times, $F_1(t),\ldots,F_L(t)$, in the tail region are invariant and described by the Weibull distribution.

Analyzing the distribution for different M_c, we find that the conditional probability distribution functions obey the Weibull distribution when the previous interoccurrence time is relatively large and that the tails of all conditional probability distribution function obey

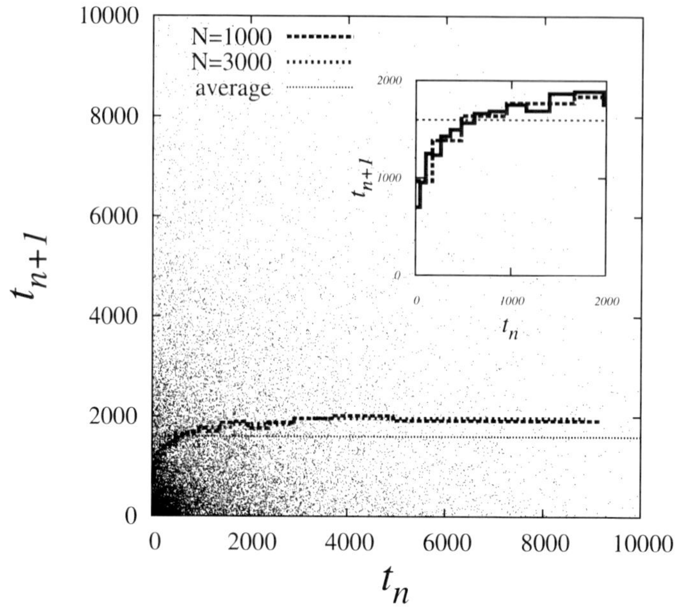

Figure 27. Return map of interoccurrence times in Japan from June 1, 2008 to November 30, 2008 ($M_c = 2.0$). The dotted lines are averages of the interoccurrence time, where averages are taken in three different bins ($N = 1000, 3000$ and 18132). Inset figure is a blow-up of the conditional mean.

the Weibull distribution. The asymptotic behavior of the constructed piecewise linear maps near the origin is the same because the conditional probability distributions in the tail region are invariant. As a result, we classify the intermittency of the occurrence of earthquakes as stationary weak intermittency. The Weibull exponent a, the intensity of intermittency η, and the Lyapunov exponent λ are summarized in Table 11. In this table, the Weibull exponent is obtained by the Weibull fitting for the tail part of the conditional probability distribution $F_L(t)$, where bins of interoccurrence times are set to be equal in logarithmic scale. In the calculation of the Lyapunov exponent, one dimensional maps are constructed from the renewal processes with the Weibull distribution obtained by the above Weibull fitting. The time unit is 1 s in the calculation of the Lyapunov exponent.

We performed the same analysis for the occurrence of earthquakes in the world to investigate the universality of the results in Japan (Akimoto et al. 2010). Surprisingly, the conditional probability distribution changes systematically, $F_i(t)$ fits a Weibull distribution asymptotically as i increases, and the conditional probability distributions for the tail part obey the Weibull distribution in the same manner as in the preceding analysis. The Weibull exponents are 0.93 and 0.92 for M_c=4.5 and 5.0, respectively, indicating the view that the intermittency of the occurrence of earthquakes in the world is classified into stationary weak intermittency.

Finally, we study the dependence of a magnitude and an interoccurrence time on the previous magnitude. The sequence of magnitudes is given as in Fig. 30. We define two

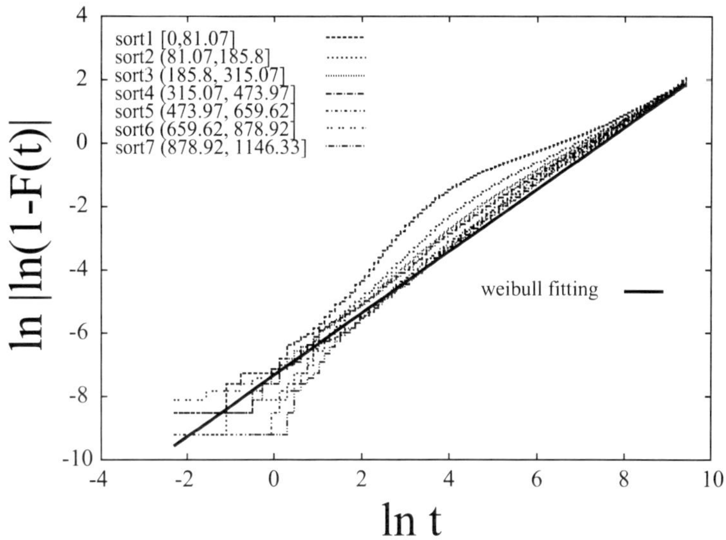

Figure 28. Weibull plot of the distribution of interoccurrence times of earthquakes in Japan from June 1, 2001 to October 31, 2007 ($M_c = 2.0$). Each legend is a time interval $[T^i_{\min}, T^i_{\max}]$, where $T^i_{\min}$ and $T^i_{\max}$ are the minimum and maximum in the ith sorted data set ($i = 1, \cdots, 13$), respectively.

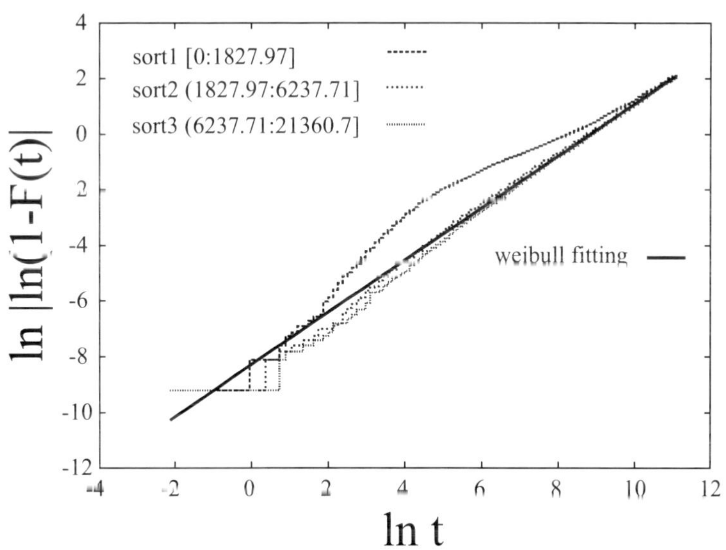

Figure 29. Weibull plot of the distribution of interoccurrence times of earthquakes in Japan from June 1, 2001 to October 31, 2007 ($M_c = 3.0$). Each legend is a time interval $[T^i_{\min}, T^i_{\max}]$, where $T^i_{\min}$ and $T^i_{\max}$ are the minimum and maximum in the ith sorted data set ($i = 1, 2, 3$), respectively.

Table 11. The Weibull exponent a, the intensity of intermittency η, and the Lyapunov exponent λ

M_c	a	η	λ	number of earthquakes
2.0	0.966 ± 0.001	0.035 ± 0.001	4.6×10^{-3}	130243
2.5	0.961 ± 0.001	0.040 ± 0.001	2.7×10^{-3}	67912
3.0	0.946 ± 0.001	0.057 ± 0.001	1.4×10^{-3}	31857

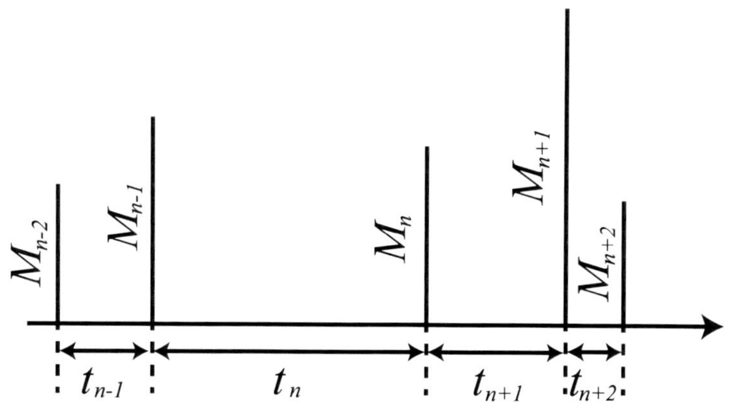

Figure 30. Schematic figure of t_n and M_n.

conditional averages as the followings:

$$\langle M_{n+1} \rangle_M \equiv \sum_{M_i \in M} \frac{M_{i+1}}{N_M} \tag{79}$$

and

$$\langle T_{n+1} \rangle_M \equiv \sum_{M_i \in M} \frac{t_{i+1}}{N_M}, \tag{80}$$

where N_M is the number of the data satisfying $M_i \in M$. The set M is determined, so that the number of M is more than 10^4. As shown in Fig. 31, magnitudes are clearly correlated to the previous magnitudes. On the other hand, interoccurrence times are not correlated to the previous magnitudes very much when the previous magnitude is not large. When the previous magnitude is large, the mean of the next interoccurrence time decreases. Figure 32 shows the conditional probability distribution:

$$F(t) \equiv F(t_{n+1} | M_n \in M), \tag{81}$$

where $F(t|t_{n+1}|M_n \in M)$ is the distribution of t_{n+1} when $M_n \in M$. Conditional probability distributions are almost the same as in the conditional average.

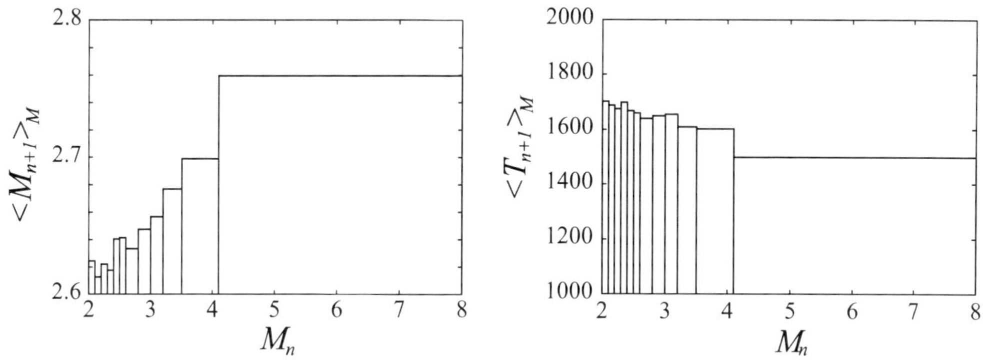

Figure 31. Conditional averages for $\langle M_{n+1}\rangle_M$ and $\langle T_{n+1}\rangle_M$.

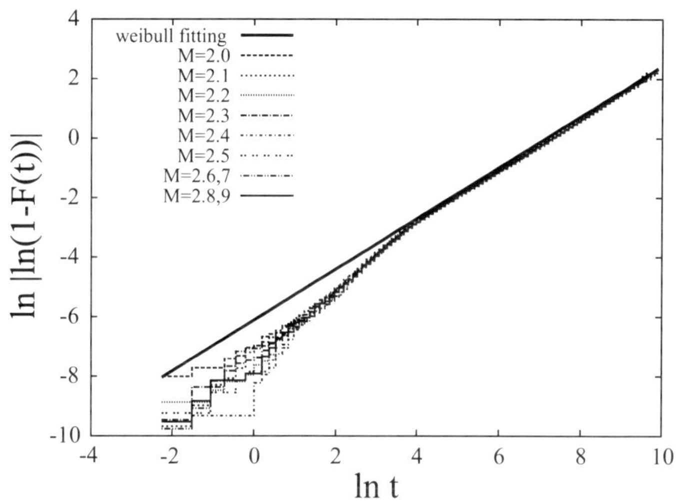

Figure 32. Weibull plot of the distribution of interoccurrence times of earthquakes in Japan from June 1, 2001 to November 30, 2008 ($M_c = 2.0$).

5.4. Summary of this Section

Intermittency of renewal processes has been studied by constructing one-dimensional piecewise linear maps from renewal processes. As a result, characterization of intermittent phenomena is extended to extremely heavy tail and stretched exponential relaxation phenomena, and we can also estimate the Lyapunov exponent of renewal processes, which measures the activity of events.

Analyzing the occurrence of earthquakes, we found that the occurrence of earthquakes is not a renewal process, which is in agreement with (Livina et al. 2005). However, interoccurrence times are i.i.d. random variables when the previous interoccurrence time is large. Moreover, the conditional probability distribution functions in the tail are described by the Weibull distribution. Corral showed that interoccurrence times of earthquakes clearly depend on the preceding interoccurrence times (Corral 2006). Compared with this study, our study gives the detailed analysis of the conditional probability distributions $F(t_{n+1}|T_n \in \mathcal{T}_{i,N})$ and $F(t_{n+1}|M_n \in M)$.

It has been known that the distribution of interoccurrence times of earthquakes shows stretched exponential decay (Corral 2004; Hasumi et al. 2009c), which is consistent with our result in that it has the same classification of intermittency. Exponential distribution or the gamma distribution can be well fitted because the exponent of the Weibull distribution is close to one. We note that the Weibull plot is the most useful to determine the decay of the distribution function. Focusing on the distribution function in the tail part, we conclude that the Weibull exponent is smaller than one. Therefore we characterize the intermittency of the occurrence of earthquakes as stationary weak intermittency. The intensity of intermittency of earthquakes depends on the threshold M_c. In particular, the intensity of intermittency increases monotonically with the threshold of magnitude, indicating the view that it is difficult to forecast the occurrence of large earthquakes. We also estimate the Lyapunov exponent of the renewal process with the Weibull distribution $F(t)$, which measures the activity of earthquakes when the previous interoccurrence time is relatively large. Generally, the occurrence of earthquakes is considered as a Markov renewal process (Cox 1965).

To quantify intermittency in point processes, Bickel (1999) proposed a clear estimation of intermittency using the correlation codimension. Although we assume that interoccurrence times are i.i.d. random variables, which does not always hold in point processes, we can study underlying dynamics of intermittent phenomena with the aid of one-dimensional maps. If we assume the integrate-and-fire model (Sauer1994), the orbit x_n in dynamical systems is considered to be the integrated value of a signal that is behind intermittent phenomena. In the occurrence of earthquakes, the integrated value would be the accumulated energy on the crust.

6. Conclusion

In this chapter, we investigate the statistical seismicity, namely, interoccurrence time statistics and intermittency of earthquakes, from the complex systems point of view. We have used the three different kinds of analysis, natural earthquake data, synthetic data produced by the 2D spring-block model, and time series of the piecewise linear intermittent map in oder to extract the new statistical properties of earthquakes. In this time, Japan Meteorological Agency (JMA) catalogue, Southern California Earthquake Data Center (SCEDC) catalogue, and Taiwan Central Weather Bureau (TCWB) catalogue are analyzed.

Firstly, we focus on the interoccurrence time statistics, statistical properties of time intervals between successive earthquakes by analyzing both natural data and the synthetic data. We have found that the probability distribution of interoccurrence time, $P(\tau)$ depends on the threshold of magnitude, m_c and can be described by the superposition of the Weibull distribution and log-Weibull distribution for all earthquake data on the basis of goodness-fit-tests. As m_c increases, the ratio of the Weibull distribution divided by the distribution, denoted here p, increases. In other words, the distribution shows the Weibull - log-Weibull transition. This result reinforces the view that the Weibull statistics and the log-Weibull statistics coexist in the interoccurrence time statistics. The difference between the results obtained from the synthetic data and from the natural data is that pure log-Weibull regime

$(p = 0)$ appears or not. For the synthetic data, the distribution exhibit the sharp Weibull - log-Weibull transition, which states that $P(\tau)$ changes from the pure log-Weibull regime $(p = 0)$ to the Weibull regime $(p = 1)$ via the superposition regime $(0 < p < 1)$. As for natural data, the distribution presents the smooth Weibull - log-Weibull transition which means that $P(\tau)$ varies from the superposition regime $(0 < p < 1)$ to the pure Weibull regime $(p = 1)$. For large earthquakes, the Weibull distribution with the exponent $\alpha_1 < 1$ is preferred, which indicates that the sequence of earthquakes is not a Possion process, but a long-term correlated process. We focus on the crossover magnitude m_c^{**} so as to figure out the geophysical interpretation of the Weibull - log-Weibull transition. We also find the region-independent relation, namely, $m_c/m_{max} = 0.54 \pm 0.04$. In addition, the crossover magnitude m_c is proportional to the plate velocity, which is consistent with an earlier observation about the velocity-dependence of the characteristic earthquake magnitude. We first propose a universal relation between the interoccurrence time statistics, namely the Weibull components, (α_1, β_1) and the b-value of the Gutenberg-Richter law. The details are presented in our forthcoming paper.

Secondly, to characterize intermittency and the correlation of interoccurrence times in earthquakes, we analyze the JMA catalog. We demonstrate that interoccurrence times are not independent and identically distributed random variables. However, we find that the tails of all conditional probability distribution functions of interoccurrence times are universal. The tail characterizes intermittency of earthquakes. It is shown that forecasting of the occurrence of large earthquakes is more difficult than that of small earthquakes.

Finally, we have investigated that the statistical seismicity for several academic region, namely, the seismology, the statistical physics, and the chaotic systems. It is useful that to understand the source mechanism and the theoretical background of earthquakes, the corroboration of different scientific filed is necessary. We hope that our research is the first step to understand fully the statistical seismology.

7. Acknowledgments

The authors thank the editor of this book and nova science publisher for giving us an opportunity to take part in the book. We would like to thank the JMA, SCEDC and TCWB for allowing us to use the earthquake data. The effort of the Taiwan Central Weather Bureau for maintaining the CWB Seismic Network is highly appreciated. T.H. is supported by the Japan Society for the Promotion of Science (JSPS) and the Earthquake Research Institute cooperative research program at the University of Tokyo. This work is partly supported by the Sasagawa Scientific Research Grant from The Japan Science Society and by the Waseda University. CCC is also grateful for research supports from the National Science Council (ROC) and the Department of Earth Sciences at National Central University (ROC).

8. References

A. Ben-Menahem and S. J. Singh, *Seismic Waves and Sources*, Springer-Verlag, New York, (1981).

A. Bunde, J. F. Eichner, J. W. Kantelhardt, and S. Havlin, *Phys. Rev. Lett.* 94 048701

(2005).
A. Corral, *Phys. Rev. E* 68 035102(R) (2003).
A. Corral, *Phys. Rev. Lett.* 92 108501 (2004).
A. Corral, *Phys. Rev. Lett.* 95 028501 (2005).
A. Corral, *Tectonophysics* 424 177 (2006).
A. Ghosh, *Computational Geosciences* 25 729 (1999).
A. H. Darooneh and C. Dadashinia, *Physica A* 387 3647 (2008).
A. Ohmura and H. Kawamura, *Europhys. Lett.* 111 690001 (2007).
A. Saichev and D. Sornette, *Phys. Rev. Lett.* 97 078501 (2006).
A. Saichev and D. Sornette, *J. Geophys. Res.* 112 B04313 (2007).
B. Enescu, Z. Struzik, and K. Kiyono, *Geophys. J. Int.* 172 395 (2008).
B. Gutenberg and C. F. Richter, Geological Society of America Special Paper, 34 1 (1941).
B. Gutenberg and C. F. Richter, *Ann. Geofis.* 9 1 (1956).
C. C. Chen, L. Y. Chiao, L. T. Lee, H. W. Cheng, Y. M. Wu, *Tectonophysics* 454 104 (2008).
C. C. Chen, Y. T. Lee, and L. Y. Chiao, *Physics Letters A* 372 4340 (2008).
C. Godrèche and J. M. Luck, *J. Stat. Phys.* 104 489 (2001).
C. H. Scholz, *The Mechanics of Earthquakes and Faulting*, Cambridge Univ Press, (2002).
C. M. R. Fowler, *The Solid Earth: An Introduction to Global Geophysics*, Cambridge University Press, New York (1990).
C. Tsallis, *J. Stat. Phys.* 52 479 (1988).
D. L. Turcotte, W. I. Newman, R. Shcherbakov, *Geophys. J. Int.*, 152 718 (2003).
D. R. Bickel, *Physica A* 265 634 (1999).
D. R. Cox and H. D. Miller, *The Theory of Stochastic Processes* Chapman and Hall, London, (1965).
D. R. Cox, *Renewal theory* Methuen, London, (1962).
E. F. Preston, J. S. Sa Martins, J. B. Rundle, M. Anghel, and W. Klein, *Computing in Science and Engineering* 2 34 (2000).
F. Omori, *J. Coll. Sci. Imp. Univ. Tokyo*, 7 111 (1894).
G. Yakovlev, D. L. Turcotte, J. B. Rundle, and P. B. Rundle, *Bull. Seism. Soc. Am.* 96 1995 (2006).
H. Haase, C. G. Hübner, E. Reuther, A. Herrmann, K. Müllen, and T. Basche, *J. Phys. Chem. B* 108 10445 (2004).
H. Kawamura, *Modelling critical and catastrophic phenomena in geoscience,* edited by P. Bhattacharyya and B. Chakrabarti Springer (2006)
H. Kumagai, Y. Fukao, S. Watanabe and Y. Baba, *Geophys. Res. Lett.* 26 2817 (1999).
H. Nakanishi, *Phys. Rev. A* 41 7086 (1990).
H. Nakanishi, *Phys. Rev. A* 43 6613 (1991).
H. Nakanishi, *Phys. Rev. A* 46 4689 (1992).
J. B. Rundle, D. L. Turcotte, R. Shcherbakov, W. Klein, and C. Sammis, *Rev. Geophys.* 41(4) 1019 (2003).
J. B. Rundle, D. L. Turcotte, W. Klein (eds), *GeoComplexity and the physics of Earthquakes*, American Geophysical Union, Washington, DC, (2000).
J. K. Gardner, L. Knopoff, *Bull. Seismol. Soc. Am.*, 64 1363 (1974).
J. M. Carlson, Phys. Rev. A 44 6226 (1991).

J. M. Carlson and J. S. Langer, *Phys. Rev. Lett.* 62 2632 (1989).
J. M. Carlson and J. S. Langer, *Phys. Rev. A* 40 6470 (1989).
J. M. Carlson, J. S. Langer, B. E. Shaw, and C. Tang, *Phys. Rev. A* 44 884 (1991).
J. M. Carlson and J. S. Langer, *J. Geophys. Res.* 96 4255 (1991).
J. M. Carlson, J. S. Langer, and B. E. Shaw, *Rev. Mod. Phys.* 66 657 (1994).
J. Schmittbuhl, J. P. Vilotte and S. Roux, *J. Geophys. Res.* 101 27741 (1996).
J. Xia, H. Gould, W. Klein, J. B. Rundle, *Phys. Rev. Lett.* 95 248501 (2005).
J. Xia, H. Gould, W. Klein, and J. B. Rundle, *Phys. Rev. E.* 77 031132 (2008).
J. Zhuang, Y. Ogata, and D. Vere-Jones, *J. Geophys. Res.* 109 (B5) B05301 (2004).
Japan Meteorological Agency Earthquake Catalog:
http://wwweic.eri.u-tokyo.ac.jp/db/jma1.
K. Christensen and Z. Olami, *J. Geophys. Res.* 97 8729 (1992).
K. Ito and M. Matsuzaki, *J. Geophys. Res.*, 95 6853 (1990).
K. Ito, *Phys. Rev. E* 52 3232 (1995).
K. Shiomi, H. Sato, and M. Ohtake, *Geophys. J. Int.* 130 57 (1997).
K. Wadati, *Journal of the Meteorological Society of Japan* (2) 10 559 (1932).
K. Z. Nanjo, D. L. Turcotte, and R. Shcherbakov, *J. Geophys. Res.* 110 B07403 (2005).
L. Ruff and H. Kanamori, *Phys. Earth Planet. Int.* 23 240 (1980).
L. Telesca, V. Cuomo, and V. Lapenna, *Geophys. Res. Lett.* 28 3765 (2001).
M. Båth, *Tectonophysics*, 2 483 (**1965**).
M. Bottiglieri and C. Godano, *Phys. Rev. E* 75 026101 (2007).
M. Enya, Report of the Imperial Earthquake Investigation Committee, 35 35 (1901).
M. Hedges and G. Takacs, physics/0505015.
M. Ishimoto and K. Iida, *Bull. Earthq. Res. Inst, Univ. Tokyo* 17 443 (1939).
M. Otsuka, *Phys. Earth Planet. Inter.* 6 311 (**1972**).
M. V. Matthews, W. L. Ellsworth, and P. A. Reasenberg, *Bull. Seism. Soc. Am.,* 92 2233 (2002).
P. A. Reasenberg, *J. Geophys. Res.,* 90 (B7) 5479 (1985).
P. Bak and C. Tang, *J. Geophys. Res.,* 94 15635 (1989).
P. Bak, C. Tang, K. Wiesenfeld, *Phys. Rev. Lett.,* 59 381 (1987).
P. Bak, C. Tang, K. Wiesenfeld, *Phys. Rev. A* 38 364 (1988).
P. Bak, K. Christensen, L. Danon, and T. Scanlon, *Phys. Rev. Lett.* 88 178501 (2002).
P. Manneville and Y. Pomeau, *Phys. Lett.* 75 1 (1979).
R. Burridge and L. Knopoff, *Bull. Seism. Soc. Am.* 57 341 (1967).
R. Sánchez D. E. Newman and B. A. Carreras, *Phys. Rev. Lett.* 88 068302 (2002).
R. Shaw, *The Dripping Faucet as a Model Chaotic System* Aerial, Santa Cruz (1984).
R. Shcherbakov, G. Yakovlev, D. L. Turcotte, and J. B. Rundle, *Phys. Rev. Lett.* 95 218501 (2005).
S. Abe and N. Suzuki, *J. Geophys. Res.* 108 2113 (2003).
S. Abe and N. Suzuki, *Physica A* 350 588 (2005).
S. B. Lowen, L. S. Liebovitch, and J. A. White, *Phys. Rev. E* 59 5970 (1999).
S. G. Abaimov, D. L. Turcotte, and J. B. Rundle, *Geophys. J. Int.* 170 1289 (2007).
S. G. Abaimov, D. L. Turcotte, R. Shcherbakov, and J. B. Rundle, *Nonlinear Processes Geophys.* 14 455 (2007).
S. G. Abaimov, D. L. Turcotte, R. Shcherbakov, J. B. Rundle, G. Yakovlev, C. Goltz, and

W. I. Newman, *Pure Appl. Geophys.* 165 777 (2008).
S. Hainzl, G. Zöller, and J. Kurths, *J. Geophys. Res.* 104 7243 (1999).
S. Hainzl, G. Zöller, and J. Kurths, *Nonlinear Processes Geophys.* 7 21 (2000).
S. Lennartz, V. N. Livina, A. Bunde, and S. Havlin, *Europhys. Lett.* 81 69001 (2008).
S. P. Nishenko and R. Buland, Bull. *Seism. Soc. Am.* 77 1382 (1987).
S. R. Brown, C. H. Scholz, J. B. Rundle, *Geophys. Res. Lett.* 18 215 (1991).
S. Shinkai and Y. Aizawa, *Prog. Ther. Phys.* 116 503 (2006).
S. Tasaki and P. Gaspard, *J. Stat. Phys.* 123 803 (2002).
S. Yoshida, M. Uyeshima and M. Nakatani, *J. Geophys. Res.* 102 14883 (1997).
Southern California Earthquake Data Center: http://www.data.scec.org.
T. Akimoto and Y. Aizawa, *Prog. Ther. Phys.* 119 849 (2003).
T. Akimoto and Y. Aizawa, *Prog. Theor. Phys.* 114 737 (2005).
T. Akimoto, T. Hasumi, and Y. Aizawa, *Phys. Rev. E,* 81 031133 (2010).
T. Hasumi, *Phys. Rev. E* 76 026117 (2007).
T. Hasumi, *Physica A* 388 477 (2009).
T. Hasumi, C. Chen, T. Akimoto, and Y. Aizawa, *Techtonophys*, 485 9 (2010).
T. Hasumi, S. Komura, M. Kamogawa, and Y. Yamazaki, *J. Phys.: Conf. Ser.* 31 237 (2006).
T. Hasumi, T. Akimoto, and Y. Aizawa, *Physica A* 388 483 (2009).
T. Hasumi, T. Akimoto, and Y. Aizawa, *Physica A* 388 491 (2009).
T. Hasumi and Y. Aizawa, *Physics of Self-organization systems*, 113 World Scientific (2008).
T. Huillet and H. F. Raynaud, *Eur. Phys. J. B.* 12 457 (1999).
T. Miyaguchi and T. Akimoto, *Phys. Rev. E* 83 031926 (2011).
T. Miyaguchi and Y. Aizawa, *Phys. Rev. E* 75 066201 (2007).
T. Mori and H. Kawamura, *Phys. Rev. Lett.* 94 058501 (2005).
T. Mori and H. Kawamura, *J. Geophys. Res.* 111 B07302 (2006).
T. Mori and H. Kawamura, *J. Geophys. Res.* 113 B06301 (2008).
T. Mori and H. Kawamura, *Phys. Rev. E* 77 051123 (2008).
T. Sauer, *Phys. Rev. Lett.* 72 3811 (1994).
T. Seno, S. Seth, and E. G. Alice, *J. Geophys. Res.* 98 17941 (1993).
T. Utsu, *Bull. Earthq. Res. Inst, Univ. Tokyo* 59 53 (1984).
T. Utsu, *Geophys. Mag.*, 30 521 (1961).
T. Wong, R. H. C. Wong, K. T. Chau, and, C. A. Tang, *Mechanics of Materials* 38 664 (2006).
Taiwan Central Weather Bureau: http://www.cwb.gov.tw/
V. B. Ryabov and K. Ito, *Pure Appl. Geophys.* 158 919 (2001).
V. Livina, S. Tuzov, S. Havlin, and A. Bunde, *Physica A* 348 591 (2005).
V. N. Livina, S. Havlin, and A. Bunde, *Phys. Rev. Lett.* 95 208501 (2005).
W. Feller, *An Introduction to Probability Theory and Its Application*, John Wileys and Son, Inc. 2 Chap. VIII. 8. (1966).
W. H. Bakun, B. Aagard, B. Dost, et al., *Nature* 437 969 (2005).
W. H. Press, S. A. Teukolsky, W. T. Vetterling, and B. P. Flannery, *Numerical Recipes in C 2nd edition*, Cambridge University Press Cambridge (1995).
W. Weibull, Ingeninörs, Vetenskaps Akademien, Handlinger, 151-3 45 (1939).

W. Weibull, *J. Appl. Mech.,* 18 293 (1951).
X. Brokmann, J.-P. Hermier, G. Messin, P. Desbiolles, J.-P. Bouchaud, and M. Dahan, *Phys. Rev. Lett.* 90 120601 (2003).
Y. Aizawa, *Prog. Ther. Phys.* 72 659 (1984).
Y. Aizawa, *Prog. Ther. Phys. Suppl.* 99 149 (1989).
Y. Aizawa, *Kagaku*, Iwanami publisher, 65 266 (1995) (in Japanese).
Y. Aizawa, *Chaos. Solitons & Fractals* 11 263 (2000).
Y. Aizawa and T. Hasumi, in preparation (2011).
Y. Aizawa and T. Kohyama, *Prog. Ther. Phys.* 71 847 (1984).
Y. Hagiwara, *Tectonophys* 23 313 (1974).
Y. T. Lee C. C. Chen, L. Y. Chiao, and Y. F. Chang, *Physica A,* 387 5263 (2008).
Y. T. Lee, C. C. Chen, T. Hasumi, and H. L. Hsu, Geophys. *Res. Lett.* 36 L02308 (2009).
Y. Y. Kagan, *Pure Appl. Geophys.* 155 537 (1999).
Y. Yamashita, *J. Phys. Earth.* 24 417 (1976).

In: Earthquakes
Editor: Kostas Konstantinou
ISBN: 978-1-62081-883-1

Chapter 6

TRIGGERING AND SYNCHRONIZATION OF SEISMICITY: LABORATORY AND FIELD DATA - A REVIEW

T. Chelidze and T. Matcharashvili[*]
M. Nodia Institute of Geophysics, Tbilisi, Georgia

ABSTRACT

Triggering and synchronization (i.e. adjustment of rhythms of oscillating objects due to their weak interaction) are encountered in various fields, from mechanics to biological and social processes. Thus, it is only natural that triggering and synchronization phenomena are observed in many geophysical fields, as the Earth is embedded in the oscillating field of different origin, with extremely wide range of frequencies, from seconds to months and years. There are a lot of observations on the significant impact of small external forcing on the seismic regime, namely on the seismicity induced by: tides, reservoir exploitation, big explosions, magnetic storms, strong electrical pulses, wave trains of remote strong earthquakes, oceanic microseisms, etc, which means that these phenomena are connected with nonlinear interactions of objects, namely with initiation of instability in systems that are close to the critical state.

Early observations of triggering/synchronization were not supported by strict validation of procedures and should be considered as mainly qualitative or even unreliable. The situation in the field changed dramatically in the last decade after implementation of dense seismic networks equipped by high sensitivity broadband devices, which allow reliable identification of dynamic triggering/synchronization of earthquakes. Besides, new analytical tools for quantitative assessment of the strength of triggering/synchronization developed in the last decades made the procedure of revealing these effects much more robust procedure.

In the chapter the results of laboratory experiments on the electromagnetic (EM) and mechanical triggering and synchronization of mechanical instability (slip) in the spring-slider system are presented. Slip events were recorded as acoustic emission bursts.

[*] M. Nodia Institute of Geophysics,1, Alexidze str, 0171, Tbilisi, Georgia.

Besides, field data on the forcing of local seismicity by strong electromagnetic pulses, water regime variations and remote earthquakes are analyzed as well as the potential of synchronization technique for earthquake prediction.

1. Introduction

Triggering and synchronization are the two faces of the same coin; both effects imply that the additional forcing causing triggering and synchronization is much smaller than the main driving force, which means that these phenomena are connected with nonlinear interactions of objects, namely with initiation of instability in systems that are close to the critical state. In seismic process the main driving component is the tectonic stress; earthquakes are considered as dynamic instabilities generated in the process of friction (stick-slip) between faces of geological faults [Brace and Byerlee, 1966; Kanamori and Brodsky, 2004; Ben-Zion, 2008]. The additional forcing is exerted by various external impacts: Coulomb stress change by previous strong earthquakes (EQ-s), mining activity and water reservoir loading [Gupta, 1992], precipitation and snow melt [Heki, 2003], magnetic storms, nuclear explosions [Nikolaev, 1974], strong electrical pulses [Tarasov et al., 1999], tides [Scholz, 2003], wave trains of remote strong earthquakes [D. Hill, S. Prejean. Earthquake Seismology, 2009], oceanic microseisms [Stich et al., 2011], solar activity, atmospheric pressure, etc.

These results show that weak quasiperiodic perturbations can trigger (synchronize) even such irregular relaxation process as seismicity, which is driven by enormous tectonic forces. The study of seismic response of the lithosphere to a weak forcing is a fundamental problem for seismic source theory as it reveals the important detail of the tectonic system, namely, how close is it to the critical state. The intensity of these superimposed periodical oscillations is as a rule much smaller (stresses of the order of 0.1-1 bar or 0.01-0.1 MPa; in some areas dynamic triggering threshold is 0.001 MPa) than that of the main driving force – tectonic stress. Nevertheless, this weak interaction (forcing) may eventually invoke the phenomenon of triggering or even synchronization, or, at least, the increased level of ordering of the system behavior in the time domain. It is evident that these phenomena cannot be understood in the framework of traditional linear approach and that such high sensitivity to weak impact implies essentially nonlinear interactions. Indeed, the onset of synchronization means that a weak periodic impact (in following – periodic forcing) may induce periodicity/order into a dynamical system driven by much stronger force. In order to be sensitive to the weak impact, the system should be close to the critical state and consequently, highly sensitive to even small external impact, which means that triggering/synchronization analysis embraces also a potential for earthquake forecast as a critical phenomenon.

Early observations of triggering/synchronization were not supported by strict validation of procedures and should be considered as mainly qualitative or even unreliable. The situation in the field changed dramatically in the last decade. Implementation of dense seismic networks equipped by high sensitivity broadband devices lead to a breakthrough, which allows reliable identification of dynamic triggering of earthquakes as spatially and temporally correlated local events following remote strong earthquakes. The last striking data on remote dynamic triggering [Hill and Prejean, Earthquake Seismology, 2009] show that this is a new important domain of earthquake seismology, where experiments leave behind the theory.

On the other hand, new tools for quantification of the strength of triggering/ synchronization between the phases of synchronized process and synchronizing impact developed last decades make the corresponding assessments much more reliable [Pikovsky et al., 2003]. For the relaxation (integrate and fire type) processes, when the signal is composed of well-defined marked events (earthquakes), the generalized phase difference, the stroboscopic technique can be used for computing and comparison of phases. The other possible measures for quantification of synchronization are: mutual information [Palus, 1997], Shannon and Tsallis entropy [Tsallis, 1988], first Fourier mode of distribution of relative phases, computation of conditional probability of phases [Rosenblum et al. 2001], assessment of flatness of stripes of synchrograms [Toledo et al., 1999], calculation of coefficients of phase diffusion [Anishchenko, 1995]. Calculation of fractal dimension, Lempel-Ziv complexity, Recurrence Quantification Analysis (RQA), mutual information, analysis of unstable periodic orbits can also be used to reveal synchronization and ordering in coupled time series [Chelidze, Matcharashvili, 2003, Tozi, 2004].

Experimentally triggering is revealed by correlation of some single external impact and single seismic event with some delay; according to this definition, statistics of triggering events is relatively small.

Synchronization, on the other hand, is defined as "the adjustment of rhythms due to an interaction" and this means that it manifests itself not as infrequent solitary acts like in triggering, but as a systematically recurrent process, which provides relatively rich statistics and allows applying rigorous methods of assessment of the strength of synchronization.

2. Synchronization of Relaxation Oscillator

We can illustrate the essence of synchronization phenomenon by consideration of relaxation or integrate-and-fire oscillator under periodic forcing. The examples are plenty from biology to electricity and mechanics. Such oscillator exhibits interchange of the epochs of slow evolution, when accumulation of energy takes place and the rapid phase, when, on reaching some critical threshold, the fast discharge of accumulated energy occur.

Relaxation oscillator can be used as a simple model of seismic cycle or, on laboratory scale, of the stick-slip motion of spring-slider system. If the force on the spring, or the shear stress on the tectonic fault, *P(t)* increases slowly and simultaneously external small oscillating stress with amplitude, frequency and initial phase correspondingly a, $\omega = 2\pi / T$, φ is applied to the system, the resulting stress is:

$$P_c(t) = P(t) + a\sin(\omega t + \varphi) \tag{1}$$

When the summary stress approaches the threshold value P_{max} the sudden discharge (stress drop) occur. The condition for stress drop is:

$$P_c(t) = P(t) + a\sin(\omega t + \varphi) = P_{\max} \tag{2}$$

Assuming that at the moment $t = t_0$ the stress $P = P_0$ and after that it increases linearly:

$$P(t) = b(t - t_0) \tag{3}$$

then from 2 and 3 we have:

$$P_0 + b(t - t_0) + a\cos(\omega t + \varphi) = P_{max} \tag{4}$$

or

$$b(t - t_0) = P_{max} - P_0 - a\cos(\omega t + \varphi) \tag{5}$$

As the initial moment of (linear) tectonic stress accumulation can be arbitrary, let us assume that this process can start any time, for certainty, at N different moments, separated by the step ε, where ε is less than period of forcing. Than we will have N equations for discharge moments:

$$b(t - t_0 + \varepsilon n) = P_{max} - P_0 - a\cos(\omega t + \varphi), \quad n = 0,1,2,...N \tag{6}$$

Solving the system (6) delivers all possible time moments of discharges caused by external forcing and located within the period of forcing. The distribution of these moments relative to the phase of forcing may reveal the synchronizing phenomenon [Lursmanashvili et al., 2010].

Figure 1 represents the family of stress evolution lines (left side of (6)) and superimposed external periodic forcing (right side of (6)). Stress lines start at different time moments in the past and stretch till intersection with forcing curve, when the sudden discharge (earthquake) occur.

Calculating the number of events n_e in different phase bins of forcing period one reveals "gap" in distribution of n_e, or a phase window, when stress lines do not intersect forcing curve, which means that discharges in this gap are forbidden (Figure 2).

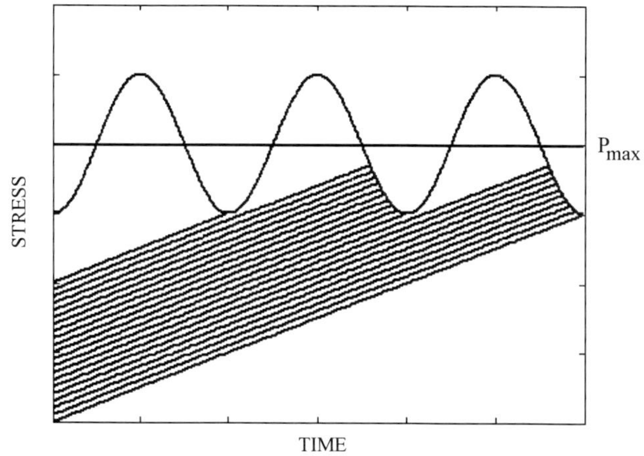

Figure 1. The diagram demonstrating the principle of stick-slip or earthquake synchronization by periodic external forcing. When the sum of linear and periodic components approaches P_{max} the stress drop (slip) occur.

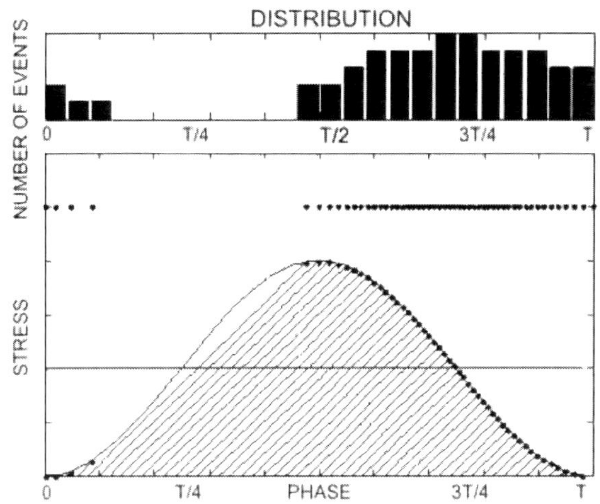

Figure 2. Distribution of relaxation oscillator discharge moments inside the forcing period *T* (upper figure) and corresponding solutions of equations (6), (dots, lower figure). [Lursmanashvili et al., 2010].

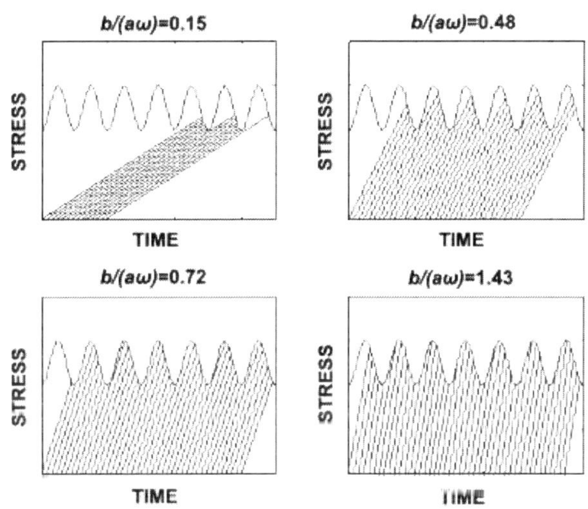

Figure 3. Four relaxation oscillator synchronization cases with different values of stress rate parameter *A*. [Lursmanashvili et al., 2010].

If we introduce the normalized variable $\tau = t/T$ the equation (6) became:

$$A(\tau + n/N) = C - \cos 2\pi\tau, \quad n = 0,1,2,...N \tag{7}$$

where $A = b/a\omega$ is the relative value of stress rate (slope of the stress line) and $C = (P_m - P_0)/a$ is a constant determining discharge moments of relaxation oscillator.

Examination of the distribution of roots of equation (7) shows that the width of gap is determined by the slope of stress line, i.e. by the stress rate relative to external factor frequency. The low stress rate with superimposed high frequency forcing leads to wider gaps. High stress rate generates narrow gaps or just modulation of distribution without gaps.

Therefore, analysis of discharge moments' distribution within the forcing periods for different forcing frequencies and looking for gaps or modulation - is the way to recognize external (synchronizing) forcing.

Figure 3 illustrates four synchronization cases with different values of A. Corresponding distributions of events within forcing period are presented in Figure 4. The increase of A causes decrease of the width of gap where oscillator discharges are forbidden and causes changes of the corresponding distribution form, but the information on synchronizing factor retains even in the case when the gap vanishes, due to modulation effect.

The general diagnostic diagram, where the roots of equation (7) are plotted for a set of values of parameter A (Figure 5) demonstrates all mentioned peculiarities of synchronization. As the stress rate parameter A increases, the gap area of diagram (the root-free area) decreases to zero and for large values of A discharge may occur for any phase of external forcing. In this case the information on forcing factor frequency remains in the root density variation, which is minimal at the phase equal to $\pi/2$ (Figure 4, case $A = b/a\omega = 1.43$).

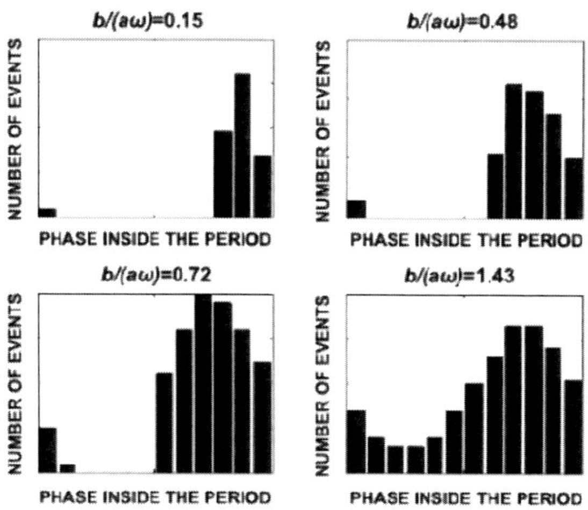

Figure 4. Distribution of discharge moments inside the period of external forcing; compare with Figs. 28-30. [Lursmanashvili et al., 2010].).

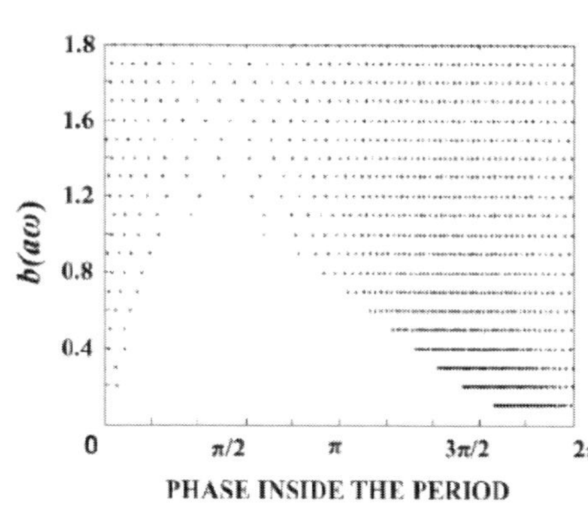

Figure 5. Diagnostic diagram of relaxation oscillator synchronization. Points correspond to solutions of equation (7). Note root-free area (gap), which is very wide at low values of A [Lursmanashvili et al., 2010].

3. Modeling of Earthquakes Triggering and Synchronization on the Laboratory Scale

Up to now the problem of triggering and synchronization of seismic process is far from being resolved and relevant publications, especially related to earthquake triggering by earth tides, are controversial due to the complexity of natural processes [Beeler and Lockner, 2003; Scholz, 2003a; Grasso, 1998; Nikolaev A.; 1994; Nikolaev V. 2003; Scholz, 2003; Kanamori and Brodsky, 2004; Ben-Zion, 2008; de Rubeis, Chechovski and Teisseyre, 2010]. Understanding of main regularities can be obtained in controllable experiments carried out on model systems. It is generally accepted that the most appropriate model of seismic process is unstable frictional movement or stick-slip [Brace andByerlee, 1966; Dietrich, 1979; Ruina, 1989].

3. 1. Electromagnetic Triggering of Slip

We carried out laboratory experiments on the slider-spring system with superimposed pulse or periodic electromagnetic (EM) forcing, which is weak in comparison with the main dragging force of the spring. The use of EM impact was prompted by experiments carried out in 1983-1988 by the Institute of High Temperatures of Russian Academy of Sciences (IHT RAS) in the Central Asia, at the Bishkek test area. After performing series of MHD soundings as well as "cold" discharges, initially aimed to finding resisitivity precursors of strong earthquakes in the upper layers of Earth crust, an unexpected effect of micro-seismicity activation by strong EM pulses has been discovered [Tarasov et al., 1999]. We reproduced the effect in laboratory conditions and it turns out that EM forcing is a flexible tool for study of triggering and synchronization phenomena in laboratory slider experiments [Chelidze et al., 2002; Chelidze and Lursmanashvili, 2003; Chelidze, Matcharashvili, 2003; Chelidze et al., 2005; Chelidze et al., 2006; Chelidze et al., 2007; Chelidze et al., 2008, Chelidze et al., 2009, Chelidze et al. 2010a, 2010b, 2010c; Matcharashvili, Chelidze. 2010].

3. 1. 1. EM Triggering

The experimental set up has been designed in such manner that the mechanical system could be easily driven to the critical state where the triggering of mechanical instability by some weak impact such as electrical pulse became more probable. The details of experiments are presented in [Chelidze et al., 2002; Chelidze et al., 2004].

The system consists of two pieces of rock; the upper piece can slip on the fixed supporting sample if a special frame tilts the system up to the critical angle α_c. In the most cases the supporting and the slipping blocks were prepared from basalt; these samples were saw-cut and roughly finished. The slipping block has 10 cm length, 10 cm width and 2 cm thick. Other samples such as granite, labradorite, and glass, which were better finished, were also tested. The height of surface protuberances was in the range 0.1-0.2 mm for basalt samples and 0.05-0.1 mm for other ones. The basalt samples were preferred because they do not contain significant quantity of piezoelectric minerals.

The electrical part consisted of EM pulse generator and acoustic signals amplifier. The signal from the standard generator of amplitude 0.5-5 V was applied to the input of the amplifier and went out from the output with the amplitude up to 1300 V. Up to 10 DC-pulses of that amplitude were applied to the sample. The duration of pulses was from 5 to 10 s; interval between pulses was also from 5 to 10 s. The scanning of the process was performed on the frequency 96 kHz, i.e. sampling rate was 1/96 000 s. It should be noted that although high voltage was applied to the outer surfaces of samples the voltage in the gap between sliding surfaces was much less – of order of 1 V.

Electrodes were applied in the following ways:

i. To the bottom of the supporting sample in a coplanar manner or to the sides of the supporting sample (the first mode).
ii. Tto the upper surface of the sliding sample and the bottom of the supporting one (the second mode). In the first case the EM field is oriented roughly parallel to the slip surface and in the last case current lines are normal to it.

Slip events were registered as acoustic bursts [Akay, 2002] by the sound card of PC.

3. 1. 2. Experimental Procedure and Case Stories

The main objective of experiments was to find out whether EM-pulse could indeed displace the rock sample, resting on the supporting sample at the slope of support, less than but close to the critical slip angle.

After finding the critical angle (Figure 6) the slope of support was decreased by 0.1 - 2°. In this state the upper sample was stable for many hours (2 days), which mean that other sources of instability such as building vibration by trucks, elevator, wind, etc were not strong enough to initiate the slip. The critical angle for the rough surface varies from one test to another because is impossible to reproduce exactly the arrangement of asperities between the support and the slipping blocks in different experiment. The scatter in values of critical angle for the same sample is of order of ± 2.5^{o} (Figure 6), so we can define only the average value of α_c, which changes from one test to another. Because of the critical angle scatter before the EM impact the sample was kept at the angle, less than the (average) critical one for 10 minutes and only after this exposure subjected to the EM-impacts also for 10 minutes.

That allows assessing correctly the statistics of EM-activation, as the probability of slip in the time intervals without EM-impact can be compared with that in the time intervals, covering the whole EM-activation period. The activation period lasts several minutes and it includes also the gaps between pulses. A sequence of pulses, applied in a single experiment and gaps between them will be related as a run. Practically the probability of slip without EM-impact at $\alpha < \alpha_c$ was zero: no slip was observed during any of 10 min. preliminary repose periods (500 min in total).

3. 1. 3. EM Triggering - the First Mode

We found that the application of EM-pulses in the first mode, i.e. to the coplanar electrodes at the bottom of support, initiates slip with probability around 0.07 either during pulse (i.e. in the active phase), either after it (i.e. in the passive phase). The last observation means that the polarization of the sample can be important for the slip initiation. The typical recording of acoustic emission generated by the slip event is shown in the Figure 7.

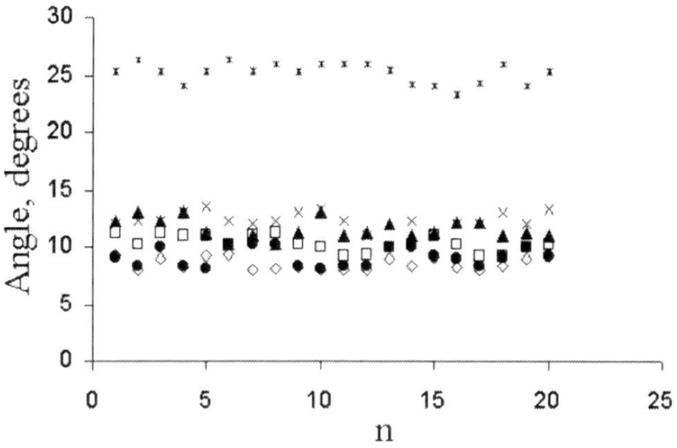

Figure 6. The critical angle of slip (degrees) for various samples on the same supporting material versus number of experiment n. ∗ - basalt, x – glass cylinder (diameter 14 cm, thickness 12 mm), □ - glass cylinder (diameter 7.5 cm, thickness 10 mm), ▲- glass cylinder (diameter 9.5 cm, thickness 15 mm), ◊- glass cylinder (diameter 2 cm, thickness 10 mm). ● - labradorite. Surfaces of all samples except basalt were finished [Varamashvili, Chelidze, 2004].

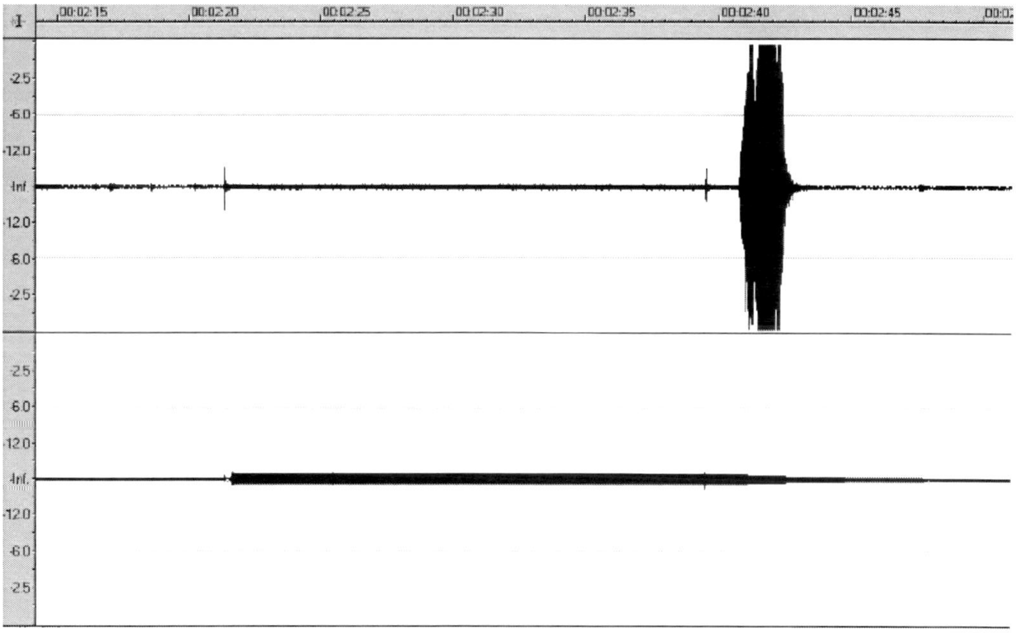

Figure 7. Recording of acoustic emission, generated by the electromagnetically initiated slip of the basalt sample (upper trace). Lower trace shows EM pulse switching on (thick line) and switching off (thin line) periods. The slip was initiated just after the active period. y-axis shows amplitude of the signal in db, x is the time axis [Chelidze and Matcharashvili, 2005].

Besides pulse generator the second source of high voltage, namely the electrical discharger TESLA OPOCMO TVI 200 has been used for initiation of slip. In this case the voltage applied was of order of 10 kV. Again series of pulses were applied to the sample in the first mode and in this case the probability of slip initiation was much higher – around 0.2.

3. 1. 4. EM Triggering - the Second Mode

In the second mode the electrodes were applied to the upper facet of the slipping block and to the bottom side of support, i.e. in this case the applied electrical field was oriented in the direction of the normal to the slip surface. That means that when the EM is applied in the second mode, it increases the friction force (EM field hampers the slip) according to electrostriction rule [Tamm, 1956].

3. 1. 5. Finding Mechanical Equivalent of EM Impact

For assessment of mechanical equivalent of electrical impact both direct and theoretical methods were used. In the first case the mechanical force, initiating slip at the same angle $\alpha<\alpha_c$ that has been set in experiments with EM-impact, was measured by spring and torsion dynamometers (accuracy correspondingly ± 0.01N and ± 0.005 N). Both methods gave comparable results. The force, equivalent to slip-initiating EM-impact is of order of 0.2 N. Another way to get mechanical equivalent is to calculate it from the general equation of balance of forces for a sample placed on the inclined plane:

$$F = mg(\mu\cos\alpha - \sin\alpha) \tag{8}$$

As far as μ is known (for basalt samples μ = 0.47) slip-initiating force can be calculated for any angle. For example, if α_c equals 25°, at $\alpha = 24^{\circ}50'$ the initiating force is 0.42 N. This value is of the same order as in direct experiments (0.2 N).

Thus in the situation, close to the critical one, even 0.2 N force can initiate slip of the sample, weighting 700 g. Our guess is that the EM impact is equivalent to the above value, i.e. it promotes slip in the first mode and hampers it in the second mode by the additional force of order of 0.2-0.4 N, which is less than the spring pulling force of order of 5-10 N.

3. 2. The Elementary Theory of EM Coupling with the Friction Force

In order to understand physics of EM-slip it is necessary to consider fundamentals of surface phenomena. Intermolecular and inter-surface forces, responsible for adhesion and friction, can be loosely divided into three categories: i. purely electrostatic, arising from the Coulomb interaction between charges; ii. polarization forces arising from the dipole moments, induced by internal (bound charges, dipoles) or external electric field; iii. quantum-mechanical forces, responsible for covalent bonding and steric interactions. All these forces can act simultaneously, resulting in some total adhesion (friction) force. For friction we have:

$$F_f = \mu F_n, \tag{9}$$

where μ is friction coefficient and F_n is the normal component of force acting on the body (gravity, compression).

From the above classification it can be deduced that in principle as far as inter-surface forces are of electromagnetic origin, external electrical field can affect the adhesion (friction) forces, changing μ and thus initiating slip of the body, placed on the inclined plane. We can rewrite (9) in the following way:

$$F_f = \mu(F_n + F_p) \tag{10}$$

F_p is the increment (decrement) of normal component of force due to the application of EM field.

Of course, coupling of EM-impact with friction for the sample, containing piezoelectric materials is a trivial phenomenon.

However, the EM-activation of slip is clearly observed on samples that are practically free of piezoelectric minerals (basalts). That is why we exclude piezoelectric effect as a principal mechanism of EM-slip.

The elementary theoretical model of EM coupling with the friction can be formulated in the following way. It is well known that application of EM field to the dielectric invokes some forces acting upon molecules of the body; the resultant of them is called the electrostriction force F_p which is affecting the whole sample. The force is proportional to the gradient of the field intensity squared and it carries away the sample in the direction of the largest intensity. The tension tensor $\vec{T}_n$ operating on the element of dielectric's surface in EM field of intensity E on the assumption that the sample of dielectric constant ε is surrounded by the immobile dielectric medium in *ESU* system is [Tamm, 1956]:

$$\vec{T}_{np} = \pm \frac{\varepsilon + \frac{\partial \varepsilon}{\partial \delta}\delta}{8\pi} E^2 \vec{n}, \tag{11}$$

where the sign (+) is applies when the field $\vec{E}$ is parallel to the external normal $\vec{n}$ to the considered surface element, and the sign (-)when the field $\vec{E}$ is normal to $\vec{n}$.

We can imagine that the elastic strings are stretched along field lines [Tamm, 1956]: in our case they pull together the surfaces of sliding and supporting samples in the second mode or build the side thrust on each other according to (11). The latter case corresponds to the first mode and the former one to the second mode of EM-triggering.

The above equations can be simplified if the dielectric increment due to the striction force is negligible, i.e. $(\partial\varepsilon/\partial\delta \to 0)$. Introducing the area of dielectric's surface S and taking into account the above assumption, the electrostriction force is:

$$\vec{F}_p = \pm \frac{\varepsilon}{8\pi} E^2 \vec{n} = \pm \frac{\varepsilon S}{8\pi}\left(\frac{\Delta V}{d}\right)^2 \vec{n}, \tag{12}$$

where ΔV is the applied voltage and d is the distance between electrodes; the sign depends on the mutual orientation of dielectric's surface and electrical field . Substituting into (12) experimental values: ΔV = 1200 V = 4 ESU, ε = 5; S = 100 cm^2 and d = 5 cm, which correspond to the capacitor, created by two electrodes we obtain:

$$\left|\vec{F}_p\right| = 0.5 \text{ dyne} = 5\ 10^{-6}\,N.$$

That is much less than the experimental values of electromagnetic pull force, which is of order of 0.2 N. Here we have to note that the value of $(\Delta V/d)$, substituted in (12), is an average value for the whole system and on the contact between two blocks the gradient can be quite different. The matter is that the most part of the voltage drop occurs in the air gap between samples, created by surface roughness and the local gradient of EM field in the gap can be much larger than for the whole system between electrodes. In order to assess the forces acting in the narrow gap between slipping and supporting samples, it is necessary to consider the gradient in the gap between the samples. The inner surfaces of slipping and supporting samples carry bound charges, due to the polarization of material and thus create the local gradient of electrical field in the gap. The opening of the gap itself is varying; we can introduce some effective value of opening d_{eff}. Then, applying again the equation (12) to the "inner" or gap capacitor we obtain for the electrostriction force F_{pi} acting on the gap surfaces:

$$\vec{F}_{pi} = \pm \frac{\varepsilon_{eff} S}{8\pi} \left(\frac{\Delta V_{eff}}{d_{eff}} \right)^2 \vec{n} \quad , \tag{13}$$

where $\Delta V_{eff} / d_{eff}$ is the effective voltage gradient in the gap and ε_{eff} is the effective dielectric constant of the gap, which is between values of ε for the air and the sample: $1 < \varepsilon < 5$. Assuming $\varepsilon_{eff} = 2.5$, $S = 100$ cm^2, we have to put in the equation (12) the gradient $(\Delta V_{eff} / d_{eff})$ = 0.07 V/cm in order to obtain the experimental values of slip-initiating electrostriction force, namely, $F_p \approx 0.2\ N$. Thus the expression (13) can be written, taking into account (12):

$$F_f = \mu\left(F_n + F_{pi}\right), \tag{14}$$

which is similar to the expression for the friction force, taking into account fluid pore pressure term [Sibson, 1994].

The above value of F_{pi} can be considered as an order of magnitude of electrostriction force that promotes the slip in the first mode and hampers it in the second mode, according to the expression (12) for the accepted set of parameters.

Both our experiments and theoretical considerations are related to the "dry" environment, namely, to the humidity of atmosphere 60±20% at the temperature (20±5)°C.

It is interesting to note that somehow similar effect of synchronization of induced acoustic vibrations (not stick-slip motion) in rock samples with the superimposed periodic EM field has been observed by G. Chernyak [1978].

In the "humid" environment, when the rock's pore space is fully or partially saturated with water, additional factors should be taken into account: i. the pore pressure increment (decrement) in the gaps, caused by the electro kinetic flow of fluid at the application of EM field; ii. the change of surface fracture energy of cracks due to electroosmotic fluid inflow into the cracks of an undersaturated rocks due to the EM impact. Both these factors can facilitate fracture process in water-bearing rocks.

3. 3. Synchronization of Stick-Slip by Electromagnetic Forcing

3. 3. 1. Stick-Slip Synchronization: Experimental Set up

Experimental set up in synchronization experiments with electromagnetic (EM) forcing represents a system of two horizontally oriented plates of roughly finished basalt. A constant pulling force F_p of the order of 10 *N* was applied to the upper (sliding) plate; in addition, the same plate was subjected to periodic mechanical or electric perturbations with variable frequency and amplitude (from 0 to 1000 V), which was much weaker (of order of 1 *N*) compared to the pulling force; the electric field was normal to the sliding plane. Acoustic bursts accompanying slip events were registered by the sound card of PC. Details of the setup and technique are given in [Chelidze and Lursmanashvili, 2003].

3. 3. 2. EM Synchronization: Results

In our experiments the following parameters were varied:

i) The stiffness of the spring K_s.
ii) The frequency, *f*, of superimposed periodical perturbation.
iii) The amplitude of the excitation (applied voltage V_a);
iv) The direction of applied electrical field;
v) The velocity of drag, v_d.
vi) The normal (nominal) stress σ_n.

The typical background behavior of the system with stiff spring during conventional stick-slip is presented in Figure 8 a. It is evident that in these conditions AE events do not manifest any appreciable periodicity at the time scale from several to hundred milliseconds. In case of soft spring the friction movement is realized by quasi periodic slips (Figure 8 b).

a)

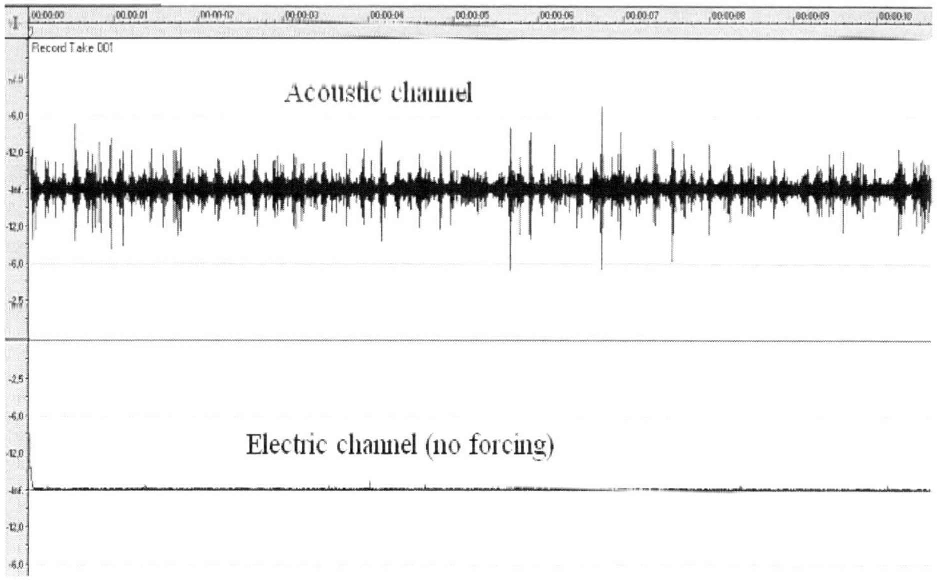

Figure 8. (Continued on next page)

b)

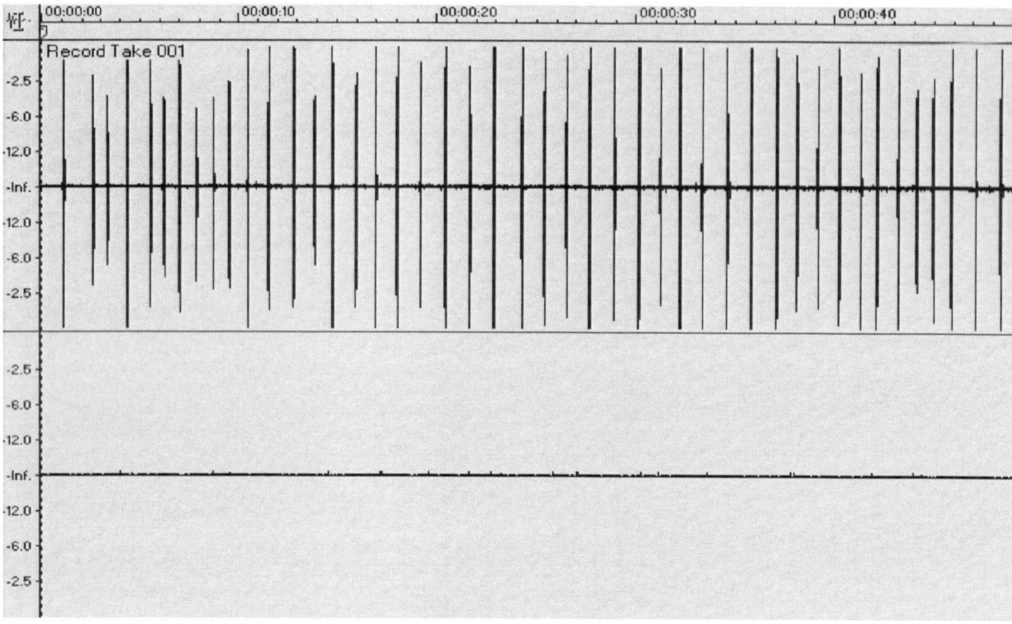

Figure 8 a, b. Typical acoustic emission during slip without external periodic impact: a – stiff spring; disordered release of acoustic bursts. The stiffness of spring K_s = 1000 N/m; σ_n = 2 kPa. ; the mean drag velocity v_d = 2.5 mm/s; dragging force F_d = 3.5 N; T = 20°C; humidity W = 40%; b – soft spring; quasi periodic acoustic bursts with the mean waiting time (natural period) 0.94 sec; v_d = 1.4 mm/s; F_d = 3 N; K_s = 150 N/m; T = 20°C; W = 50%. Here and in Figs. 9, 11, 13; 24; 25 a; 26 a; 27 a; 28 a, b the amplitude of AE during slips (upper channel) and that of synchronizing field (lower channel) is given in dB.

a)

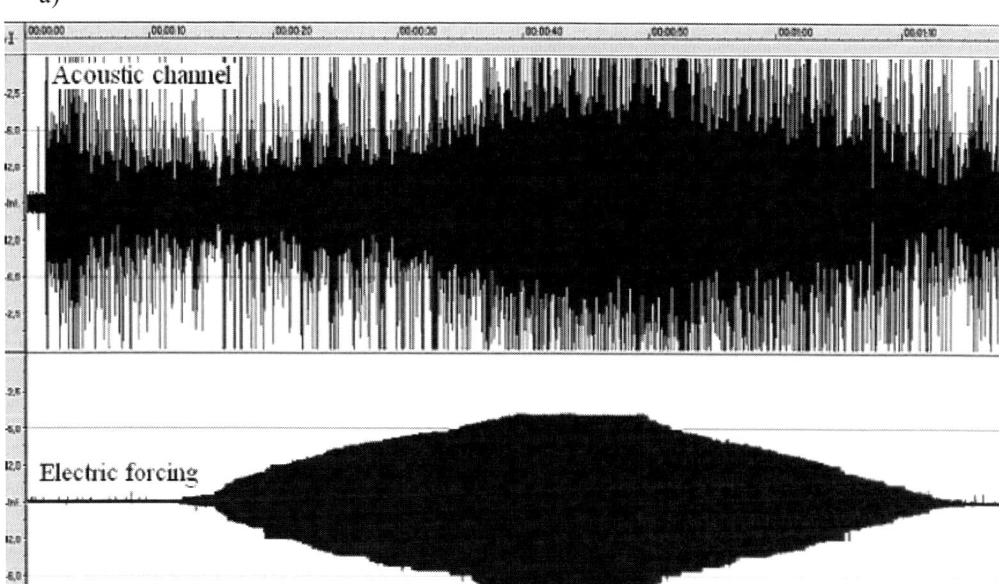

b)

c)

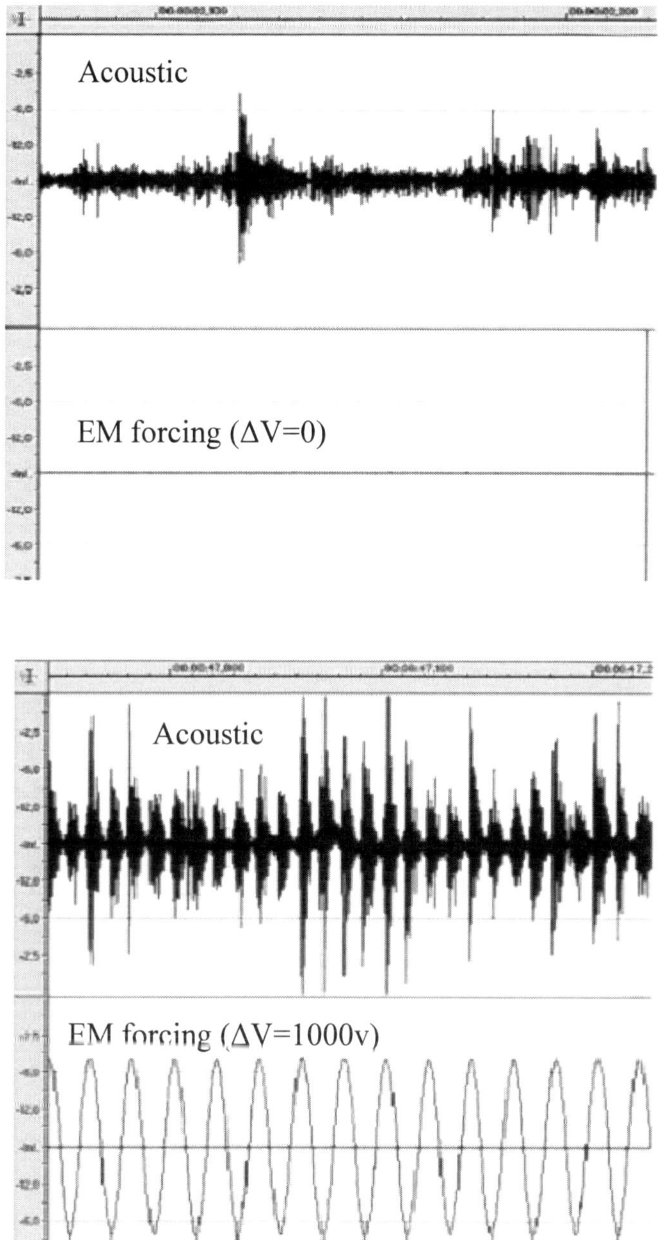

Figure 9. a. Acoustic emission during slip under variable from zero to 1000 V external periodical voltage: b. the extended part of record with zero EM forcing; c – the extended middle part of record under maximal EM forcing. Note complete phase synchronization (PS).

Slip with superimposed periodic low-frequency EM field ($f \approx 60$ Hz) of varying intensity, oriented normally to the slip surface is presented on Figure 9 a, b, c [Chelidze et al., 2008c]. The significant synchronization at this frequency occurs at $V_a \geq 500$V. Under EM excitation the AE events (micro slips) occur twice per period (Figure 9 c). The maxima of AE coincide (approximately) with the extreme points of oscillation.

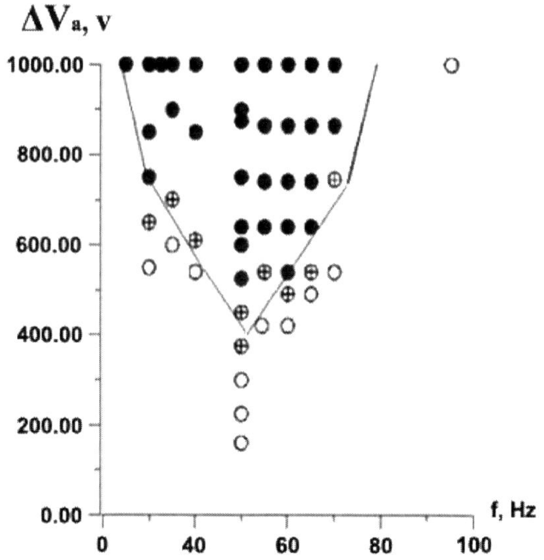

Figure 10. Stick-slip synchronization area (Arnold's tongue) for various intensities *(V_a)* and frequencies *(f)* of the external EM forcing. Filled circles – perfect, circles with crosses – intermittent and empty circles – absence of synchronization [Chelidze et al., 2008c].

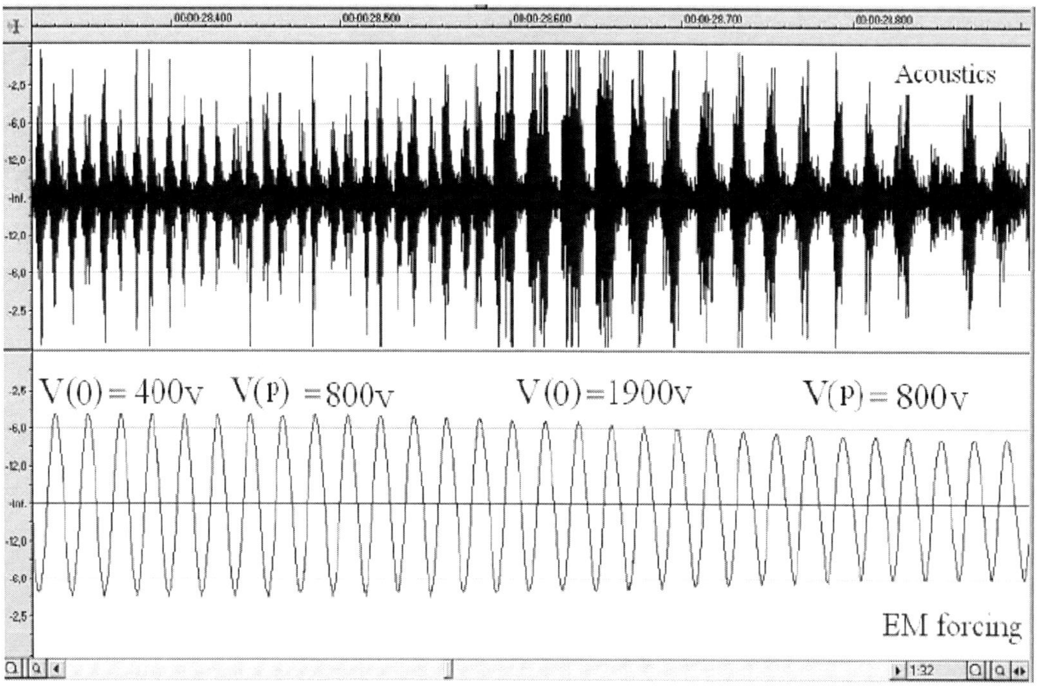

Figure 11. Transition (bifurcation) in stick-slip from 1:2 synchronization (period doubling) to 1:1 synchronization at simultaneous action of direct V(0) and periodic V(p) voltages; transition occur at V(0) > V(p), [Chelidze et al., 2008c].

Synchronization was observed only at some definite values of set of parameters (K_s, f, V_a). The "phase diagram" for variables f, and V_a or so-called Arnold's tongue [Pikovsky et al., 2003] is presented on the Figure 10.

It should be noted that the phenomenon of synchronization was observed only with the EM field directed normally to the slip surface. When the EM field was applied to the vertical sides of the slipping block, that is, roughly parallel to the slip plane, we were not able to observe the effect of synchronization.

We conclude that the synchronization is related to "electromagnetic braking" of slip at passing the extreme values of sinusoidal EM forcing and a sudden slip after accumulation of enough stress provided by spring pull.

Moistening of the slip surface by wetted blotting paper does not affect the "mechanical' synchronization, but practically kills the "electromagnetic" synchronization.

We observe transition (bifurcation) in stick-slip from 1:2 or period doubling synchronization, when two slip events occur per a period of EM forcing, to 1:1 synchronization, when one slip event occur per a period of forcing (Figure 11) at simultaneous action of direct $V(0)$ and periodic $V(p)$ voltages; transition occur at $V(0) > V(p)$.

The simple explanation of the period doubling effect follows from the expression of electrostriction at simultaneous application of direct and periodic voltages.

As the mechanical instabilities synchronize with both positive and negative sections of sinusoid (the respond is symmetric) we can postulate that the additional elastic strain u, induced by forcing, has a quadratic dependence on the intensity of electrical field E:

$$u = kE^2 \tag{14}$$

which is the well known expression of electrostriction in solids; here k is some proportionality constant depending on the forcing frequency and physical properties of rock [Chernyak, 1978]. If the electromagnetic forcing contains besides periodic the constant component as well, that is:

$$E = E_c + E_m \sin \omega t\,, \tag{15}$$

after inserting (15) into (14) the elastic response is:

$$u = kl(E_m)^2 \{1 + 2(E_c / E_m)^2 + 4(E_c / E_m)\sin \omega t - \cos 2\omega t\}, \tag{16}$$

where $k1$ is proportionality constant, which depends on the forcing frequency, physical properties of rock and constant component intensity E_c. It is evident that the stick-slip response to forcing in the latter case depends on the value of ratio E_c/E_m. At $E_c/E_m \ll 1$ the $cos2\omega t$ term of (16) is dominant, which means that the slip events will occur with the double frequency of forcing, but at $E_c/E_m \gg 1$ the slip regime is governed by the $sin\ \omega t$ term that is only one slip event occur per period of forcing. These conclusions are confirmed by our experiments (Figure11).

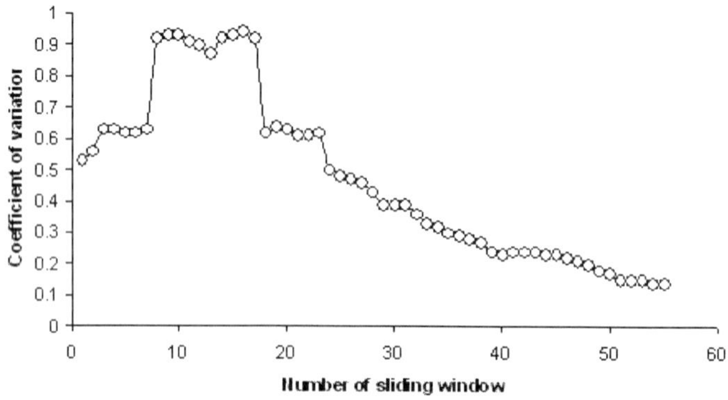

Figure 12. Coefficient of variation of power of acoustic emission time series at increased external forcing for 500 data length sliding window with 50 data shift. Here only half of record (Figure5 a) is used from the start of test to the maximal synchronization state [Chelidze et al., 2008c].

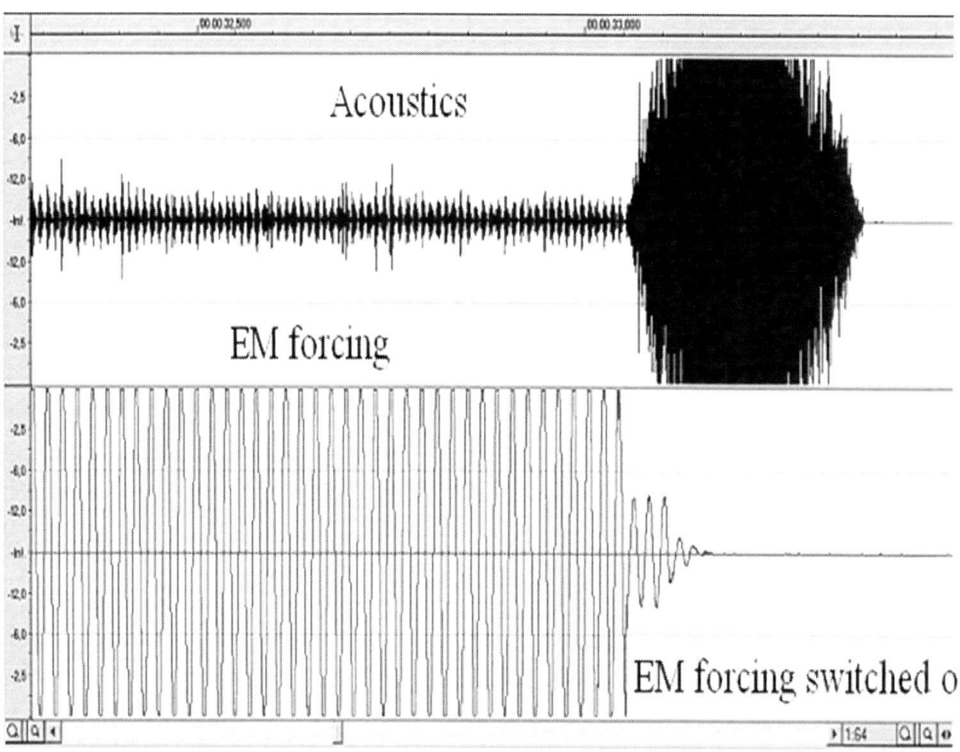

Figure 13. Increased acoustic energy release after total cessation of periodic EM forcing, which means that synchronization limits the energy release associated with a single event (quantization effect).

3. 3. 4. Synchronization: Amplitude Quantization

Synchronization affects not only waiting times, but also frequency-energy distribution: the amplitudes of bursts are much less scattered at application of periodic forcing.

Decrease of contribution of extreme events to synchronized AE activity is confirmed by calculation of the coefficient of variation CV (CV=standard deviation/mean). As it is shown in Figure 12, the extent of the deviation from the mean value of released AE power calculated for consecutive sliding windows, decreases at synchronization.

That means that synchronization limits the energy release associated with individual events (quantization effect). Sudden decrease or total cessation of synchronizing forcing is followed by acoustic burst of much larger energy than during periodic forcing (Figure 13).

3. 4. Synchronization: Quantitative Analysis Tools

3. 4. 1. Nonlinear Dynamics Methods for Measuring Complexity/Ordering of Natural Processes

Synchronization is one of manifestations of nonlinearity/complexity of a system [Meyers, 2009]. That is why the tools of nonlinear dynamics are main methods for revealing synchronization and measuring its strength. One of very popular models of complexity is a well-known theory of Self-Organized Criticality [Bak , Tang and Wiesenfeld. 1988]. Natural systems and/or processes are complex [Rundle, Turcotte, Klein, 2000] mainly due to their nonlinearity, an intrinsic property of the underlying laws conditioning absence of determinism of universe. This property (complexity) incorporates phenomena with the very broad diversity of dynamical features. Generally this diversity manifests itself in a certain kind of hierarchy of dynamical behavior ranging from strict determinism to total randomness. The most important is that between these extremes there are many intermediate states that reveal different degree of orderliness such e.g. periodicity, quasiperiodicity, deterministic chaos, low and high dimensional dynamics, hyperchaos etc. [Kantz, Schreiber, 1997]. The fundamental problem is how to measure the complexity from the observed time series. One of the most often used is application of time delay phase space reconstruction method to original time series [Takens, 1981; Kantz, 1998; Sprott, 2003, etc.]. After phase space reconstruction, phase space structure (attractor) can be analyzed using both qualitative and quantitative methods [e.g. Packard et. al [1980], Takens, [1981], [Abarbanel, 1993; Kantz, 1998; Sprott, 2003]. We will shortly describe here some of them.

Since phase space structure attractor-image of dynamics is formed, two most popular ways for the quantitative evaluation of complexity of analyzed dynamics are: quantification of the average evolution patterns of neighboring trajectories in the state space and/or quantification of the geometric patterns of the state space object.

Evolution of phase space trajectories could be analyzed by calculation of spectrum of Lyapunov exponents or, as it is often done, by calculation of maximal Lyapunov exponent $\lambda_{\max}$. Generally Lyapunov exponents quantify the average exponential rate of divergence of neighbouring trajectories in the state space, and thus provide a measure of the system's response to the local perturbations [Rosenstein, 1993; Kantz, 1997]. For measured data sets the maximum Lyapunov exponent $\lambda_{\max}$ of a dynamical system can be determined from the equation: $d(t) = d_0 e^{\lambda_{\max} t}$, where $d(t)$ is the mean divergence between neighboring trajectories in the state space at time t and d_0 is the initial separation between neighboring points. Generally, if λ<0, phase trajectories are drawing together and considered dynamical system has an attractor in the form of a fixed point. When λ=0, the system tends to the stable limit cycle. $\lambda>0$ means that phase trajectories are mowing away and such system may be chaotic or random [Rosenstein, 1993].

In order to characterize unknown dynamics by geometry of their reconstructed phase structures the algorithm for calculation of fractal dimensions of phase space point sets should be used. It is known that the fractal dimension of an attractor roughly characterizes the complexity and gives a lower bound for the number of equations or variables needed for modeling the underlying dynamical process. There are several such measures based on quantification of self-similar properties of phase space objects. These measures are information dimension (d_i), Hausdorf dimension d_H, etc.[Abarbanel,1993; Abarbanel et al. 1996; Kantz, 1997;Sprott, 2003; Sornette, 2000; Strogatz, 2000]. We shortly describe here only the (GPA) method of computing correlation dimension or fractal dimension as proposed by Grassberger and Procaccia [1983]. In spite of difficulties in dealing with real data sets GPA remains the most popular and often used method for quantifying geometrical features of phase space objects. This is probably due to the simplicity of the algorithm [Bhattacharya, 1999] and the fact that the same intermediate calculations are used to estimate both dimension and entropy. The correlation sum, $C(r, N)$, quantifies the way in which the density of points in the state space scales with the size of the volume containing those points. This approach is based on the idea of correlation sum. Correlation sum $C(r)$ of set of points in the vector space is defined as the fraction of all possible pairs of points which are closer than a given distance r. The basic formula useful for practical application is

$$C(r,N)=\frac{2}{(N-w)(N-w+1)}\sum_{i=1}^{N}\sum_{j=i+w}^{N}\Theta(r-\|x_i-x_j\|) \tag{17}$$

where $\Theta(x)$ is the Heaviside step function, $\Theta(x)=0$ if $x<0$ and $\Theta(x)=1$ if $x\geq 0$. $\|x_i-x_j\|$ is the Euclidian norm. Points with $i=j$ are excluded. w is the Theiler's window for fractal systems for long enough time series. For small r, $C(r)\propto r^{\nu}$ relationship is correct. Commonly such dependence is correct only for the restricted range of r values, so called scaling region. Correlation dimension ν or d_2 is defined as

$$\nu=d_2=\lim_{r\to 0}\frac{\log C(r)}{\log(r)}. \tag{18}$$

In practice d_2 value is found from the slopes of $\log C(r,N)-\log r$ curves for different phase space dimensions. In order to achieve appropriate linearity of the relationship between $\log(C(r, N))$ and $\log r$ one has to determine local slopes, or so called ''local scaling exponents''[Kantz, 1997].

True correlation dimension of an unknown process is the saturation value of d_2, which do not change by increasing phase space dimension. If saturation do not takes place, correlation dimension is infinitely large, which is typical for random processes.

For a correct analysis it is necessary to have long enough data sequences, et least $N\geq 10^{d/2}$ where N is a length of time series and d is the dimension of attractor [Abarbanel,

1993]. Since linear correlations lead to many spurious conclusions in nonlinear time series analyses, it is important to verify obtained results using the so called surrogate data approach. This is a method to test the null hypothesis that analyzed time series are generated by a specific process with the known linear properties [Theiler et al., 1992]. The mentioned method of surrogate data equips us for testing the nonlinear structure of complex dynamics [Theiler, 1992]. The surrogate data is inherently a stochastic signal which mimics certain statistical properties such as temporal autocorrelation or Fourier power spectra of the original signal. The surrogates can be constructed from the original time series on the basis of different null hypotheses. The three types of mostly used surrogates address the three main hypotheses: temporally independent noise, linearly filtered noise, and nonlinear transformation of linear filtered noise. So whenever we try to quantify the degree of nonlinearity the results of calculation of above measures should be compared with the similar quantities for surrogate data sets. Phase randomized surrogate sets (obtained by destroying the nonlinear structure through randomization of the phases of a Fourier transform of the original time series and following invert transformation) is often used to test the null hypothesis that the time series are linearly correlated with Gaussian noise [Theiler et al., 1992]. Also Gaussian scaled random phase (GSRP) surrogate set can be generated to address a null hypothesis that the original time series is linearly correlated noise that has been transformed by a static, monotone nonlinearity [Rapp et al., 1993; 1994].

Generally these two methods of generation of surrogates are based on shuffling of the original data set but, in the case of Gaussian scaled random phase surrogates, the controlled shuffles [Rapp et al., 1994] can give more precise and reliable results than the unstructured shuffles of the random phase surrogates.

The most commonly used measure of the significance of the difference between the original time series and the surrogate data is given by the criterion: $S = |\langle M_{surr}\rangle - M_{orig}| / \sigma_{surr}$, where σ_{surr} denotes standard deviation of M_{surr}. The details of the procedure, as well as an analytic expression for ΔS, the uncertainty in S, are described in Theiler et al., [1992].

For rejecting the null hypothesis, the Barnard and Hope nonparametric test can be used [Rapp et al., 1994]. With this criterion, the null hypothesis is rejected at a confidence level $p_c = 1/(N_{surr} + 1)$, if $M_{orig} < M_{surr}$ for all the surrogates.

One of the serious real data analysis problems is the influence of noises. It is preferable to use the so called nonlinear noise reduction (which in fact is phase space nonlinear filtering) instead of common linear filtering procedures. The latter, as it is well known, may destroy the original nonlinear structure of analyzed complex processes [Schreiber, 2000]. Nonlinear noise reduction relies on the exploration of reconstructed phase space of considered dynamical process instead of frequency information of linear filters [Hegger and Kantz. 1999; Schreiber, 1993; Kantz, and Schreiber, 1997].

As the most methods of analysis need rather long and stationary data sets what is commonly not typical features of measured time series a strong impetus was given to development of new techniques for analysis of not too long and noisy time series. Here we will focus on the approach, based on testing of topology of phase space images of unknown dynamics. This new technique oriented on exploration of phase space structure-image of dynamics, is the method of recurrence plots (RP) [Eckmann et al., 1987]. Let us recall here that if the dynamical system has any deterministic structure, an attractor appears in the state space. As it was already mentioned, attractor is the set of points in phase space, towards which a dynamical path will converge. Again, recurrence is a fundamental property of nonrandom dynamical systems, the state of which although exponentially diverges under small disturbances, but after some time the system will come back to a state that is arbitrarily close to a former state. Recurrence plots visualize such a recurrent behavior of dynamical system. RP enables to investigate structure in these high dimensional phase spaces through two dimensional representation of its recurrences. It is important that recurrence plot method is effective for nonstationary and rather short time series.

Generally recurrence plots are designed to locate hidden recurring patterns and structure in time series and is defined as $N \times N$ symmetric matrix:

$$R_{i,j} = \Theta(\varepsilon_i - \|\vec{x}_i - \vec{x}_j\|),\ i, j = 1,..., N, \tag{19}$$

here $\vec{x}_{i,j}$ are phase space vectors reconstructed using Taken's time delay method. Insofar as RP is based on Takens delay-coordinate embedding, when this last is correctly chosen, the dynamical invariants of the true and reconstructed dynamics are identical. Therefore it is natural to assume that the RP of a correctly reconstructed trajectory bears similarity to a RP of the true dynamics. In fact $\vec{x}_i$ stands for the point in phase space at which the system is situated at time i, ε_i is a predefined cut-off distance, $\Theta(x)$ is the Heaviside function. The cut-off distance defines a sphere centered at $\vec{x}_i$. As far as the trajectory in the reconstructed phase space returns at time i into the ε - neighborhood of where it was at time j (i.e. if $\vec{x}_i$ is closer to $\vec{x}_j$ than the cut-off distance) $R_{i,j} = 1$ and these two vectors are considered to be recurrent. Otherwise $R_{i,j} = 0$. According to Eckman, one can visualize the $R_{i,j}$ values by black and white dots [Eckman et al. 1987], but often the recurrence plot relates $R_{i,j}$ distances to a color, e.g. the larger is the distance, the "cooler" is the color. Thus, the recurrence plot is a solid rectangular plot consisting of pixels whose colors correspond to the magnitude of data values in a two-dimensional array and whose coordinates correspond to the locations of the data values in the array.

The black points indicate the recurrences of the investigated dynamical system revealing their hidden regular and clustering properties. By the definition RP is symmetric and has black main diagonal (the line of identity) formed by distances in matrix. In order to understand RP it should be stressed that it visualizes distance matrix, which represents autocorrelation in the series at all possible time (distance) scales. As far as distances are computed for all possible pairs, on the RP plots elements near the diagonal correspond to short range correlation, whereas the long range correlations are revealed by the points distant from the diagonal. Hence if the analyzed dynamics (time series) is deterministic (ordered, regular), then the recurrence plot shows short line segments parallel to the main upward diagonal. At the same time if dynamics is purely random, the RP will not present any structure at all. One of the crucial points in RP analysis is selection of cutoff distance ε or radius. If ε is selected too low no recurrent point will be found. At the same time it cannot be set too high as then every point will be assumed as recurrent. Exhaustive overview on this subject can be found in Zbilut [1998], Marwan [2003]. etc.

The initial aim of RP testing was the visual inspection of structures located in high dimensional phase spaces where other methods are useless especially when we deal with real data sets.

Zbilut and Webber [Zbilut, 1992] have developed a tool, which quantifies the structures in RPs, namely, the Recurrence Quantitative Analysis (RQA). They define measures using the recurrence point density, the length of diagonal and vertical structures in the recurrence plot, the recurrence rate, the entropy of recurrent points' distribution etc. Presently at least 8 different statistical RQA values are known [Zbilut, 1992; Marwan, 2003]. Computation of these measures in small windows moving along the main diagonal of the RP reveals the time-dependent behavior of these variables making possible the identification of unknown dynamical patterns in time series [Zbilut, 1992; Marwan, 2002].

Here we will briefly touch only main RQA statistical parameters. The first of these statistics, termed % *recurrence* (% REC), is simply the percentage of points on the RP that are darkened or in other words those pairs of points whose spacing is below ε_i - predefined cut-off distance. It quantifies the number of time instants characterized by a recurrence in the signals' interaction: the more periodic is the signal dynamics, the higher is the (% REC) value.

The second RQA statistic is called % *determinism* (%DET); it measures the percentage of recurrent points in a RP that are contained in lines parallel to the main diagonal. The main diagonal itself is excluded from these calculations because points there are trivially recurrent. Intuitively, %DET measures how ''organized'' a RP is. This variable discriminates between the isolated recurrent points and those forming diagonals. Since a diagonal represents points close to each other successively forward in time, DET also contains the information about the duration of a stable interaction: the longer the interactions, the higher the DET value.

The third often used RQA statistics, called entropy, is closely related to %DET. Entropy (ENT) is Shannon information entropy of line distribution measured in bits and is calculated by binning the diagonal lines according to their lengths and using the following formula:

$$ENT = -\sum_{k=1}^{N} P_k \log_2 P_k$$

where N is the number of bins and Pk is the percentage of all lines that fall into bin k. In other words Pk is defined as the ratio between the number of k-point long diagonals, and the total number of diagonals. ENT is measured in bits of information, because of the base-2 logarithm. Thus, whereas DET accounts for the number of the diagonals, ENT quantifies the distribution of the diagonal line lengths. The more different are the lengths of the diagonals, the more complex is the deterministic structure of the RP. A more complex dynamic will require a larger number of bits (*ENT*) to be represented.

It is reasonable to assume that the existence of some determinism (i.e. recurrence of definite states) in the forced system, dependent on the forcing time structure should be a signature of synchronization. That is why we widely used tools of nonlinear dynamics for quantitative assessment of the complexity of system and synchronization strength.

Besides of phase space testing there are also other qualitative tests which can be used for complex data analysis, such as Iterated Function Systems (IFS) [Jeffrey, 1992], Lempel-Ziv complexity (LZC) measure (Lempel, A., Ziv, 1976), mutual information (Sprott, 2006), nonextensivity analysis [Tsallis, 2010] etc.

In following we present examples of calculation of some mentioned parameters for real time series. The main idea is that the decrease of complexity under weak forcing means increase of ordering, i.e. increase of synchronization strength.

3. 4. 2. Measuring Synchronization Strength

There are several kinds of synchronization between oscillating system with natural frequency ω_0 and forcing frequency ω. As a result of interaction the natural frequency oscillating system can be modified to some another frequency Ω. We are looking for the phase synchronization (PS) – when amplitudes of oscillator are irregular and uncorrelated, but the frequencies ω and Ω are adjusted, i.e. there is a regular phase shift between ω and Ω. Pikovsky et al [2004] describe different approaches to quantifying the synchronization of processes by analysis of measured time series. Some of them have roots in theories of dynamical systems and fractals and includes Lyapunov exponents and fractal (mostly correlation) dimensions. These methods are presented in the previous section.

Several tools for analysis were tested on the recordings of stick-slip process, where the superimposed periodic EM field intensity was raised monotonously from zero to 1000V and then decreased in the same way to zero (Figure 9 a). At first we used the easiest approach for estimating phases of acoustic signal: digitized waveforms were transformed to sharp spikes to have well pronounced markers. Then time series (catalogs) of time intervals between consecutive maximums (waiting times) ($\Delta t=t_i-t_{i-1}$) in AE wave trains for π periods of external sinusoidal forcing were composed (time scale in Figs. 14-18 corresponds to sequential values of onsets of AE events t_i). Significant reduction of Δt scatter during EM forcing is evident.

In order to assess synchronization in quantitative manner various tools of nonlinear dynamics (synchronization) theory described in [Zbilut, 1992; Rosenblum et al., 1996; Kantz, 1997; Pikovsky et al., 2003; Marwan, 2003] were applied to experimental data (Figure 9a) obtained under variable intensity of forcing; the results are shown in Figs 14-18. All these approaches yield similar results, namely, strong increase in the ordering of Δt in AE time series, which we consider as the signature of stick-slip synchronization by EM forcing. Figure 14 shows that increments of consecutive waiting times for experiment presented in Figure 9a between max amplitudes of AE is minimal at strong EM forcing, i.e. in the middle part of the graph.

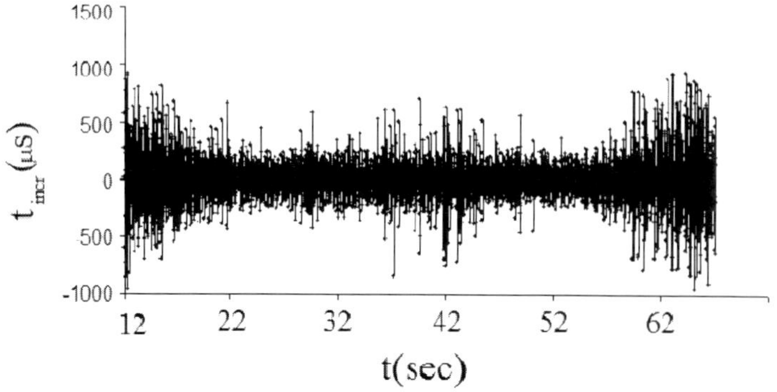

Figure14. Time series of time increments of consecutive waiting times between max amplitudes of AE in consecutive π– periods of external forcing for a whole record presented in Figure 9a.

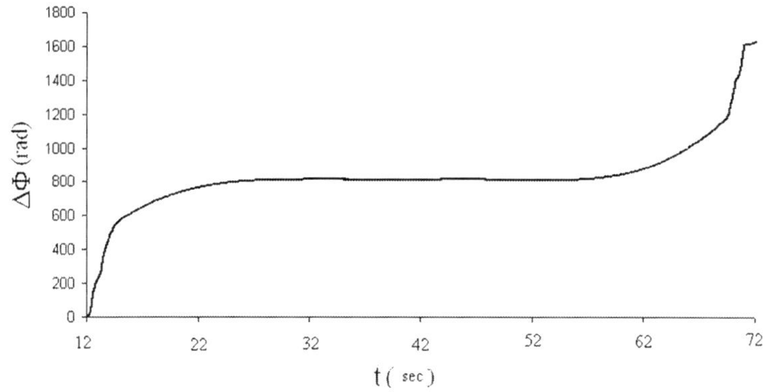

Figure15. Phase difference $\Delta\Phi$ between the sequence of maximums of acoustic emissions' bursts (AE catalogue) and external sinusoidal forcing for a whole record (Figure 9a).

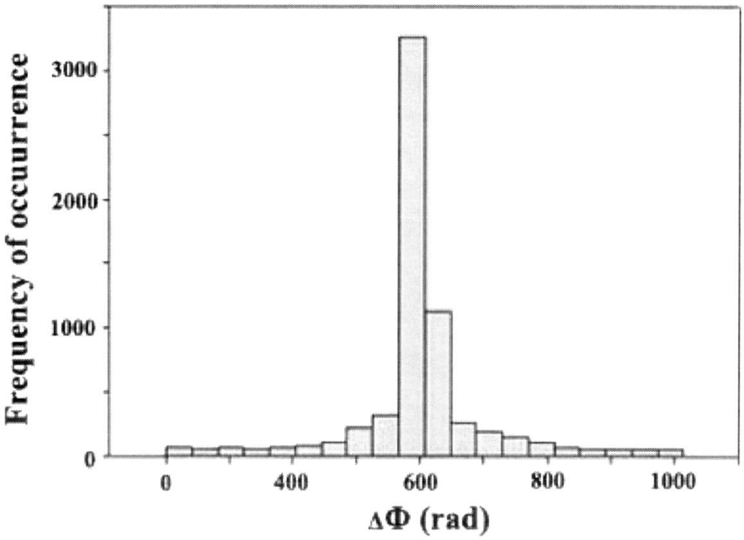

Figure16. Frequency distribution of phase difference $\Delta\Phi$ between the sequence of maximums of AE and external sinusoidal forcing. The phase difference $\Delta\Phi$ is practically constant ($\Delta\Phi = 800$ rad) at maximal forcing for data presented in Figure 9a.

In Figure 15 we present the temporal evolution of phase difference $\Delta\phi$ obtained from Hilbert transform of waiting times time series. Well-defined horizontal part of $\Delta\phi$ versus t (Figure 15) represents the time, during which the acoustic emission becomes phase synchronized to the external sinusoidal forcing in the wide range of forcing amplitudes (from approximately 500 V to 1000 V). The most synchronized part of the plot phase difference variation $\Delta\phi$ does not exceed 10 radian (compare with the total $\Delta\phi$ increment of 1800 radian during the whole experiment).

It is known that the probability density distribution must be narrower for the synchronized signal compared to a non-synchronized one. As it follows from Figure 16, the phase difference $\Delta\Phi$ between the sequence of maximums of AE and external sinusoidal forcing is practically constant ($\Delta\Phi = 800$ rad) at maximal forcing for the data presented in Figure 9a.

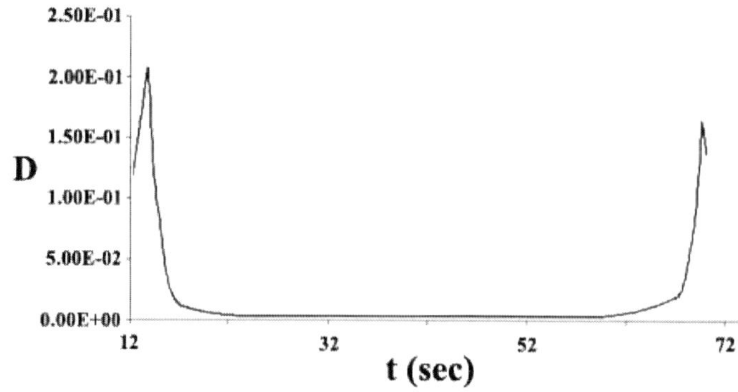

Figure17. Variation of phase diffusion coefficient *D* of phase differences $\Delta\Phi$, calculated for consecutive sliding windows, containing 500 events for data presented in Figure 9a.

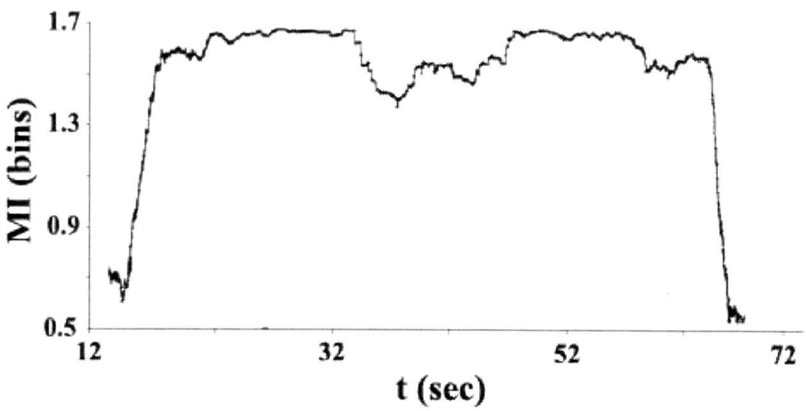

Figure18. Variation of Mutual Information between phase of forcing and AE bursts calculated for consecutive sliding windows, containing 500 events (for the data presented in Figure 9a).

Frequency locking, expressed as a minimum of the phase diffusion coefficient D [Pikovsky et al., 2003] is also a quantitative measure of the phase synchronization: Figure 17 confirms results, obtained by other methods.

Mutual Information (*MI)* is a quantitative assessment of strength of correlation between two variables: X (here phase of forcing) and Y (here, AE onset). Clear increase of Mutual Information (*MI)* between the phase of forcing and AE bursts indicates that dynamics of acoustic emission becomes much more correlated in the synchronized part of acoustic emission data set (Figure 18).

One of interesting methods of revealing synchronization in relaxation-type processes is a "gap" technique [Lursmanashvili, 2001], which is described in detail in the [de Rubeis et al., 2010] and which has an analogue in the analysis of radio technical systems [Blekhman, 1988]. It is founded on the effect of concentration of slips (and associated seismic/acoustic events) in the definite phases of forcing period. Naturally, this means that occurrence of dynamic events in the remaining phases of forcing is less probable (prohibited). The width of the prohibited zone is larger for stronger synchronization. Of course, some intervals of silence can occur just randomly, but these intervals are relatively short and are distributed randomly relative to the forcing phase. These short gaps can be considered as a noise; the gaps, due to synchronization are much wider. The method has been tested on the laboratory spring-slider model with mechanical forcing.

Consider stick-slip as a nearly periodic relaxation process, where the stress is accumulated for a long time in order to release it in a fast slip. As this last phase is very short, the duration of a whole cycle is practically equal to that of accumulation stage. Assume that a periodic relaxation process with period τ is exposed to also periodic forcing of period T. Let us suppose that the forcing period is unknown exactly, but some preliminary guess on the range of periods can be made. Is it possible to retrieve the unknown forcing period accurately from the observed modified relaxation process? So the objective is to find exact forcing period from a given sequence of onsets of slips $d(j)$, $j = 1,2,3, \ldots m$. Here m is the total number of slips. We are looking for the forcing period inside the interval $(T_{min}; T_{max})$ of a width $\Delta T = T_{max} - T_{min}$. For this we use a scanning window with the optimal length dT, which is much less than ΔT. The window length dT is calculated by the following empirical formula: $dT = T^2_{min}/ 50t$, where t is the duration of experiment. As a result we will have $n = \Delta T/ dT$ (virtual) periods to test distribution of slips inside the chosen interval of periods ΔT. For the $T(i)$-th trial (virtual) forcing period we have:

$$T(i) = T_{min} + i \cdot dT \,,\; i = 0,1,2.3\ldots\ldots n \tag{20}$$

The times of slip occurrences $d(j)$ are normalized by dividing on the corresponding period $T(i)$:

$$d(j)/T(i);\; j = 1,2,3\ldots\ldots m;\; i = 1,2,3\ldots\ldots n. \tag{21}$$

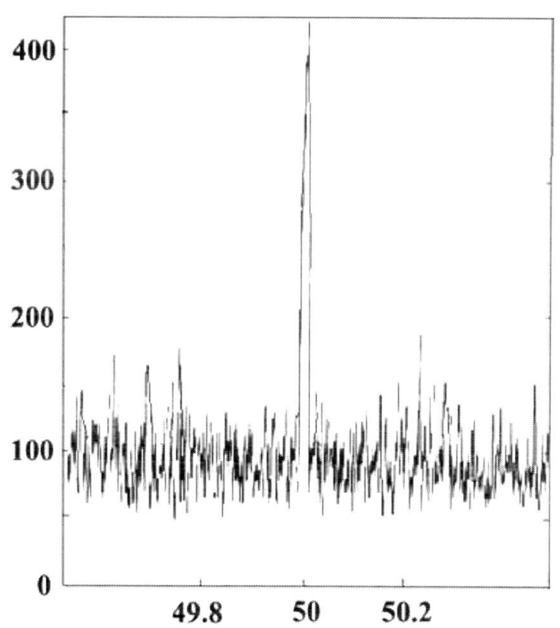

Figure 19. Recurrence spectrum of local maxima of gaps widths for a chosen range of trial forcing periods spectrum of slips: Y axis - the local maxima of the gap widths $dF(j,i)^{loc}_{\max}$; X axis - trial periods $T_i \cdot 1000$; Statistics – 20 events.

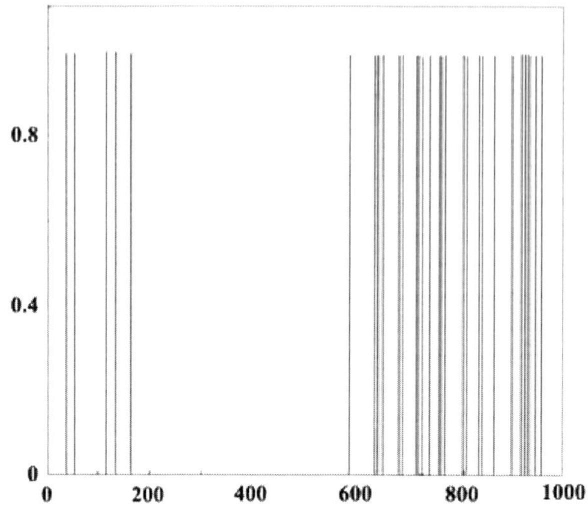

Figure 20. The distribution of slip moments inside the genuine forcing period T_{ig} divided in 1000 intervals. Statistics - 20 events.

Let us consider for the each period *T(i)* the remainder of the above division *F(j,i):*

$F(j,i) = remainder(d(j)/T(i));$ $\qquad j = 1,2,3.......m;\ i = 1,2,3......n.$

For revealing phase synchronization just this non integer part is essential as it characterizes the distribution of phases of "discharges" inside the trial forcing period. For convenience these values are normalized in the following way:

$$F_n(j,i) = F(j,i)T(i) \cdot 1000 \qquad (22)$$

After such normalization considered values of $F_n(j,i)$ do not depend on the absolute value of the trial period *T(i)* and are distributed inside the range (0;1000). The normalized phases are arranged according to their ascending values and the largest free of slips interval between neighboring phases or the local maximal "gap" width $dF(j,i)_{\max}^{loc}$ is found for each trial period *T(i)*. As a result, two numerical sequences for *T(i)* and $dF(j,i)_{\max}^{loc}$ are obtained. The graph of $dF(j,i)_{\max}^{loc}$ versus *T(i)* actually forms a spectrum of local maxima of gaps widths for a chosen range of trial forcing periods. The largest gap (or the absolute maximum in the set of local ones) returns the value of the sought forcing period.

The method has been tested on the laboratory spring-slider model with mechanical forcing. The results are shown in Figures 19-20. In these experiments the forcing frequency (50 Hz) was known beforehand and the objective was to retrieve it from the observed data as if it was unknown. So the known forcing frequency was used only for validation of the gap method. Figure 19 shows the hidden periodicity of slips (in particular, recurrence spectra of local maxima $dF(j,i)_{\max}^{loc}$ for a trial period T_i) and Figure 20 present gaps in distribution of slip moments inside the genuine (T_{ig}) period of forcing, (retrieved from the value of absolute maximum of the gap width $dF(j,i)_{\max}^{abs}$), here T_{ig} = 50 s. From the analysis of the gap spectra we can conclude that synchronization is present at the forcing frequency (50 Hz) and the genuine forcing period T_{ig} = 1/f = 0.02 s can be extracted with no less than 0.01 Hz accuracy, which means that the forcing frequency can be determined accurately from the observed synchronized slip recurrence spectra. It is worth to note that such a clear distinction of the hidden periodicity was obtained on the very restricted statistics, namely for 20 events.

The method can be used for detection and quantitative assessment of seismic process synchronization strength induced by a weak external impact, such as tides, reservoir loading, etc.

The above approach has been tested on the Catalogue of Caucasian earthquakes of M. Nodia Institute of Geophysics and some significant gaps related to tidal effects of various periods were revealed [Chapter 18 in de Rubeis et al., 2010].

3. 5. Phase Time Delay

The acoustic response lags behind the periodic forcing phase; the lag is inversely proportional to the forcing intensity (Figure 21 a, b, c, d). The delay is quite similar for both AE burst onsets and AE wave train maxima. The dependence of phase delay on the intensity of pulse-like (Figure 21c) or periodic (Figure.21.d) forcing points to a nonlinear response of a system to weak external impact.

3. 6. Synchronization by Mechanical Forcing

Of course, synchronization of stick-slip can be achieved by a weak mechanical forcing also. Ponomarev et al (2010) studied spatial and temporal patterns of acoustic emission (AE) on the pre-fractured sample of rock loaded in a triaxial machine at 160 MPa confining pressure under constant strain rate 10^{-7} s^{-1} modulated by weak mechanical periodic forcing with periods 175 and 570 seconds (Figure 22). The log-likelihood function *lnL* of acoustic emission was calculated: maximum of the *lnL* function shows which frequencies provide the largest contribution to AE compared to purely random model. Dark vertical strips correspond to broadband acoustic emission during strong slip events. Correlation of AE with applied oscillating load after the mark of 300000 s is revealed by dark horizontal strips of AE. The intensity of correlated AE is most pronounced just after strong AE events, i.e. in the "aftershock" period.

Friction experiments with mechanical forcing are quite numerous as the phenomenon of stick-slip is ubiquitous in friction processes and it is considered as a negative factor, disturbing stable displacement of contacting surfaces. In order to achieve smooth functioning of technical systems with friction, application of weak mechanical vibrations of relatively high frequency is studied mainly as a tool for stabilization of friction process [Bureau et al , 2000; Perfettini et al., 2001; Boissou et al., 1998]. It has been found that at some intensity (but still much smaller than the main driving force) of mechanical forcing with the frequency 120 Hz the stick-slip phenomenon is strongly reduced and the frictional movement became more stable: the amplitude of slips is reduced strongly (Figure 23).

a)

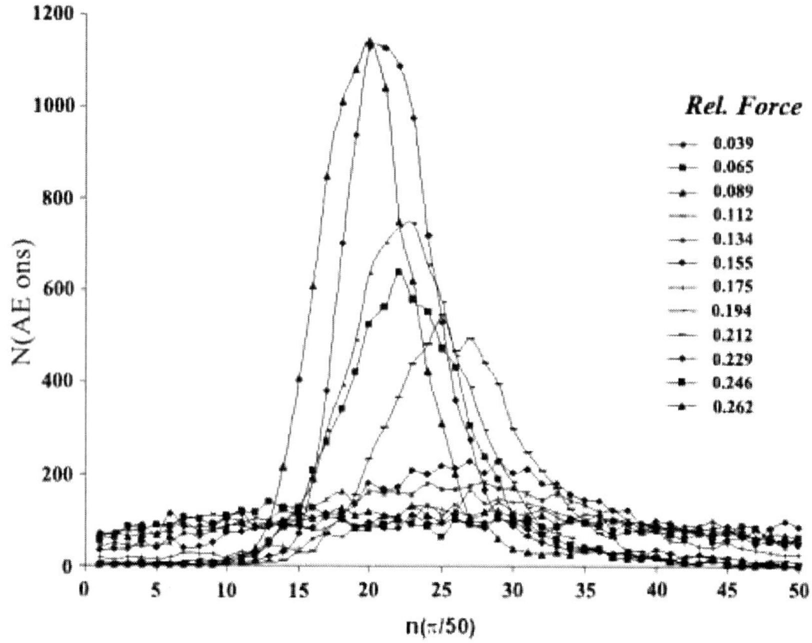

Phase distribution of onsets of stick slip AE

b)

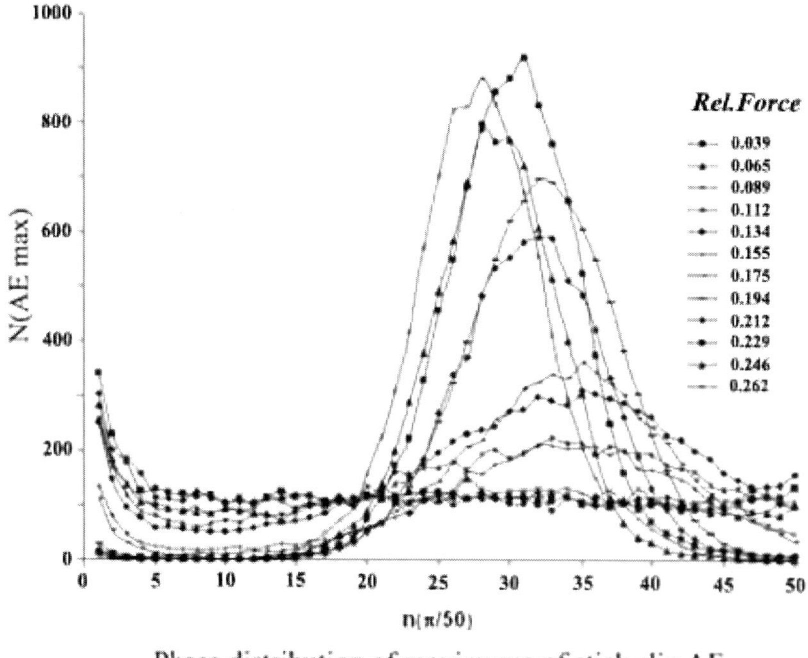

Phase distribution of maximums of stick slip AE

c)

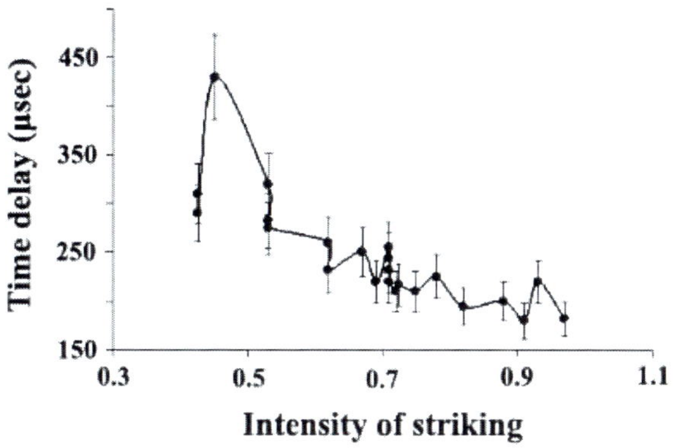

d)

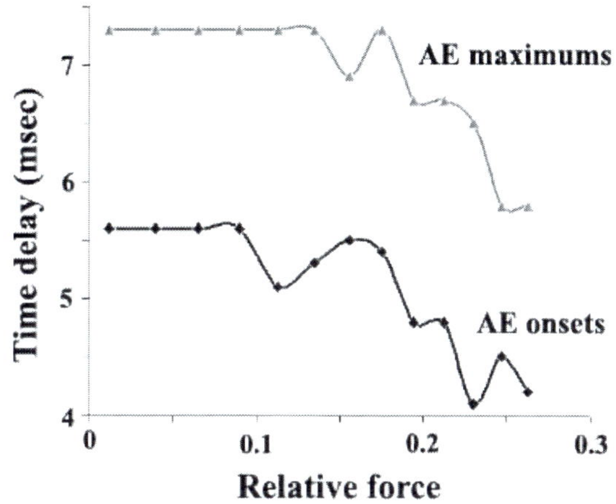

Figure21 a, b, c, d. AE phase delay relative to forcing phase: distributions of AE number versus delays for: onsets (a) and maximums (b) of acoustic signal at periodic EM forcing; c – time delays for slips triggered by mechanical strikes (Sobolev et al., 1993), d - phase time delays at synchronization by periodic EM forcing for Figure 21a ,b.

At the same time it is clear from analysis of recordings that what is considered as stabilization of friction by Bureau et al [2000] is in fact small amplitude stick-slip, synchronized with the (high) frequency of forcing – 120 Hz. Thus the stabilization of motion is achieved not by total elimination of stick-slip, but by drastic decrease of amplitude of slips, which occur much more frequently than in natural stick-slip motion without forcing and are synchronized with the forcing frequency.

The phenomenon can be considered as a "quantization" of slips, which has been observed also in our experiments with EM forcing (Figure 12). Thus according to Figure 23 a single dynamic instability (slip) of magnitude of the order of 3.5 μm in a non-modified friction process is released in the same time interval by 12 small-scale events of magnitude 0.2 μm in synchronized friction as result of quantization. In total the cumulative slip rate of 12 small displacements was 2.4 μm/s, which is close to the one large slip rate 3.5 μm/s.

It is quite unrealistic at present, but the quantization effect of periodic forcing can in principle be considered as a tool for control/reducing the magnitude of impending strong earthquake.

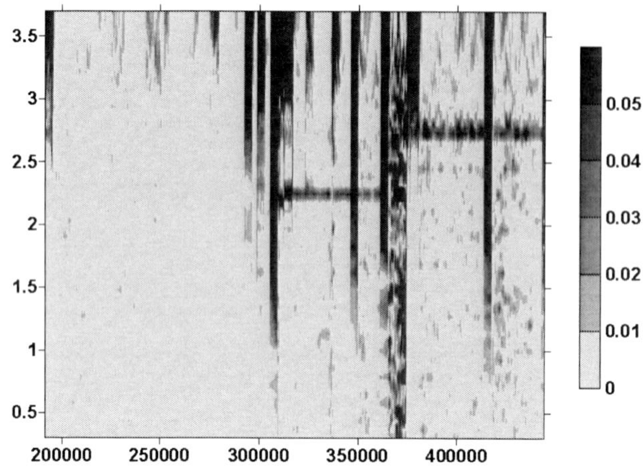

Figure 22. Plot of log-likelihood function *lnL* of acoustic emission intensity versus time in seconds for a sample of granite loaded by constant strain till the mark of 300000 s and by combination of constant and oscillating strain after this mark. Darker colors correspond to larger values of *lnL*. Dark vertical strips correspond to broadband acoustic emission during strong slip events. Note correlation of AE with applied oscillating load after the mark of 300000 s, revealed by two dark horizontal strips of AE, corresponding to two forcing frequencies with periods 175 and 570 seconds [Ponomarev et al., 2010].

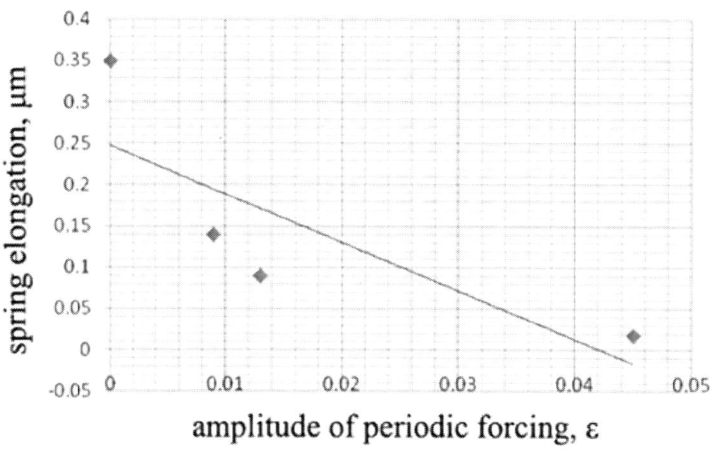

Figure23. The loading spring's maximal elongation during slips experienced in stick-slip motion for drag velocity $v = 8$ μm/s versus periodic mechanical modulation amplitude ε (plot compiled using data of Bureau, 2000). Slips amplitude are maximal at zero forcing. At the maximal forcing sliding seems to be much more stable, but the blow up of the trace shows the remaining oscillating slips exactly at the frequency of the load modulation, $f = 120$ Hz, which is much higher than the natural stick-slip frequency, approximately 2 Hz. It seems that what is considered as stabilization is indeed synchronization of slips with quantization of slip amplitudes.

In our experiments the mechanical forcing was exerted by vibrator, which was glued to the sliding plate and whose impact was directed in either normal or tangential direction to the slip plane. The intensity of vibration forcing was varied, but it always was much weaker (approximately 10^{-4} less) than the pulling force, [Varamashvili et al., 2008]. In our experiments with mechanical forcing synchronization phenomenon was mainly observed as high order synchronization [Pikovsky et al. 2003] in contrast with results of [Ponomarev et al., 2010] shown in Figure 22. The difference can be related to different conditions of experiment (triaxial loading versus free stick-slip in our experiments) and intensity of forcing, which in Ponomarev et al [2010] experiments was a few percent of applied load.

3. 7. High Order Synchronization of Stick-Slip Process: Experiments on Spring-Slider System

3. 7. 1. High Order Synchronization - Definitions

In the previous chapters we considered relatively simple synchronization regimes. At the same time it is well known that when the processes of various timescales are coupled there is a possibility of high order synchronization (HOS). Namely, if ω_n is the natural frequency of the first (driven) oscillator, ω is that of the second (forcing) one and Ω is the resulting frequency of the driven system under forcing, the systems tends to synchronize at different integer ratios of ω/Ω. Such effects are observed, for example, in biological systems, lasers, electronic relaxation generators, etc [Pikovsky, Rosenblum and Kurths, 2003; Kantz, Schreiber, 1997].

Our experiments were aimed to reveal HOS in the frictional system, namely in the stick-slip process of spring-slider set up, subjected to weak electromagnetic (EM) or mechanical forcing.

In the chapter the following definitions are used: ω_0 – natural frequency of autonomous oscillator (here spring-slider system); T_0 – corresponding natural period; ω, T, ϕ - forcing frequency, period and phase accordingly; Ω, T_{obs}, ϕ_{obs} – frequency, period and phase accordingly of autonomous oscillator observed after application of forcing.

We again are looking for the phase synchronization (PS) – when amplitudes are irregular and uncorrelated, but the frequencies ω and Ω are adjusted, i.e. there is a regular phase shift between ω and Ω.

High-order or $(n{:}m)$ synchronization means that observed and forcing frequencies satisfy equation [Pikovsky et al.2003]:

$$n\,\omega = m\,\Omega\text{; or } nT_{obs} = mT_f\text{;} \qquad (21)$$

where n and m are some integer numbers. In our case ω is the EM or mechanical forcing frequency and Ω is the observed frequency of AE bursts under forcing. The ratio n/m is called the winding or rotation number ρ and is defined as:

$$\rho = \Omega / \omega = T_f/T_{obs}. \qquad (22)$$

This condition of high-order synchronization can also be incorporated in the general framework of the frequency-locking model, using equation (21) ; in this case the ratio $n{:}m$ is the winding number: $\rho = n/m$. The phase-locking can be also expressed in terms of the oscillators' phases:

$$\left| n\phi - m\phi_{obs} \right| < const \qquad (23)$$

where ϕ is the phase of the forcing and *Ω* is that of the kicked oscillator.

3. 7. 2. HOS Synchronization by Electromagnetic Forcing

The experiment shows that addition of the strong enough component of the constant electric field to the high frequency EM forcing signal (40 Hz) invokes transition from 1:2 to 1:1 synchronization; thus Figure 11 represents actually the example of high order (n/m = 1:2) synchronization.

The 1:1 regime was obtained also at application of relatively low frequency (T_0 = 0.5 *s*) signal (the left part of Figure 23). At still lower forcing frequencies, for example, T_0 = 4.5 *s* the stick-slip process demonstrates swarm-like behavior: one EM forcing generates dozens of AE bursts (the right part of Figure 23).

Figure 25 represents the HOS of stick-slip at EM forcing by alternating pulses of different duration.

Both the onsets of swarms and these of the individual events within sequences (swarms) turn out to be very well organized, which is evident from synchrograms. Synchrogram or stroboscopic technique implies observing the phase of slow (forcing) signal at the moments of spiking, here acoustic pulses (Pikovsky, Rosenblum and Kurths, 2003). If the moments of n-th spiking of the swarm (n = 1, 2, 3, etc) always occur in the definite phase of the forcing than both oscillators are synchronized and the plot of delays of n-th spike relative to the onset of forcing signal versus time (or number of cycle for periodic forcing) reveals n stripes parallel to the time axis.

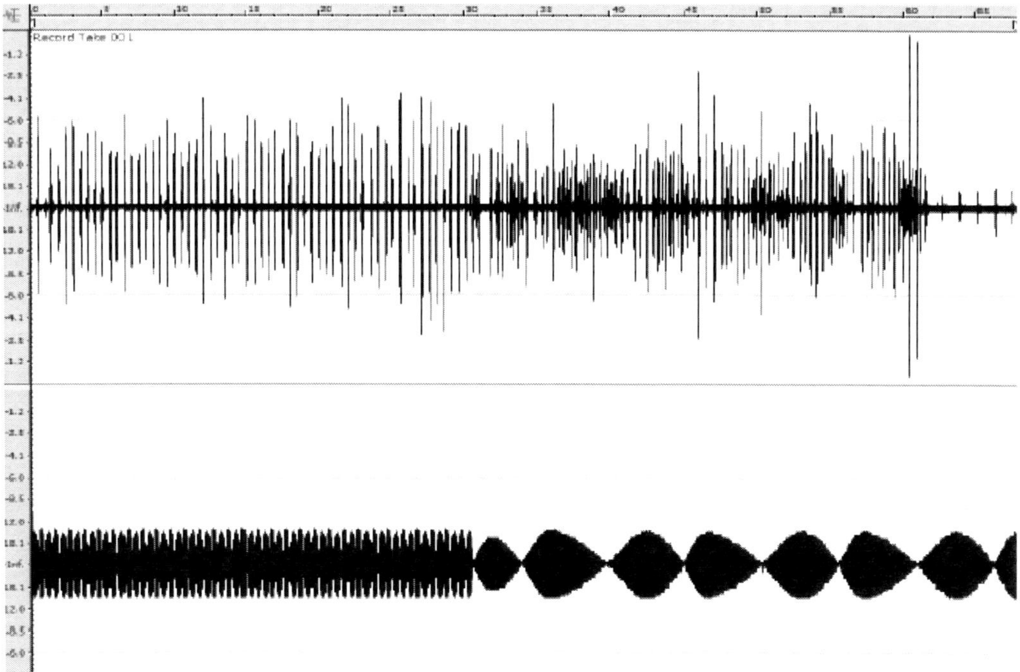

Figure24. Transition (bifurcation) in stick-slip from 1:1 synchronization to high order synchronization at increasing the period of forcing from 0.5 s to 4.5 s (exp: Zura5). Note that in Figs. 24, 24 a and 26 a the forcing signal was filled by high frequency (HF) oscillations in order to visualize the low frequency forcing on the computer screen; the HF signal was applied only to computer and not to the rock plates.

In our case the onsets of n-th AE-pulses of swarms have almost the same delay relative to the times of onsets of identical forcing (t_{if0}; $i = 1,2 \ldots k$, where i is the number of forcing pulse) and the events within swarms manifest regular phase difference. Besides, even small difference in the duration of the forcing pulses invokes regular changes in the delays, number of AE bursts in the induced swarm and in phase differences. Here are characteristics of EM forcing and AE response. Short forcing pulses: $n{:}m = 1/3$; mean duration $t = 0.618$ s; mean delay $t_{if0}= 0.295$ s; the AE response consists of three consecutive bursts with_mean phase differences: correspondingly 0.258; 0.411; 0.537 s and St. Dev. correspondingly 0.037; 0.048; 0.034. Long forcing pulses: $n{:}m = 1/4$ or $1/5$; mean duration $t = 0.82$ s; mean delay $t_{if0}= 0.358$ s; the AE response consists of four consecutive pulses with mean phase differences: correspondingly 0.268; 0.467; 0.601; 0.736 and St.Dev. correspondingly 0.046; 0.05; 0.07; 0.055; 0.046.

Figure 26 a presents even more populated AE swarms generated by longer EM pulses; in this case the swarms contain up to 40 AE events, so $n{:}m \approx 1/40$. Again, the AE bursts within the swarm are well organized. Corresponding synchrogram (Figure 26 b) shows that the first ten bursts show almost constant phase shift relative to the EM forcing onsets. The following bursts demonstrate regular small increase of the phase delay in sequential swarms [Chelidze et al., 2008b]. It is striking how Figs. 26 a, b are similar to synchronization pattern of neuron spiking under external electrical field forcing [Pikovsky et al. 2003, Figure 6.7]; this is a vivid illustration of universality of synchronization phenomenon.

So, it seems that lower is the frequency of forcing larger is the number of triggered synchronized events in the forcing-generated AE swarms. Probably the same mechanism can lead to formation of seismic swarms.

a)

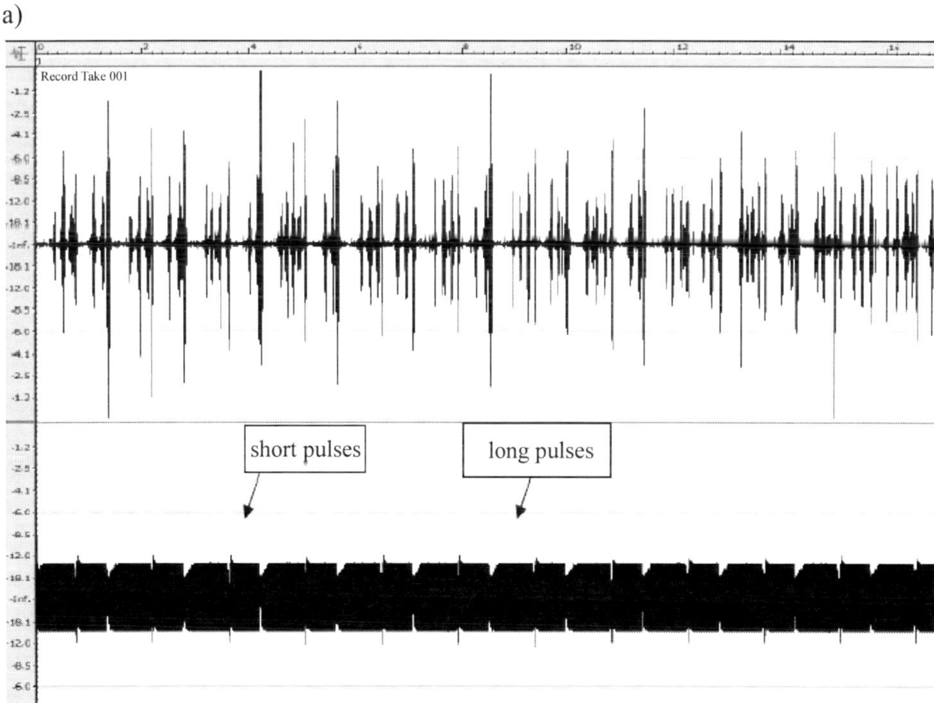

Figure 25. (Continued on next page)

b)

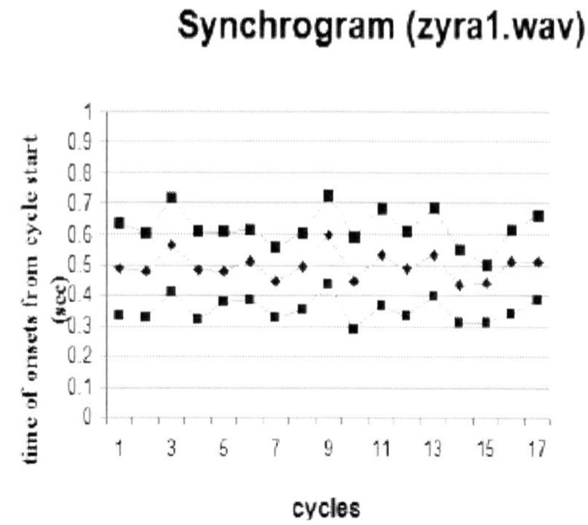

c)

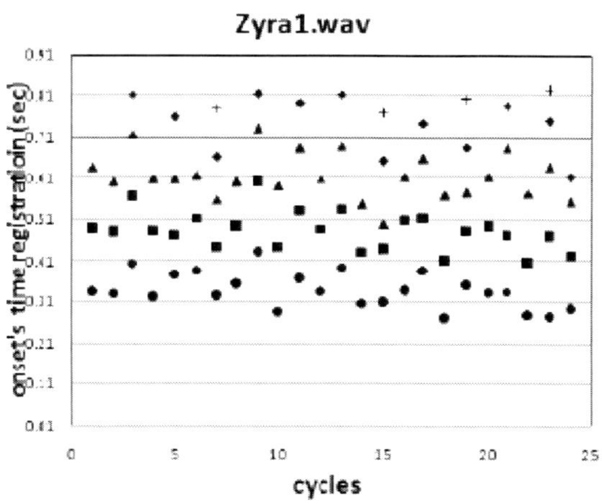

Figure 25. High order synchronization: $n<m$ (exp. Zyra1). Note that the onsets of swarms and these of the individual events within sequences (swarms) turn out to be very well organized (see text). a) swarms of AE pulses generated during forcing by EM pulses; b) synchrogram for shorter forcing pulses $n:m = 1/3$; c) synchrogram for longer forcing pulses $n:m = 1/4$ or1/5. Note that the onsets of swarms and these of the individual events within sequences (swarms) turn out to be very well organized (see text).

Besides $n<m$ coupling, we observe also $n>m$ HOS (Figure 27 a, b). Here the AE bursts occur seldom in comparison with EM forcing pulses. The forcing pulse (repetition) rate was $T_f = 1.8$ sec, AE delay relative to the pulse onset – 0.24 s. Observed period of AE recurrence was $T_{obs} = 25$ s. Thus in this case $n>m$, namely the winding number is 14:1. Figure 26 b presents the AE distribution relative to the forcing period phase, which is very sharp though slips wait for quite a long time (hundreds of forcing periods) to occur.

3. 7. 3. HOS during Mechanical Forcing.

We also investigated synchronization of the same spring-slider system under weak periodic mechanical forcing [Chelidze et al. 2007; Varamashvili et al. 2007; Varamashvili et al. 2008; de Rubeis, ChechovskiandTeisseyre, 2010].

a)

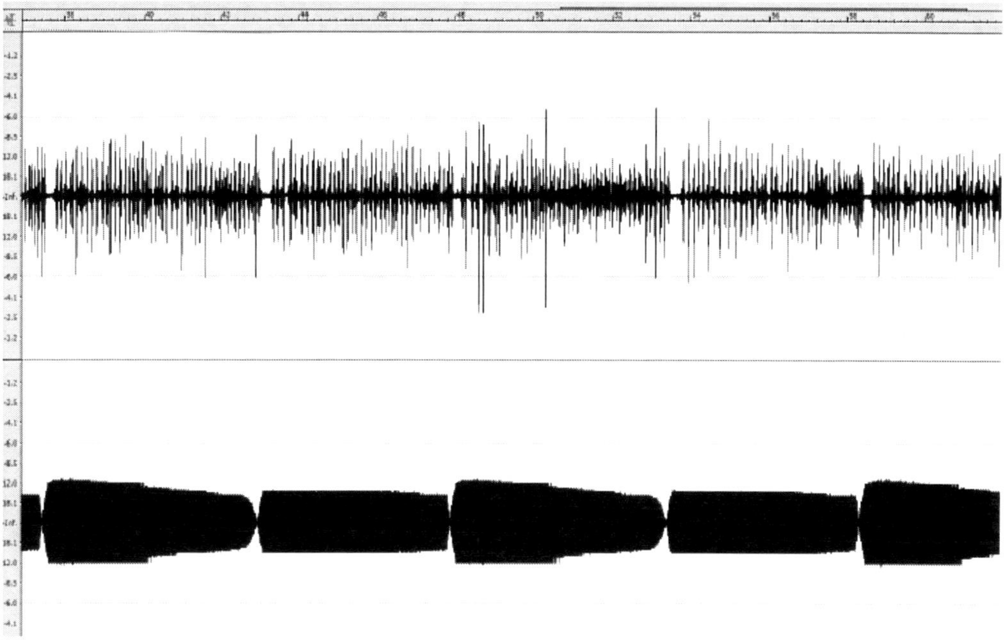

b)

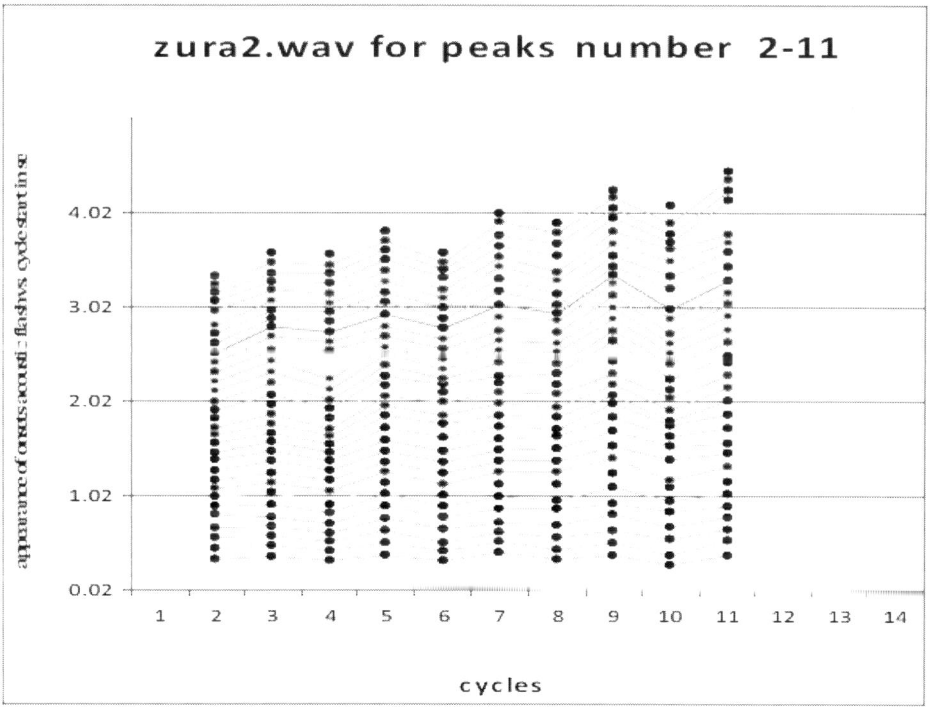

Figure 26. High order synchronization: $n{:}m \approx 1/40$ (exp. Zura2); a) swarms of AE pulses generated during forcing by very long EM pulses; b) stroboscopic diagram (synchrogram) for the first 31 AE bursts in the swarms number 2-11 generated by 11 forcing EM pulses; note stripe structure of synchrogram, which shows that the phase shift between EM forcing and AE pulses is almost constant for the first 10 slips [Chelidze et al., 2008b].

a)

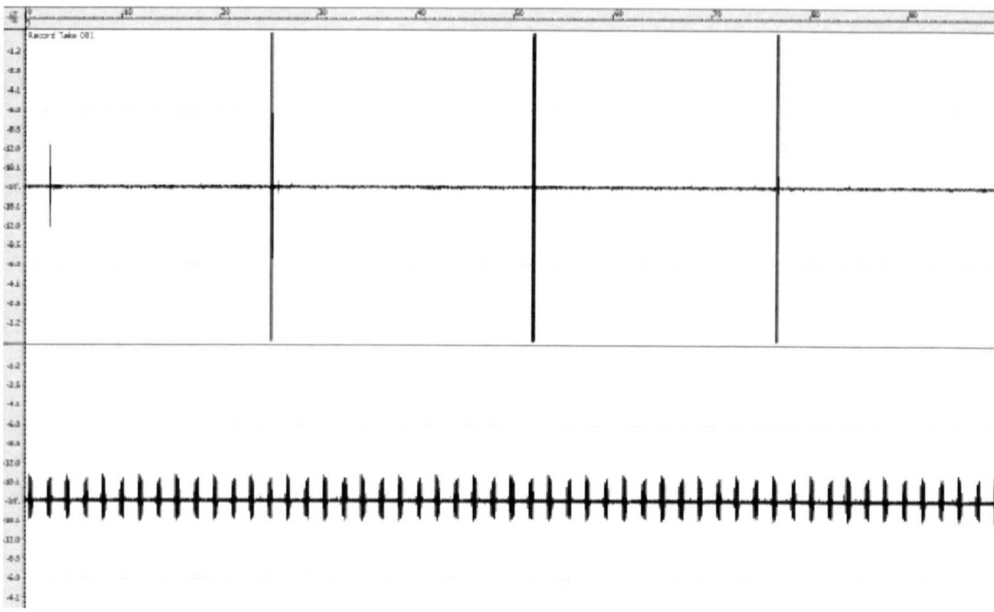

b)

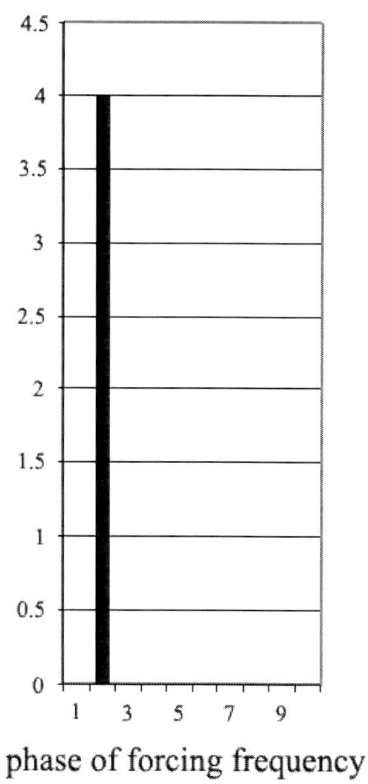

Figure 27. High order EM synchronization: $n>m$; (exp.Sopo1). $T_f = 1.815$ sec; delay time = 0.240 ms; $T_{ob} = 25$ sec; $n\ \omega = m\ \Omega$; $14n \cong m$. a. recording of AE at synchronization by EM pulses; b. histogram of AE bursts distribution in phases of EM forcing.

The experiments were conducted for two modes of mechanical forcing: i. the forcing is applied normal to the slip surface and ii. forcing is applied parallel to the slip surface; for brevity we will refer to them as normal and tangential forcing accordingly. In the case of normal forcing we calculated the maximum value of mechanical forcing, which corresponds to the maximum measured voltage applied to mechanical vibrator (i.e. when the voltage applied to the vibrator equals 6.5 V). The mass of the oscillating element of the vibrator m is ≈ 20 g, so we obtain for the natural frequency f of the vibrator's oscillating element: $f = \sqrt{k/m} = 5\,\text{Hz}$, where k is the stiffness of the vibrator spring. Substituting $k = 25\, m = 0.5$ N/m and the value of the maximum deflection x_{max} of the oscillating element at the 6.5 V applied voltage, $x_{max} \approx 4\text{x}10^{-3}$ m, we obtain for the corresponding (maximal) intensity of mechanical forcing F_{max}:

$$F_{max} = k\, x_{max} \approx 2\text{x}10^{-3}\, N$$

At lesser voltages the forcing is much weaker: our assessment for $1V$ intensity of vibrator input is approximately $5\text{x}10^{-4}\, N$. Thus the forcing was always much less than the driving force $F = 4N$.

Similar numbers were obtained for tangential mechanical forcing. The forcing frequencies were 30 Hz for the tangential and 20 Hz for the normal loading cases. In both cases the forcing rate was larger than the dragging rate, which means that the synchronization of the process is possible.

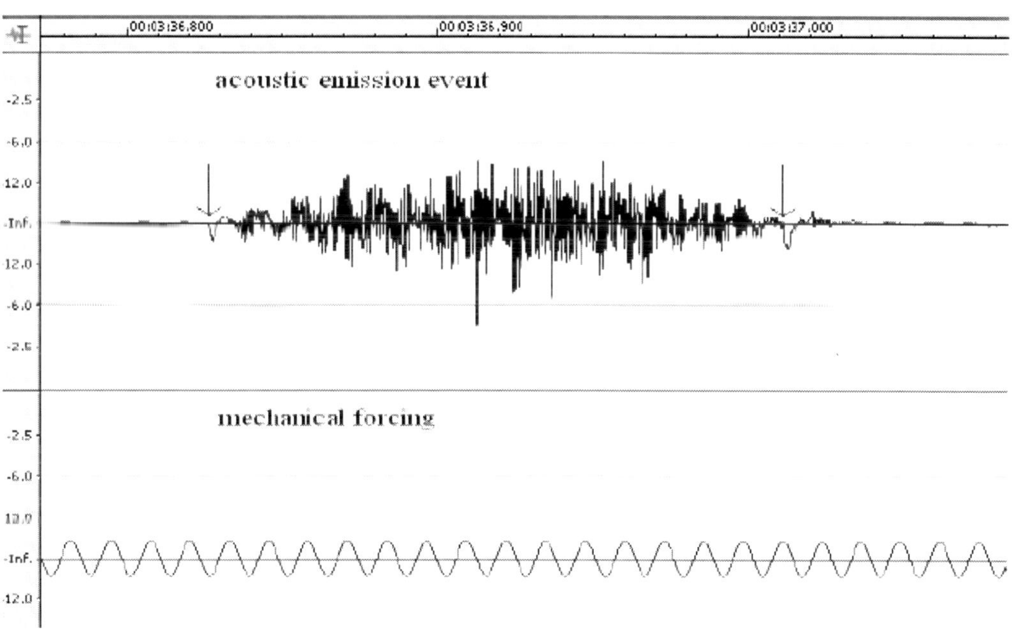

Figure 28. Record of mechanical forcing: a single acoustic pulse and corresponding tangential mechanical forcing on the expanded scale example corresponds to vibration intensity generated by application of 4 V voltage to the vibrator. The initial slow deviation from the background line is considered as the onset of the AE pulse and the start of slow terminal phase as a pulse termination; these moments are marked by arrows.

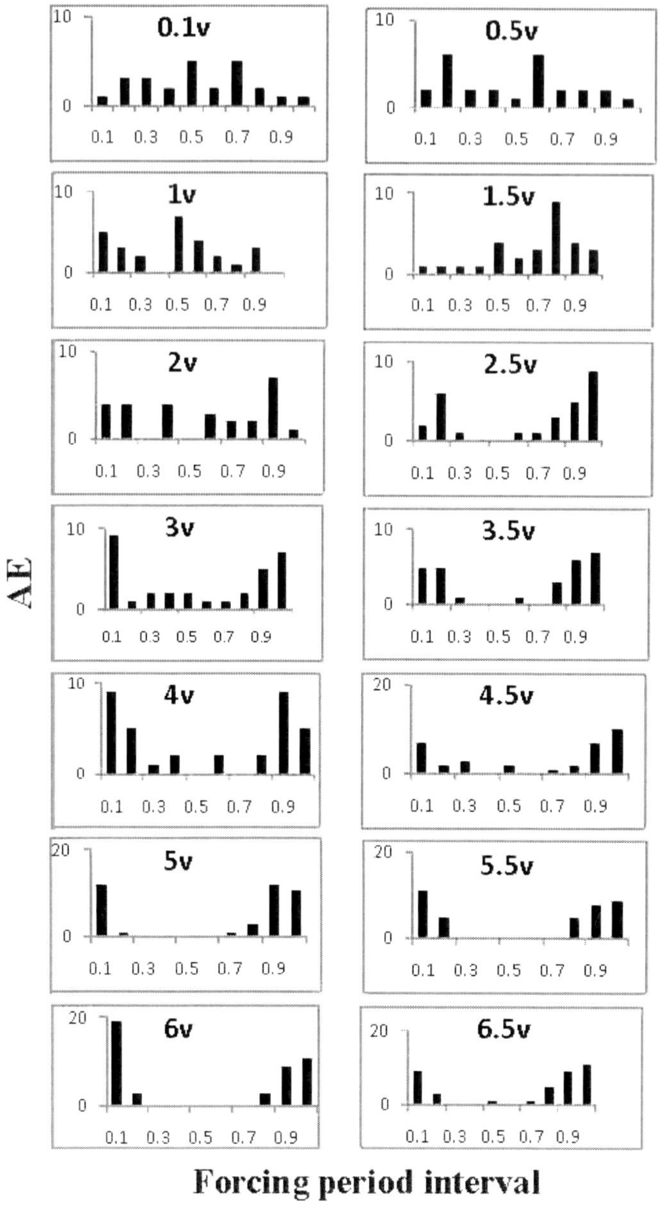

Figure 29. Distribution of acoustic emission onsets relative to mechanical forcing period phases (in decimals) for different intensities of normal forcing. Forcing frequency – 20 Hz.

Figure 28 a, b presents experimental records, when the mechanical forcing is in the range $5x10^{-5}$- $2x10^{-3}$ *N*.

On Figure29 are presented distributions of the AE burst onsets relative to the phase (in decimals) of mechanical forcing period for the normal forcing.

At low voltages (up to 1*V*) the onsets are more or less randomly distributed in the decimals of the forcing period. Voltage increase results in concentration of the offsets at a definite part of forcing period, namely in the first and the last decimals of forcing phase. Evidently, increasing of voltage applied to mechanical vibrator promotes synchronization of

AE offsets with external forcing. The same behavior is observed for tangential mechanical forcing (Figure 30), namely, at low voltages (up to 2*V*) the onsets are more or less randomly distributed in the decimals of the forcing period (Figure 29, 30). Voltage increase results in concentration of the offsets at a definite part of forcing period, namely in the first and the last decimals of forcing period for normal forcing; in the case of tangential forcing synchronization is most pronounced in the phase interval 0.25-0.5 of the forcing period (Chelidze et al., 2008a) and fills the gap, observed for normal forcing (Figure 29).

Alteration of the forcing frequency affects the phase distribution of AE. Figure 31 (the left column) shows the AE distribution for tangential forcing at frequency 80 Hz. It is evident that the maximum of AE shifts to other phases of forcing period compared to the distribution for 30 Hz (Figure 30).

3. 7. 4. Synchronization of AE Signal Terminations

It was a surprise to discover that not only AE onsets can be synchronized by a weak mechanical forcing; Figure 31 (right column) illustrates that the terminal parts of the signal also are synchronized by forcing with the same strength as the onsets.

We suppose that the forcing (here 80 Hz) can affect not only the phases of onsets but also the phases of terminations of AE bursts, but their influence is realized in quite different phases of forcing: presumably, the onsets are triggered by tangential forcing around the minimum area of forcing period and suppressed by forcing in the maximum area of forcing period.

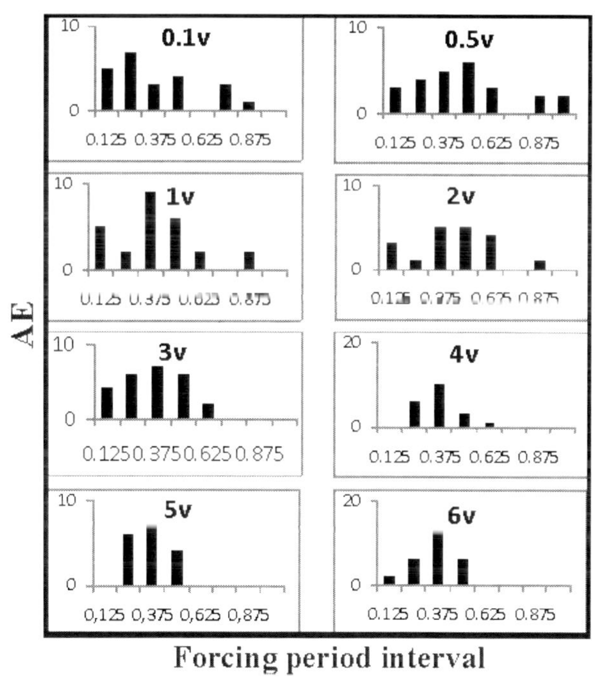

Figure 30. Distribution of acoustic emission onsets (the left column) and terminations (the right column) relative to the (mechanical) forcing period phase (in twelfths of the forcing period) for different intensities of tangential forcing. Forcing frequency – 80 Hz [Chelidze et al., 2008a].

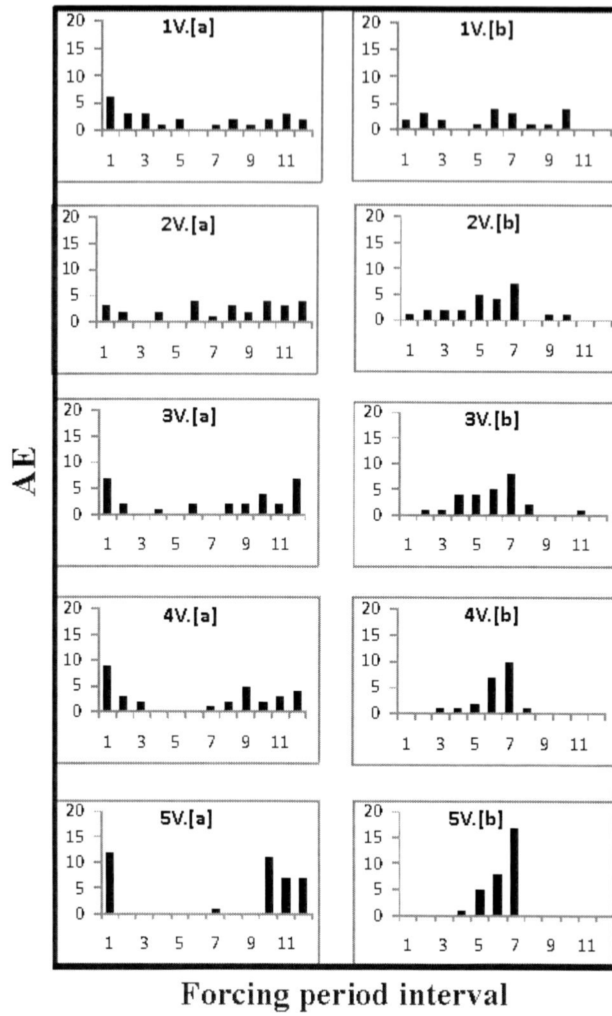

Figure 31. Distribution of acoustic emission onsets (the left column) and terminations (the right column) relative to the (mechanical) forcing period phase (in twelfths of the forcing period) for different intensities of tangential forcing. Experiment shows also that AE onsets can be synchronized both by the onsets and the terminal parts of the mechanical forcing signal (right column). Forcing frequency – 80 Hz.

3. 7. 5. Statistics of Synchronization

The standard deviation of onset and termination times regularly decreases with increase of forcing intensity (Figure 32).

Increase in the forcing intensity besides better synchronization of onsets and terminations, also brings on regular shortening of duration of AE bursts (Figure 33).

It is evident that increasing of voltage applied to vibrator promotes synchronization of AE offsets with external forcing.

It is striking that the AE bursts are so well synchronized though the waiting interval of bursts varies between 100-200 periods of forcing, i.e. to initiate a single phase-synchronized slip event the forcing oscillator should pump the energy of hundreds of oscillations to the slider-spring system.

a)

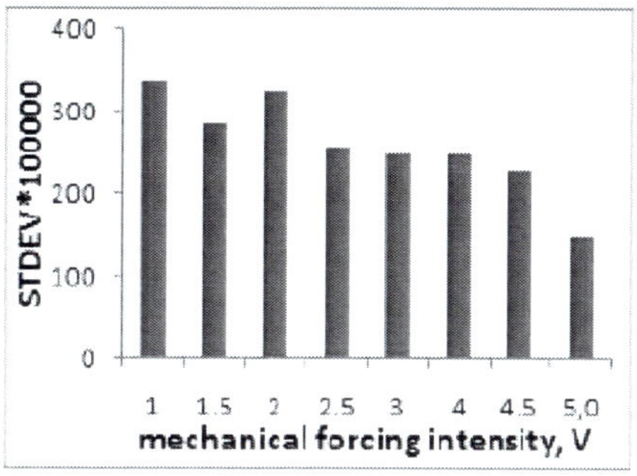

b)

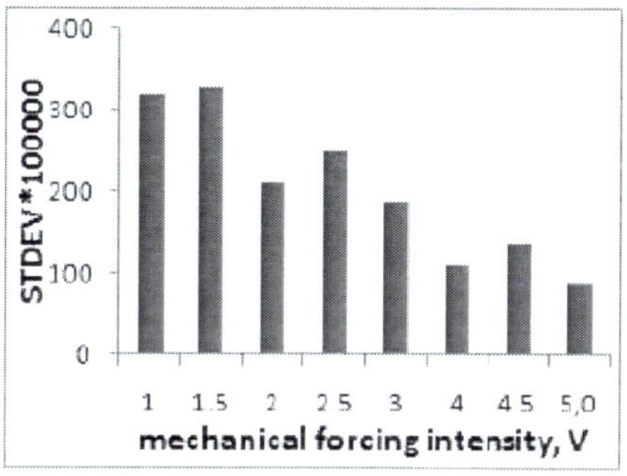

Figure 32. Standard deviation of onsets (a) and terminations (b) distributions for different intensities of external applied voltages, which are proportional to intensity of (tangential) mechanical forcing.

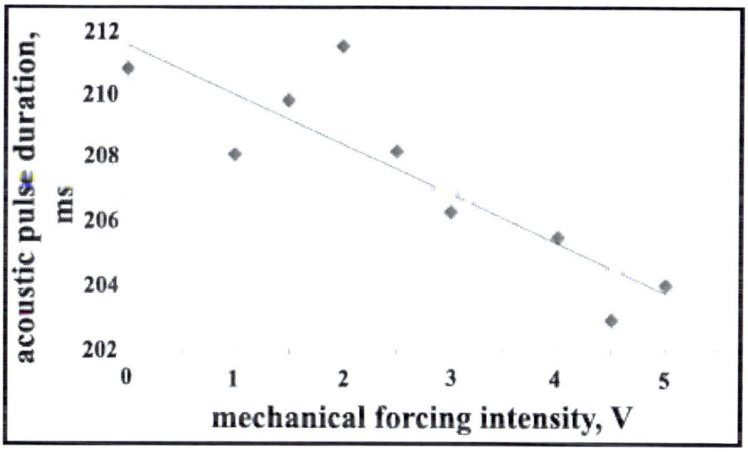

Figure 33. Mean duration of stick slip generated acoustic pulses for different intensity of normal mechanical forcing, with a trend line.

The important conclusion from the presented data for the much discussed interaction between tidal deformation and EQ-s [Beeler, Lockner, 2003; Scholz, 2003] is that in order to find it we should not only look for direct 1:1 correlation between events recurrence frequency and forcing, namely for the increase of seismicity exactly at the tidal frequencies (say, 12 h etc). The high order phase synchronization can occur at multiples of tide frequency like it was shown in laboratory experiments. It is important to note that HOS can be responsible for phase synchronization of AE or seismic events irrelative to the duration of their nucleation; of course, the optimal condition of synchronization (minimal forcing) should correspond to the forcing period, close to natural event nucleation time. Additional complication arouses from the phenomenon of delay; the response can be shifted quite significantly for weak forcing. Thus the question of tidal forcing of earthquakes should be reconsidered taking into account the new experimental evidence.

3. 8. Dynamic Triggering/Synchronization of Earthquakes and Tremors

The study of seismic response of the lithosphere to a weak forcing is a fundamental problem for seismic source theory as it reveals the important detail of the tectonic system, namely, how close is it to the critical state. Last years introduction of new sensitive broadband seismographs, new dense seismic networks and new methods of signal processing lead to the breakthrough in triggering and synchronization studies and formation of a new important domain of earthquake seismology, related to dynamic triggering (DT) of local seismicity by wave trains from remote strong earthquakes [Hill, Prejean, 2009; Prejean, Hill, 2009; Hill, 2010; Chao et al., 2012]. The trivial aftershocks' area is delineated mainly by static stress generated by earthquake and decay rapidly with distance d as d^{-3}, whereas the dynamically triggered stresses decay much slower (as $d^{-1.5}$ for surface waves). That means that dynamic stresses generated by seismic wave trains can induce local seismicity quite far from the epicenter; they can be defined as remote aftershocks. The first well documented DT episode is connected with 1992 Landers earthquake, when the sudden increase of seismicity above background value (calculated as β-statistic of Matthews and Reasenberg, 1988) after the main event was observed by many seismic stations at distances up to 1250 km with delays ranged from seconds to days. Later on DT was observed in different remote areas after Denali Fault 2002, Hector Mine 1999, Kurile 2007, Sumatra, 2004 and many other EQ, though most clearly the effect is expressed in active extensional regime areas, as well as in volcanic and geothermal regions.

3. 8. 1. Characteristics of DT Process (Stresses, Periods, Repeat Time)

The main characteristic of DT events are peak dynamic values of stress (T_p) or strain (ε_p); for shear waves $T_p \approx G\,(u_p/v_s)$ and $\varepsilon_p \approx u_p/v_s$; here G is the shear modulus, u_p is particle' peak velocity and v_s is velocity of the shear wave. Calculated from the field data give values of T_p from 0.01MPa to 1MPa (ε_p from 0.03 to 3 microstrain). Such large scatter is due to the impact of another important factor, namely, the local (site) strength of earth material, which is highly heterogeneous. Thus what matters is not the absolute value of T_p or ε_p, but the difference between local stress and local strength (resistance to failure). This is why in some areas high T_p do not trigger local seismicity and, on contrary, some areas manifest DT even at low peak stresses (Hill, Prejean, 2009;). One of main factors reducing local strength is the pore pressure of fluids, which is the scope of relatively new direction, so called hydroseismology [Costain and Bollinger, 2010].

The stresses imparted by teleseismic wave trains according to assessments of D. Hill [2008] are 10^{5} times smaller than confining stresses at the depth, where the tremors are generated. This is not surprising as the synchronization theory predicts that even smallest forcing is able to adjust the rhythms of oscillating systems [Pikovsky et al., 2003].

In most cases triggering is observed during surface waves, especially during Rayleigh wave arrivals, i.e. long periods and large intensity of shacking are favorable for exciting remote triggered events. Periods in the range 20-30 sec are considered as most effective in producing triggered events for the same wave amplitude. In principle the optimal period of DT should depend on the earthquake preparation characteristic time (consider Arnold's tongue phenomenon, Figure 10) and can change from dozens of seconds for microearthquakes to hours and days for moderate events. For tidal stresses with periods 12-24 h the threshold can be as low as 0.001 MPa.

Timing of triggered events is very variable: they can be excited immediately by the some phase of the wave train (say, Rayleigh) or delayed by quite a long time, hours or days. Duration of triggered activity period is also variable - from minutes to a month.

Magnitude of reported triggered events varies between M = 0.2 or less to M = 5.6. It is likely that most of triggered seismicity are just ignored due to their small intensity and are not included in seismic catalogs. Small (local) triggered events in a given area are revealed using very simple method: the original broadband record of the strong (remote) earthquake are filtered in order to separate low-frequency component (0.01-1 Hz), i.e the dominant component of passing wave train, which can be considered as a forcing and high-frequency component (1-20 Hz), where local triggered events can be recognized.

The triggered events belong to one of two classes: regular earthquakes with sudden onset and so called non-volcanic tremors or tectonic tremors (TT) with emergent onset.

Tectonic tremors are considered as a new class of seismic events related to recently discovered phenomena of low frequency earthquakes and very low frequency earthquakes [Obara, 2003; Richardson et al., 2010]. As a rule individual tremor has dominant frequencies in the range 1-10 Hz, lasts for tens of minutes and propagates with shear wave velocity, which means that they are composed by S body waves. Spatially the phenomenon of triggering is most frequently encountered in hydrothermal areas [Hill, Prejean, 2009].

It is necessary to mention that some seismogram processing methods may introduce high frequency (HF) artifacts similar to triggered events. Artificial HF signals appear in Fourier spectra and spectrograms when predominant periods of the input seismic waves are longer than the window length [Peng et al., 2011]. These artificial HF energy emissions correlate very well with arrivals of intensive surface waves and thus mimic the real triggering. The simplest method to avoid appearance of false triggering signals is application of high pass filter [Peng et al., 2011] like this has been done in [Hill, Prejean, 2009; Prejean, Hill, 2009; Hill, 2010]. In our analysis of remote triggering in Georgia we also used this approach.

3. 8. 2. Examples of Triggering and Synchronization

At present a lot of instances of triggering and synchronization are documented using statistical β-approach, but the most informative technique is the above mentioned double filtering method. As a rule, triggered events belong to the class of triggered tremors. Tremor's signatures are: emergent onset, lack of energy at frequencies higher than 10 Hz, long duration from dozens of seconds to several days, irregular time history of oscillations' amplitude, close correlation with large-amplitude surface waves.

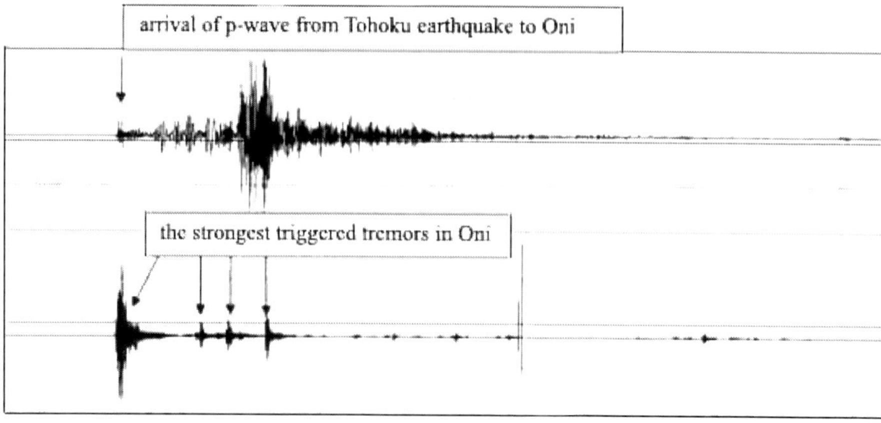

Figure 34. Broadband record of M= 9 Tohoku EQ, Japan (11.03.2011) wave train z-component (upper channel) and the same high-pass band (0.5-20 Hz) filtered record (lower channel) at the Oni seismic statsion.

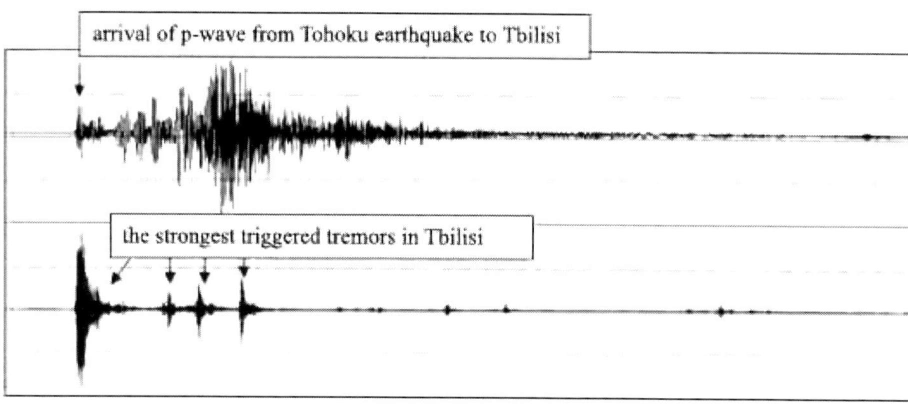

Figure 35. Broadband record of M= 9 Tohoku EQ, Japan (11.03.2011) wave train z-component (upper channel) and the same high-pass band (0.5-20 Hz) filtered record (lower channel). The lower channel shows local triggered events; the strongest event corresponds to arrival of p-wave at Tbilisi seismic station.

Of course, different patterns can be observed also. For example great Tohoku M= 9 earthquake, Japan, triggered local seismic events in Georgia (Caucasus), separated from Japan by 7800 km. Caucasus is a continental collision area, i.e. the region is under compression, which means that triggering here should be less expectable. High pass (0.5-20 Hz) filtered records at two broadband seismic stations located in Oni (South slope of Greater Caucasus) and Tbilisi (valley of river Kura) show that in this case the strongest triggered event at both sites corresponds to arrival of *p*-wave instead of surface waves, which again contradicts the idea about dominant role of Rayleigh waves in triggering TT (Figure 34, 35). Tbilisi is a hydrothermal area and so it falls into general class of triggering-prone regions, but Oni is not a hydrothermal area. Here the fracture can be promoted just by pore fluid pressure.

The triggered events have a form of asymmetric tremor with a long tail (coda).

The comparison of three components of records (N, E and Z) shows that (Figure 36): i. on horizontal components (E and H) triggered events, besides p-arrival are also generated by Love and more intensively by Rayleigh waves; ii. vertical component (Z) generates tremors only at arrival of *p*- and Rayleigh waves, as it could be expected (Figure 34, 35).

The counting of tremors' rate (number of local events per hour) before, during and after Tohoku event both in Oni and Tbilisi reveals clear maximum just during the strong earthquake wave train passage, including coda (Figure 37 a, b), which confirms the reality of triggering phenomenon. The duration of anomalously high tremor rate is of order of 6-8 hours.

Power spectrum of the triggered tremors shows that the maximal energy is released in the frequency range 0.4-0.8 Hz, i.e. these event are deficient at relatively high frequencies (Figure 38 a). Tremor spectrum differs very much from the power spectrum of the broadband recording of Tohoku earthquake, which indicates that maximal power in Georgia was relieved at much lower frequencies, in the range 0.01-0.1 Hz (Figure 38 b). That means that very low-frequency forcing is necessary for triggering tremors. In other words, forcing of a period 100-10 sec is the time, necessary for tremor area activation.

The reliability of local events in Georgia triggered by Tohoku earthquake is confirmed by the analysis of seismic noise records in Tbilisi (device used was BAZALT 4X, Kinemetrics): the number of weak local events increased immensely during passage of wavetrains from Tohoku. The effect can be seen directly on the ground acceleration record [Kartvelishvili et al., in press], i.e. without any processing/filtering of seismogram records, which in principle can introduce artifacts, i.e. without even application of such safe technique as the high pass filter [Peng et al., 2011]. This could be expected as the acceleration manifests better response to fast changes than velocity records. Thus the seismic noise devices can be successfully used for direct identification of triggered events.

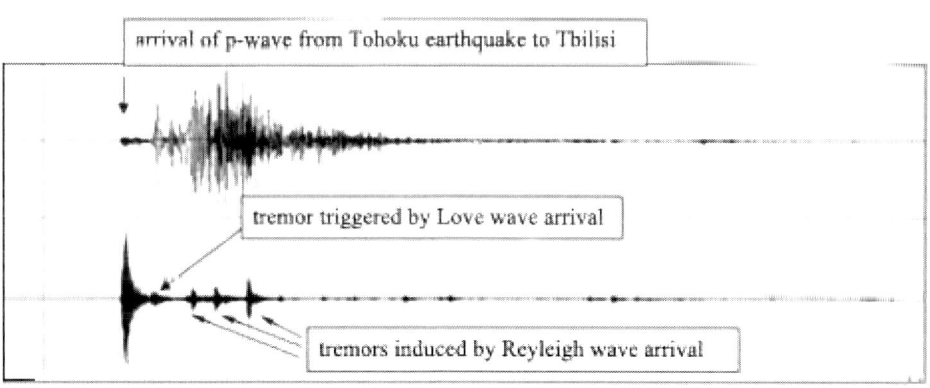

Figure 36. Broadband record of M= 9 Tohoku EQ, Japan (11.03.2011) wave train N-component (upper channel) and the same high-pass band (0.5-20 Hz) filtered record (lower channel) in Tbilisi. Arrows mark p-wave arrival. The lower channel shows local triggered events; the strongest event corresponds to arrival of p-wave. Here the Love wave also generates relatively weak tremor.

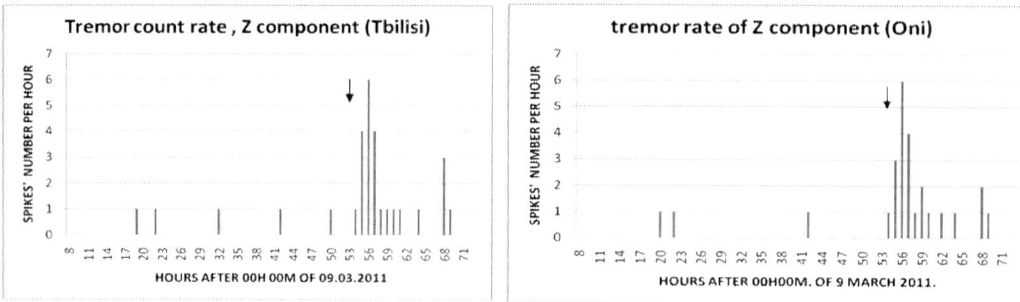

Figure 37. Tremor rate (number of local events per hour) before, during and after Tohoku event in Tbilisi and Oni. Tohoku earthquake arrival time is marked by arrow.

a)

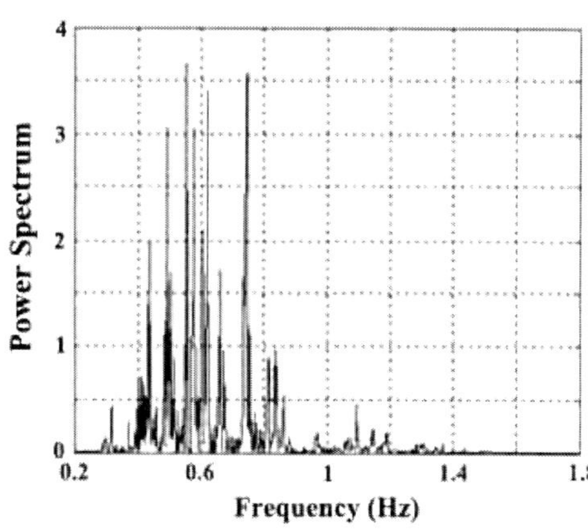

b)

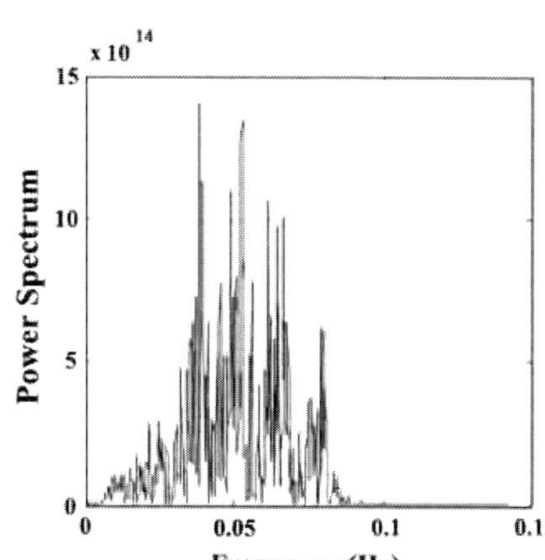

Figure 38. a. Spectrum of the largest (first) triggered tremor in Tbilisi. Bandpass Butterworth filter was used to filter data in a range 0.5-20 Hz. b. spectrum of the broadband recording of Tohoku earthquake in Oni.

a)

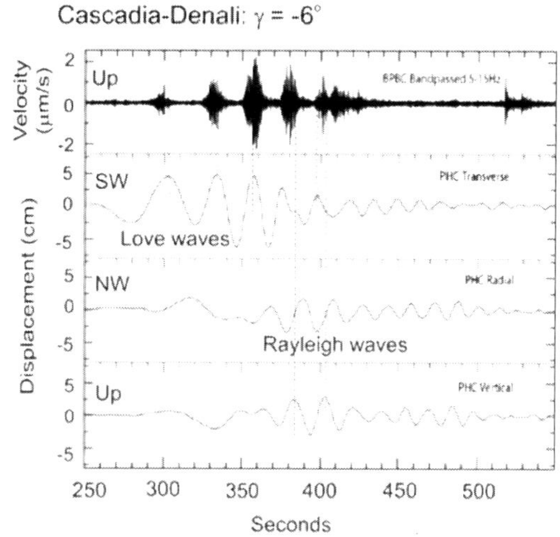

b)

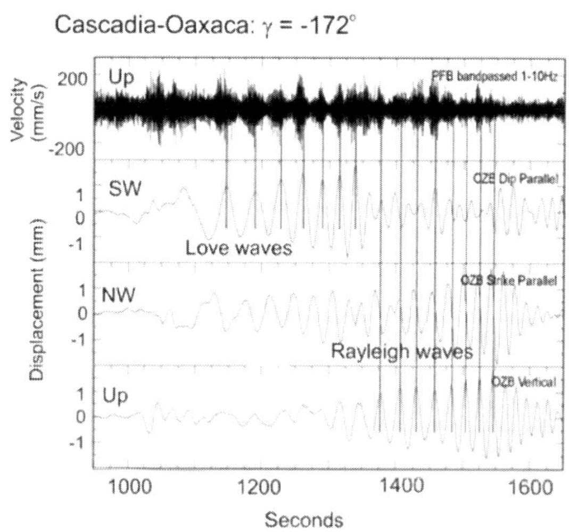

Figure 39. Examples of tremor triggered on the Cascadia megathrust beneath Vancouver Island, B.C., by surfaces waves from four Mw >7. 5 earthquake with incidence angles γ [Rubinstein et al. 2009; Hill, 2010]. The top panel in each example shows broadband displacement waveforms for the incident surface waves (bottom three traces) and the high-frequency (5 to 15 Hz) traces for the triggered tremor (upper trace). (a) The Mw 7.9 Denali fault earthquake of 2002, tremor depth ~15 km; (b) the Mw 7.5 Oaxaca earthquake of 1999, tremor depth ~35 km.

Rubinstein et al. [2009] and Hill [2010] show clearly that the weak forcing by wave train of remote strong earthquake can not only trigger, but also induce phase synchronization of induced events with surface waves (Figures 39, 40).

The strong resemblance between our experimental results on EM or mechanical synchronization of stick-slip (Figures 9-26, 28-30) and large scale natural events (Figures 39, 40) shows that the phenomenon of synchronization has universal character and it can be successfully modeled in laboratory.

Another example of possible synchronization is the discovery of periodic activation of deep (20-30 km) tremors in some active areas [Schwartz. 2009]: at Cascadia the periods of

TT activation occur every 13-16 month the phenomenon named as episodic tremor and slip events (ETS). Incidentally, this periodicity is close to Chandler's wobble period, which can be considered as a week quasi-periodic forcing affecting roots of fault zones (Figure 5 in Schwartz, 2009). Every ~14 months the trend of the GPS locations reverses direction for ~2 weeks away from the secular trend. These 2 week excursions represent the slow slip events. There is a remarkable correlation between the times when the GPS displacement has a reversal and periods of strong tremor activity.

3. 8. 3. Mechanisms of Triggering.

The physical mechanism of remote triggering is not clear. The mechanism should be different for triggered events closely correlated to wave train phase (direct triggering) and for delayed response.

a)

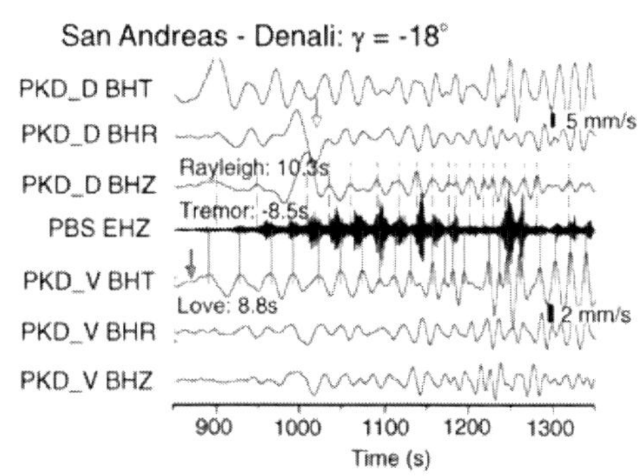

b)

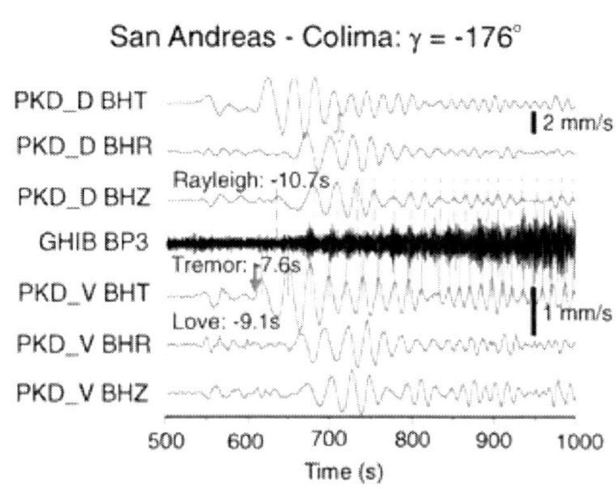

Figure 40. Examples of tremor triggered on the SAF in central California by surfaces waves from four Mw > 7:5 earthquakes [Peng et al.2009; Hill, 2010]. The top panels in (a), (b), and (c) show the broadband displacement (top three traces) and velocity (bottom three traces) waveforms for the incident surface waves at station PKD with the thick, vertical bars indicating displacement and velocity amplitude scales. The middle trace is the high-frequency (2 to 8 Hz) velocity waveforms for triggered tremor from a nearby short-period station. In (a) and (b), time shifts given for broadband traces BHZ and BHT and the short-period (tremor) traces correct for estimated traveltimes for Rayleigh, Love, and tremor waves from the tremor source to the respective stations. (a) The Mw 7.6 Denali fault earthquake of 2002, tremor depth 15 km; (b) the Mw 7.6 Colima earthquake of 2003, tremor depth 20 km.

a)

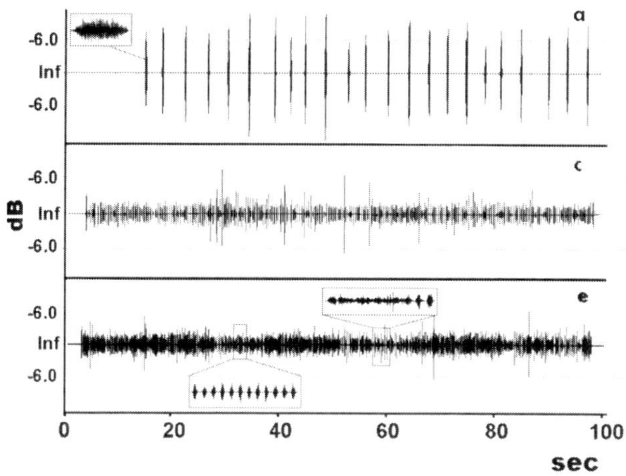

b)

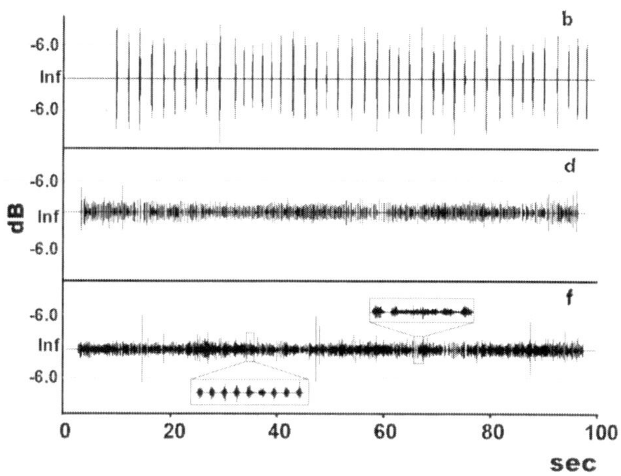

Figure 41. Typical examples of AE recordings at different values of dragging spring stiffness: a) K= 78.4 N/m, b) K= 143.1 N/m , c) K= 1068 N/m, d) 1705 N/m, e) K= 2000 N/m , f) K= 2371.6 N/m. Insets show AE wave train on extended time scales.

Hill (2010) assessed (direct) triggering potential of wave trains from the fracture mechanics point of view, using Mohr and Coulomb-Griffits failure criteria. In general, Love waves incident on vertical strike-slip faults have greater potential than Rayleigh waves, but the potential of Rayleigh waves incident on dip-slip faults dominates over Love wave potential. At the same time, the fault geometry and frictional strength are variable. Such heterogeneity leads to deviations from the above simple rule.

For large delays frictional failure, subcritical crack growth and excitation of crustal fluids are suggested as appropriate models [Hill, Prejean, 2009; Prejean, Hill, 2009; Hill, 2010].

We can stress close resemblance of typical recordings of ETS (Figure 7 in Schwartz, 2009). with our laboratory stick-slip experiments (Figure 41); it seems that different morphology of the ETS signals can be explained by the various conditions of frictional motion, in particular, by different stiffness of dynamical system (spring-slider system, fault).

Obara [2002] and Rubinstein et al [2010] note that periods of tremor activity turn on and off by local or teleseismic earthquakes and remark that ‘no satisfactory model has been proposed to explain how teleseismic event might stop a period of active tremor". The general explanation of how small-amplitude teleseismic wave can start of stop a period of tremor activity is the extremely high sensitivity of nonlinear systems to a weak forcing. The physical (laboratory) model of mentioned tremor arrest effect has been realized in our experiments with mechanical synchronization of stick-slip (Figures 29-31). This remarkable result shows that even very small mechanical forcing, 10^5 times smaller than the main driving force can affect both onsets and terminations of stick-slip generated acoustic wave train.

It seems that further development of sensitive devices, dense networks and processing methods will develop a new avenue in seismology, which can be defined as microseismology and which will study systematically small earthquakes and tremors, especially triggered and synchronized events. These events at present are ignored by routine seismological processing and are not included in traditional catalogues. At the same time, microseismic events contain very important information on geodynamical processes and can give clues to understanding fine mechanism of nonlinear seismic process and may be, even contribute to the problem of earthquake prediction. Role of microseismicity in geodynamics/geophysics can be compared by its importance to studies of elementary particles in physics.

3. 9. Hydrological Triggering/Synchronization of Seismicity

During the last decades seismologists discovered that besides plate tectonics, earthquakes, and especially induced and slow (silent) earthquakes are also strongly affected by underground waters regime,. The term coined for the phenomenon is hydroseismicity. The effect is explained by the contribution of fluids to pore pressure, which reduces normal stress and thus facilitates fracture/slip process.

The most convincing proofs come from the well-known Rocky Mountain Arsenal (Colorado), where close correlation between controlled fluid injection in the borehole and induced seismicity was convincingly documented [Evans, 1966]. Shapiro et al [2006] suggested mathematical framework for linking migration of misroseismicity with diffusion of pore pressure from the borehole, using data from the deep KTB borehole in Germany.

Besides borehole fluid injections, many other hydrological phenomena are linked to seismic process, such as reservoir loading, oil/gas production, precipitation, snow melt, hurricanes, etc [Gupta, 1992; Talwani, 1997; Grasso and Sornette, 1998; Heki, 2003; Hainzl et al., 2006; Costain and Bollinger, 2010; Wang and Manga, 2010].

3. 9. 1. Reservoir-Induced Seismicity

For the last two decades it has been documented that large water reservoirs can cause so called reservoir-induced seismicity or RIS [Sibson, 1994; Talwani, 1997]. We have investigated the character of dynamical changes in the local seismic activity under quasi periodic forcing caused by water level variation around a large Enguri dam reservoir (Western Georgia, Caucasus; 42.030 N, 42.775 E), where Enguri Dam International Test Area – EDITA is operating in the framework of the Open Partial Agreement on Major Disasters of the Council of Europe [Chelidze et al., 2006].

The height of the Enguri dam is 272 m, the (average) volume of water in the reservoir 1.1 $x10^{9}$ m^3. Enguri reservoir was built in 1971-1983. Preliminary flooding of the territory started at the end of December 1977; since 15. 04. 1978 reservoir was filled step by step to a 510 m mark (above the sea level) in 1987. Since 1987 the water level in reservoir changes seasonally, almost periodically.

Accordingly, we have defined three distinct periods for local seismicity dynamics analysis, namely,

i) Before impoundment.
ii) Flooding and reservoir filling, and
iii) Periodic change of water level.

Figure 42 a,b shows the daily record of the water level in the Enguri dam reservoir for the period 1978–1993 as well as seismic activity in this period, where some RIS activation is evident after beginning of reservoir filling.

The effect of reservoir on is rather small compared to the regional tectonic stress field at the seismogenic depths. The level of water in the lake changes by 100 m, which means that the pressure at the bottom changes by 10 bar. At the depth of 10 km, where the most of hypocentres of EQ are located, the pressure will be much less, as the reservoir is finite. We use for the crude assessment of the order of load decrease a well known expression for the stress field of a defect $\Delta\sigma$ at the distance r from it: $\Delta\sigma/\sigma \approx (a/r)^{1/2}$, where σ is applied stress, a is the size of defect. Substituting $\sigma = 10\ bar$, a = 100 m (the depth of lake) and r = 10 000 m (the average depth of hypocenters) we obtain for $\Delta\sigma$ the value of order of 1 bar. This is much less than the tectonic stress on this depth, which is of the order of several kilobars.

RQA, often used to detect changes in the dynamics of complex systems [Marwan, 2003], is the most convenient data analysis tool for revealing ordering in time series, which can be considered as a hallmark of synchronization. As follows from our RQA results, when the external influence on the earth crust caused by a water reservoir becomes periodic the extent of regularity of earthquake daily distribution (evaluated as REC and %DET) essentially increases (Figure 43 a, bold line). This result was tested by comparing with the surrogate data. Averaged results derived from RQA of 20 shuffled (asterisks) and phase randomized (triangles) surrogates (Figure 43a), assure that the above-mentioned increase of regularity in earthquakes distribution should not be an artifact [Matcharashvili et al., 2010].

The influence of increasing amount of water and its subsequent periodic variation essentially affects the character of both earthquakes temporal and magnitude distribution (Figure 43 b). The extent of order in earthquakes temporal (black columns) and magnitude (grey columns) distribution, calculated as value of %DET, substantially increases when the reservoir forcing becomes periodic.

The dynamics of temporal and energetic distributions of earthquakes became ordered even under not quite regular variation of water level, though not so much as under periodic variation. The above conclusions regarding the increase of the order in seismic process at periodic variation of water level using %DET measurements are confirmed by calculation of other RQA measures (% REC, Entropy, Laminarity) not shown here.

We conclude that the order in dynamics of earthquakes' daily occurrence, as well as in earthquakes' temporal and energetic distributions increases when water level variations become periodic.

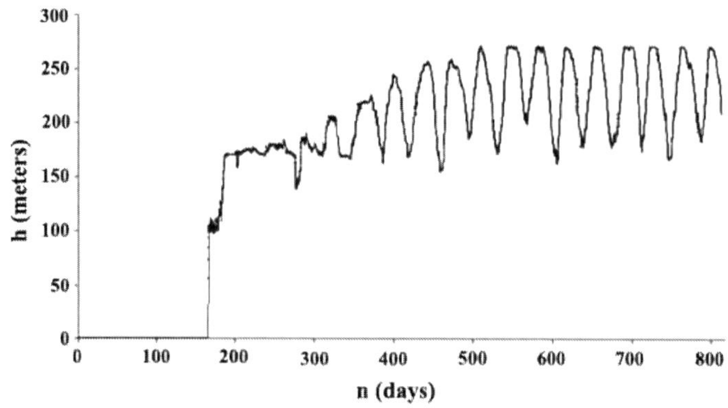

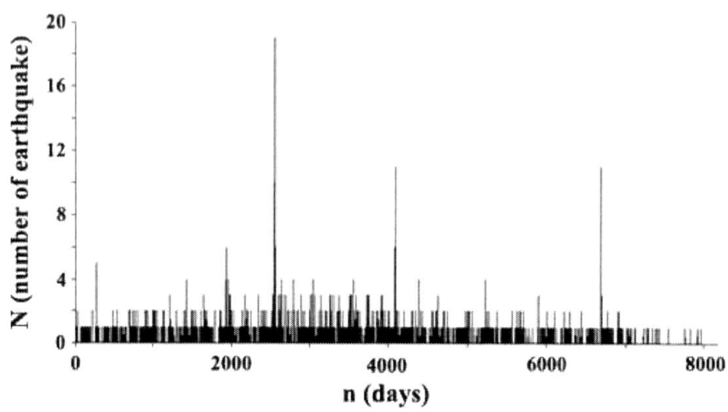

Figure 42. (a) Record of the daily water level in the lake of Enguri dam from 1975 to 1993, (b) daily number of earthquakes.

a)

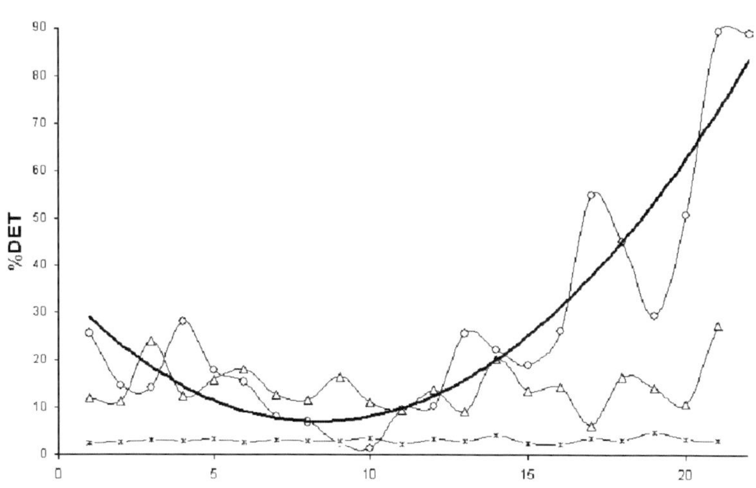

b)

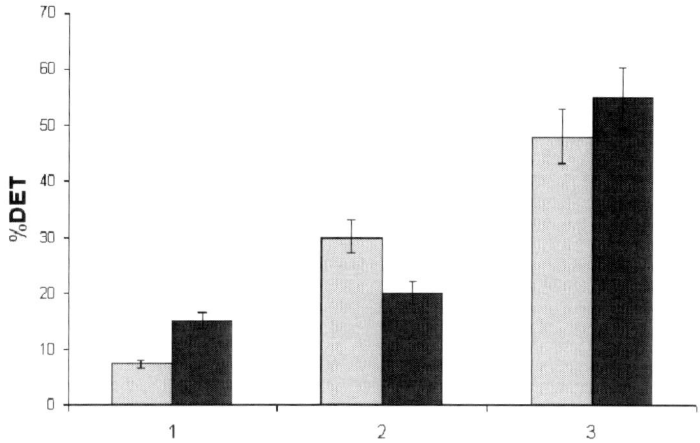

Figure 43. (a) RQA %DET of daily number of earthquakes calculated for consecutive non overlapping one-year sliding windows (circles). Averaged results of RQA %DET for 20 shuffled (asterisks) and phase randomized (triangles) surrogates of daily number of earthquakes in consecutive one-year sliding windows; (b) RQA %DET of magnitude (black columns) and waiting time (grey columns) sequences: (1) before impoundment, (2) during flooding and reservoir filling, and (3) periodic change of water level in reservoir [Chelidze et al., 2006, Matcharashvili et al 2010].

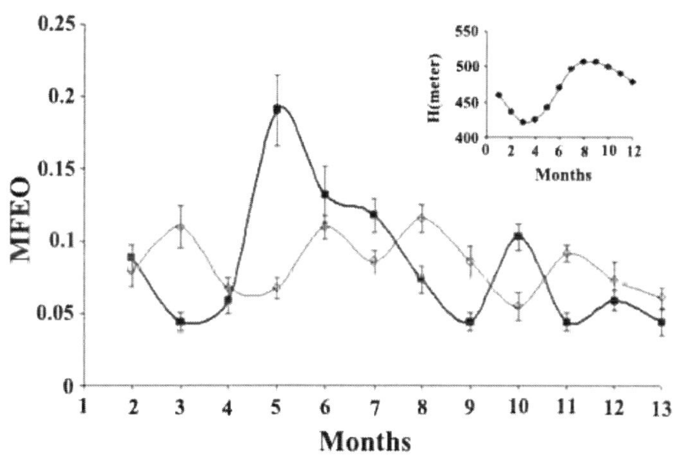

Figure 44. Monthly frequency of earthquake occurrence (MFEO) before (open circles, thin line) and during water level periodic variation (dark squares, bold line).

Our results allow presenting concept of reservoir induced seismicity synchronization (RISS) leading to decrease of seismic activity around the Enguri dam after initial period of increased reservoir induced seismicity (RIS) as a result of phase synchronization between water level variation in the lake and regional seismic activity. We guess that under external weak periodic influence (during RISS) the existing tectonic energy may be released via series of relatively small earthquakes due to regularization of tectonic process by external forcing (compare with Figure 12, 13), which excludes accumulation and quick release of extreme stresses [Chelidze et al.2005, 2006; Peinke et al., 2006].

Our results point to existence of three maxima in the annual seismic activity (Figure 44); as the water level has only one (seasonal) maximum per year, this fact can be considered as manifestation of high order synchronization.

Telesca [2010] carried out the joint analysis of water level and seismicity recorded at Koyna (India) over the last 50 years. Cross-correlation coefficients (CCC) between the daily water level W_d and the daily number of earthquakes n_d (with magnitudes greater than or equal to 2.2) have been calculated. Cross-correlation between W_d and n_d is very weak for standard method of calculation, but after using power spectrum of the standardized water level the CCC value for the lower frequency part of the spectrum became quite high. In particular, clear12, 6 and 4 months periodicities were revealed. The 12 months cycle is linked with the seasonal load/unload of the reservoir; the 6 and 4 months cycles can be considered as high order synchronization phenomenon (see section 2.7). From the time lag of 223 days between reservoir-induced enhanced seismic activity and drawdown phase of the reservoir and data on the mean earthquake depth provide a vertical hydraulic diffusivity of $D \approx 0.2$ m^2/s. This illustrates the potential of RIS and borehole injections for assessing hydraulic properties of rocks in the depth.

3. 9. 2. Precipitation and Seismicity

The idea has a long history: in 1912 N. Drake and in 1813 R. Sayles published papers on correlation of seismic activity and precipitation. Costain and Bollinger (2010) reviewed results of earthquake-rainfall connections for 22 last years and show that there really is a correlation. The hydroseismicity is explained by pore-fluid pressure diffusion and its delay relative to discharge moment by slowness of fluid diffusion process.

Hydroseismicity manifests itself by slow (silent) earthquakes, which last many hours or even days compared to seconds for regular earthquakes. Biot's [1941, 1962] theory gives wave equation for coupling fluid flow in pores with the stress field; the solution predicts two fast (P and S) and one slow (P) body waves. The slow one describes pore fluid pressure diffusion and it is related to generation of slow earthquakes.

The most impressive results on rainfall-induces seismicity are obtained by Hainzl et al [2006]. The authors analyzed seismic activity and precipitation at Mount Hochstaufen in South Germany and found that the in summer, when the precipitation is the highest, the number of earthquakes is maximal. The spatiotemporal distribution of pore pressure (one-dimensional diffusion with hydraulic diffusivity 3.3 m/s^2) and the rate of pore pressure variations with depth were calculated using recorded rainfall data and the Biot's model of pore pressure diffusion. The data on recorded precipitation were compared to recorded number of earthquakes per day and to seismic rate predicted by the Biot model. The close correlation of recorded and predicted from the Biot's slow wave propagation (spatiotemporal pore-pressure variation caused by rainwater diffusion) seismic activity was found.

3. 9. 3. Synchronization by Oceanic Microseisms

The recent observation of long-period (LP) events with dominant frequencies around 2 Hz and waiting times between events (individual swarms) ranging from 10 s to 20 s at Deception Island volcano in Antarctica (Stich et al., 2011) can be an example of high-order synchronization. The matter is that inter-event times of swarms are close to integer multiples of the dominant periods of the oceanic microseism – approximately 5 s. Authors attribute the effect to control of LP activity in hydrothermal system of volcano by weak periodic external forcing generated by ocean noise. The volumetric strain variation due to the action of oceanic microseisms is estimated as low as 10^{-7} using the vertical component of Rayleigh wave amplitude. The property of periodicity of swarms is more evident at increasing strain, which points to nonlinear nature of effect.

3. 10. Triggering (Ordering) of Seismicity by Strong Electromagnetic (MHD) Pulses

Experiments on MHD (magnetohydrodynamic) soundings of the Bishkek test area (Central Asia) have been performed in 1983 –1996 by Institute of High Temperatures of Russian Academy of Sciences - IVTAN [Tarasov, et al., 1999; Jones, 2001]. During these experiments, deep electrical soundings of the crust were regularly carried out. The source of electrical energy was MHD generator with electrodes separated by 4.5 km from each other. When the generator was fired, the load current was 0.28 - 2.8 kA, the sounding pulses were emitted in regular series of 10-14 pulses, which had durations of 1.7 to 12.1 s, and the energy generated was mostly in the range of 1.2 - 23.1 MJ [Volykhin, 1993]. This area is seismically active and a catalogue with about 14100 events from 1975 to 1996 has been analyzed.

Statistically significant evidence of increased seismic activity with delay of 2-3 days after EM discharges has been obtained by Tarasov [1997].

In order to reveal possible ordering influence of MHD pulses non linear analysis tools were applied to waiting times series of earthquakes in the following three sub-catalogue periods: the first, 1975-83, before EM experiments; the second, 1983-90, during EM experiments; and the third part, 1988-96, after the experiments. Calculations for these three periods were performed both on the complete catalogs (threshold $M \geq 1.7$) and the whole catalog [Chelidze et al., 2010a].

Investigation was performed according to general scheme of time series nonlinear analysis [Abarbanel et al. 1993; Sprott and Rowlands, 1995; Kantz and Schreiber, 1997; Goltz, 1998; Hegger and Kantz 1999], i.e data analysis was performed firstly in a qualitative and visual manner and successively a more quantitative methodology was applied.

Qualitative approach included a visual inspection of the reconstructed phase space. Namely, *p*-dimensional phase space from the scalar time sequences was reconstructed by the method of time delay [Packard et. al, 1980, Takens, 1981]. Other qualitative tools have also been used, such as Iterated Function Systems (IFS) [Jeffrey, 1992] and Recurrence Plots (RP) (Eckman et al. 1987). The Tsallis entropy calculation according to [Tsallis ,1988, 1998) has also been used as a measure of the complexity in earthquakes time distribution:

$$S_q = k\ (1/(q-1))\ \ (1- \Sigma_{i=1}^{W} p_i^{\,q})$$

where p_i are the probabilities of the separate configurations (W) and q is intrinsic parameter with a value greater than zero, which demonstrates the correlation between subsystems.

In Figure 45, the results of qualitative analysis of waiting times sequences for the complete catalog are presented [Chelidze and Matcharashvili, 2005]. The results of IFS-clumpiness test presented in Figure 45 a, c, e, [Jeffrey, 1992; Sprott and Rowlands 1995] and the recurrence plot analysis in Figure 45 b, d, f [Zbilut and Webber, 1992] reveal that after the beginning of the experiments some structure in the corresponding plots is visible, namely, diagonal lines in IFS plot and compact structure in RP during experiments, which points to the increased amount of functional interdependence in earthquake temporal distribution.

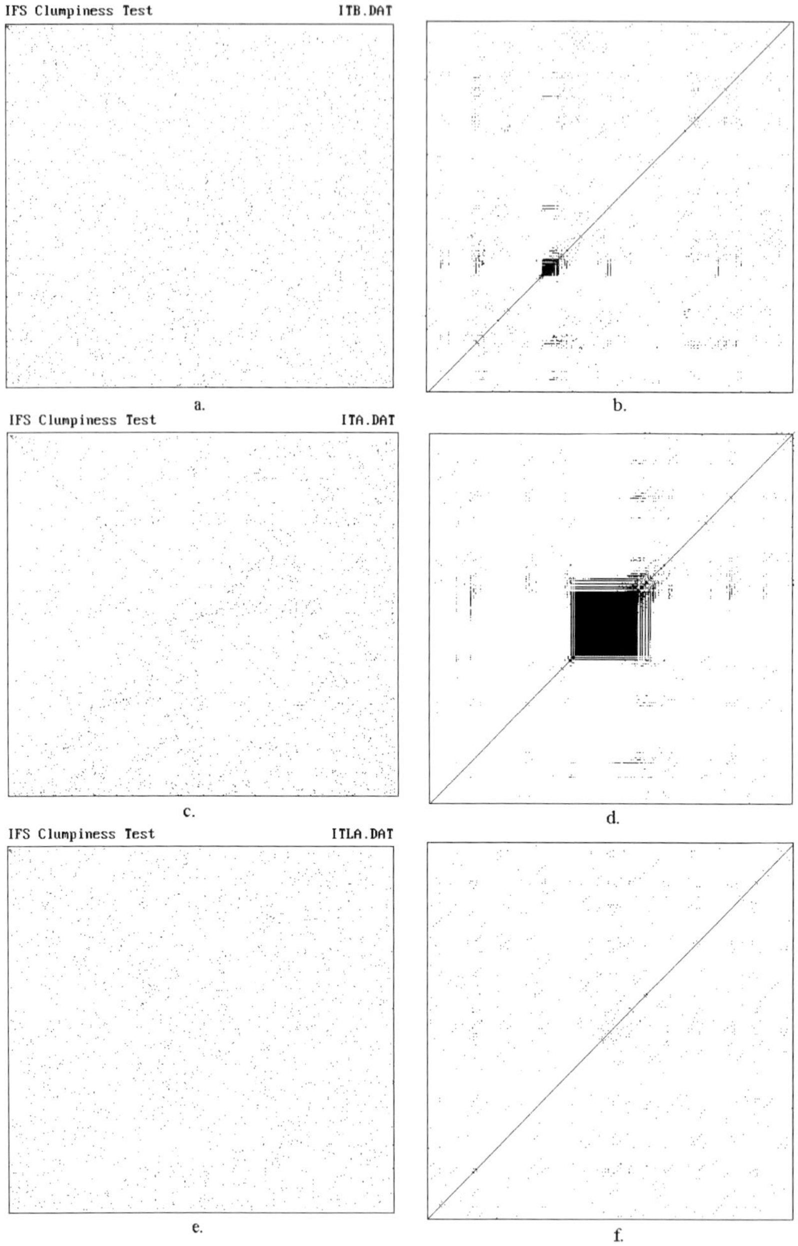

Figure 45. Qualitative analysis of temporal distribution of earthquakes (complete catalogue, $M \geq 1.7$) before the beginning of EM experiments (1975-1983), during experiments (1983-1988) and after accomplishing of experiments (1988-1992). IFS-clumpiness and Recurrence plots (RP) analysis test of waiting times sequences before experiments (a and b respectively); during experiments (c and d respectively); after experiments (e and f respectively). Note diagonal lines in IFS plot (c) and compact structure in RP (d) during experiments.

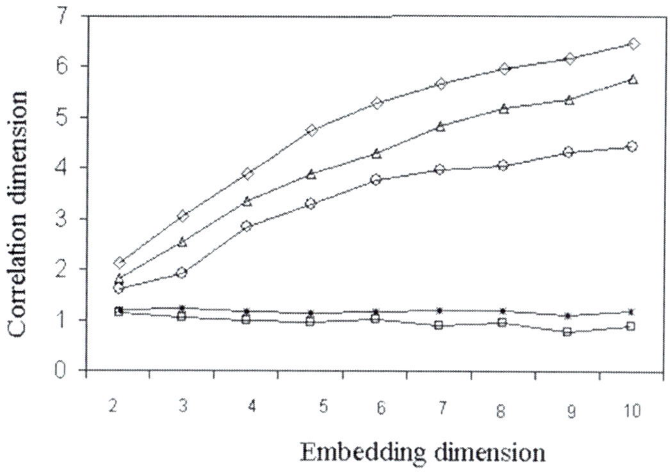

Figure 46. Correlation dimension versus embedding dimension of waiting times sequences (complete catalogue, $M \geq 1.7$): (a) asterisks - integral time series (1975-1996), (b) circles - before the beginning of experiment (1975-1983), (c) squares - during experiments (1983-1988), (d) triangles – after experiments (1988-1992), (e) diamonds correspond to random number sequence.

As to the quantitative approach, the variation of correlation dimension vs. dimension of phase space, where the reconstructed dynamics is embedded (embedding dimension) is presented in Figure 46 [Chelidze and Matcharashvili, 2005]. The integral time series (5297 time intervals) for the whole period of observation (1975-1996) containing time intervals sequences between all events above the threshold reveals clear low correlation dimension ($d2 = 1.22 \pm 0.43$) (asterisks). The three characteristic time series were considered, namely, 1760 waiting times data before (1975-1983), 1953 waiting times during MHD experiments (1983-1988) and 1584 waiting times of the period after experiments (1988-1992). Time series before have relatively low ($d2 = 3.83 \pm 0.80$) dimension, but during MHD experiments correlation dimension have especially low values: $d2 = 1.04 \pm 0.35$. On the other hand, after cessation of experiments (Figure 46, triangles) correlation dimension of waiting times sequences noticeably increases ($d2 > 5.0$), exceeding low dimensional threshold ($d2 = 5.0$). This means that after termination of experiments the extent of regularity or extent of determinism in earthquake temporal distribution decreases: this is confirmed by both qualitative (Figure 45) and quantitatively (Figure 46) assessments.

The main results of the analysis, the increase of ordering in earthquake time series during MHD experiments do not change after application of non-linear noise reducing and surrogate testing: they have confirmed the above results. The same results have been obtained with the use of the whole catalogue, regardless of the completeness criteria [Chelidze et al., 2010a].

This analysis is certainly not exhaustive: the seismic catalogue covers a broad area and all data were used, with no distinction for space location of seismic events. The energy aspect has not been fully considered: all events were considered equal, regardless of their magnitude. These are strong simplifications and the results must be considered under these constrains.

However, the results appear to be consistent: EM experiments influence seismic time dynamics to some extent, increasing the regularity of waiting times. The 2-3 days lag of seismic activity behind EM pulse injections point to involvement of pore fluids into EM-induced earthquakes.

3. 11. Synchronization Technique in Earthquake Prediction

In Chelidze et al [2006] the idea about connection of synchronization phenomenon with closeness of the system to the critical state (earthquake) has been suggested. A. Lyubushin systematically explored variation of different synchronization parameters of low-frequency microseismic noise field obtained during tens of years on the seismic network of Japan. As a result, several months before great Japan (Tohoku) earthquake of 11 March 2011 (M=9) he officially issued prediction on the possibility of strong earthquake in Japan starting from June 2011 [Lyubushin, 2009, 2011]. By the way, the alert on the increased probability of magnitude 8 (and higher) earthquake in Japan was issued several months before Tohoku earthquake (and communicated to many seismologists around the world) by V. Kossobokov using M8-MSc algorithm, but ironically it has been withdrawn in the beginning of 2010.

The vertical component of seismic records with sampling rate 1 Hz from global IRIS (1988-2009, 131 stations) and Japan F-net (1997-2010, 83 stations) networks was analyzed. Several fractal parameters were integrated, namely, the width of singularity spectra $F(\alpha)$ support $\Delta\alpha = \alpha_{max} - \alpha_{min}$, generalized Hurst's exponent α^*, coefficient of singularity spectra asymmetry $\gamma = \alpha^* - (\alpha_{max} - \alpha_{min})/2$, linear predictability index of waveforms, spectral exponent and logarithm of variance. Combinations of these parameters were calculated within 30 minutes time windows for initial 1Hz data and after down-sampling, for 1 day time windows for 1 minute averaged data. The seismic stations were grouped in 7 spatial clusters and the median values of all parameters were taken from stations in each cluster. Lastly a multiple correlation measures for different combinations of parameters were estimated for 1 and 2 years time windows with uniform 1 day sampling rate.

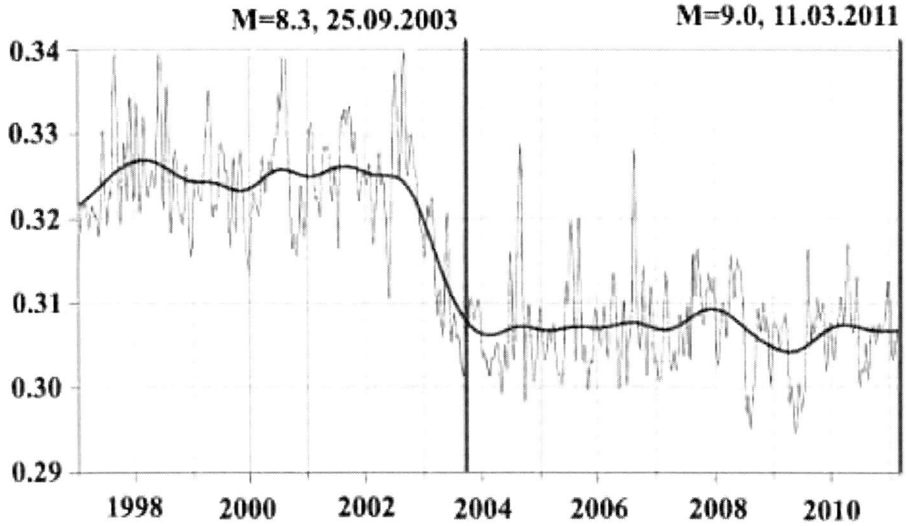

Figure 47. Smoothing of multi-fractal singularity spectra support width estimates $\Delta\alpha$ for microseismic noise within adjacent time windows of the length 30 minutes (median value over all stations). Thin grey line – Gaussian kernel smoothing in a window of the length 27 days. Solid black line – Gaussian kernel smoothing in a window of the 1 year. Parameter $\Delta\alpha$ describes a variety of stochastic oscillations and indirectly their synchronization: less is the $\Delta\alpha$ value – more is the synchronization [Lyubushin, 2011].

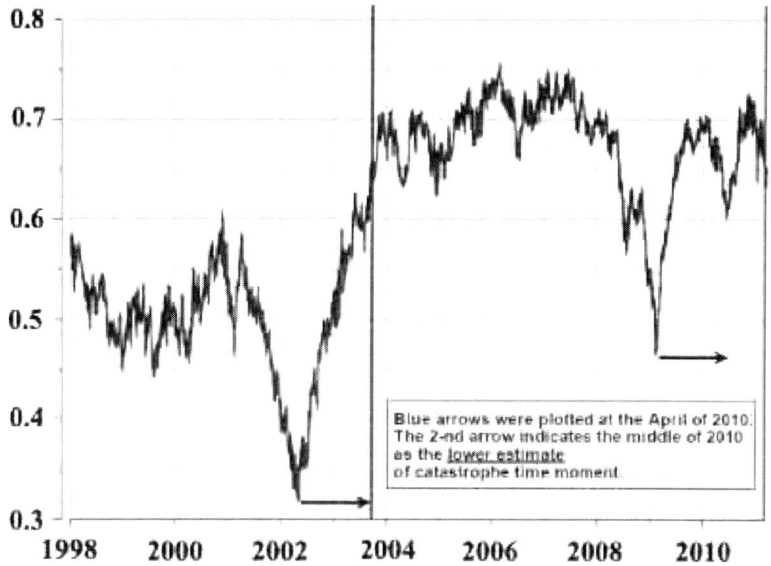

Figure 48. Squared robust correlation between $\Delta\alpha$ and α^*. Note two strong minima, one before Hokkaido and second before Tohoku earthquake [Lyubushin, 2011].

As a result of integrated synchronization analysis the periods of increased synchronization were revealed: first in 2003-2007 and the second from 2008 till 2010 (Figures 47, 50).

According to Figs. 47, 48 the high level of synchronization of microseisms' field at Japan islands started in the middle of 2002, one year before the Hokkaido (25 September 2003, M=8.3) earthquake. This strong synchronization of microseismic noise last up to present. In the plot of squared robust correlation between $\Delta\alpha$ and α^* (Figure 48) there are two large minima, one before the Hokkaido event and other in mid-2010. Lyubushin's guess was that 2003 event was only a foreshock of a still stronger earthquake (Tohoku event) in the region, as after the Hokkaido earthquake of September 25, 2003 (M = 8.3) a strong synchronization in the multi-fractal parameters of the low-frequency microseisms was still preserved at the same level up to the Tohoku catastrophe of March 11, 2011 (M = 9.0). As the synchronization level remains high after Tohoku event also, possibility of another strong earthquake cannot be excluded.

Of course it is too prematurely to claim that the synchronization analysis is universal tool for earthquake prediction. It is necessary to analyze microseimic noise before many other events in order to check reliability of these results

CONCLUSION

The phase synchronization of stick-slip process induced by a weak electromagnetic or mechanical periodic forcing is investigated on the laboratory scale. Acoustic emission bursts were used as markers of slip. For quantitative measuring of synchronization strength several modern tools of nonlinear dynamics were used, such as mean effective phase diffusion coefficient, Shannon entropy based characteristic phase synchronization measure *(γH-Sh)*, Mutual Information, recurrence plots RP and recurrence quantitative analysis RQA, namely, percent of determinism %DET, etc. Application of varying frequencies and intensities of forcing allow compiling of Arnold's tongue for EM forcing. We found that not only the onsets/maxima of a definite kind of AE signals are synchronized with forcing, but also AE wave train terminations.

The effect of high order synchronization of stick-slip events by weak electromagnetic or mechanical periodic forcing is discovered. There are two kinds of high order synchronization: i. one or more AE bursts during one forcing period and ii. a single AE burst during many forcing periods.

It was found that the onset time of the synchronized slip events delay behind the forcing phase; the delay being smaller for stronger forcing.

Acoustic emission energy distribution (analogue of frequency-magnitude plot) undergoes significant changes at variation of excitation intensity, namely, the slope of the plot is maximal in synchronization area, which means that synchronization limits the energy release associated with individual events (quantization effect). Switching off the synchronizing forcing provokes drastic increase of energy emitted by following slip event.

The review of recent results on dynamic triggering of local seismicity by remote earthquakes is presented, as well as other examples of triggering by strong electromagnetic pulses, reservoir exploitation and precipitation. The first attempt to use microseismic noise synchronization analysis for earthquake prediction is reviewed. Further development of sensitive devices, dense networks and processing methods will develop a new avenue in seismology, which can be defined as microseismology and which will study systematically small earthquakes and tremors, especially triggered and synchronized events. These events at present are ignored by routine seismological processing and are not included in traditional catalogues. At the same time, microseismic events contain very important information on geodynamics of processes and can give clues to understanding fine mechanism of nonlinear seismic process and may be, even contribute to the problem of earthquake prediction. The role of microseismicity can be compared by its importance for geophysics to studies of elementary particles in physics.

The results obtained point to possibility of revealing some new fine details in the stick-slip process which can be very useful for refining the physical mechanism of frictional motion in general. These findings can also help to find new regularities in seismic time series.

ACKNOWLEDGMENTS

The authors express their gratitude to the Georgian National Science Foundation (Grant № GNSF/ST06/5-028) and INTAS foundation (Ref. № 05-100008-7889) for financial support.

Authors are grateful for the kind permission of World Scientific and Engineering Academy and Society (WSEAS), Journal of Georgian Geophysical Society and Bulletin of Georgian National Academy of Sciences as well as to authors of official reports to reuse Figures from published in these editions papers in the present review.

REFERENCES

Akay, A. 2002. Acoustics of Friction. *J. Acoust. Soc. Am.*, 111, 1525-1548.

Abarbanel, H. and Tsimring, L. S. 1993. The Analysis of Observed Chaotic data in Physical Systems, *Rev. Mod. Phys*, 65, 1331-1392.

Abarbanel, H., Rabinovich, M., Sushchik, M. 1996. Introduction to nonlinear dynamics for physicists. World Scientific. Singapur.

Anishchenko, V. 1995. Dynamical Chaos. Models and Experiments. World Scientific. Singapure.

Bak , P., Tang, C. and Wiesenfeld, K. 1988. Self-organized criticality. *Phys. Rev.* A38, 364-374.

Bhattacharya, J. 1999. In Search of Regularity in Irregular and Complex Signals, Ph. D Thesis.

Beeler, N. M. and Lockner, D. A. 2003. Why earthquakes correlate weakly with the solid Earth tides: Effects of periodic stress on the rate and probability of earthquake occurrence. *Journal of Geophysical Research*, B108, 2391-2405.

Ben-Zion, Y. 2008. Collective Behavior of Earthquakes and Faults: Continuum-Discrete Transitions, Evolutionary Changes and Corresponding Dynamic Regimes. *Rev. Geophysics*, 46, RG4006, doi:10.1029/2008RG000260, 2008.

Biot, M. 1941. General Theory of three-dimensional consolidation. *J. Appl. Phys.* 12, 155-164.

Biot, M. 1962. Mechanics of deformation and acoustic propagation in porous media. *J. Appl. Phys.* 33, 1482- 1498.

Blekhman I. I. 1988. Synchronization in Science and Technology. A. S. M. E. Press, New York.

Bouissou, S., Petit, J., Barquins, M. 1998. Experimental evidence of contact loss during stick-slip: possible implications for seismic behavior. *Tectonophysics*, 295, 341-350.

Brace, W. E., and Byerlee, I. D. 1966. Stick slip as a mechanism for earthquakes. *Science*, 153, 990-992.

Bureau, L., Baumberger, T. and Caroli, C. 2000, Shear response of a frictional influence to a normal load modulation. *Phys. Rev.* E, 62, 6810-6820

Chao, K., Peng, Z; Wu, C., Tang, C.-C., Lin, C.-H. 2012. Remote triggering of non-volcanic tremor aroundTaiwan. *Geophys. J. Int.* 188, 301-324.

Chelidze, T., Varamashvili, N., Devidze, M., Chelidze, Z., Chikhladze, V. and Matcharashvili, T. 2002. Laboratory study of electromagnetic initiation of slip. *Annals of Geophysics*, 45, 587-599.

Chelidze, T., Lursmanashvili, O. 2003. Electromagnetic and mechanical control of slip: laboratory experiments with slider system. *Nonlinear Processes in Geophysics*, 20, 1-8.

Chelidze, T., Matcharashvili, T. 2003. Electromagnetic control of earthquake dynamics? *Computersand Geosciences*, 29, 587-593.

Chelidze, T., Gvelesiani, A., Varamashvili, N., Devidze, M., Chikchladze, V., Chelidze, Z. and Elashvili, M. 2004. Electromagnetic initiation of slip: laboratory model. *Acta Geophysica Polonica*, 52, 49-62.

Chelidze, T., Matcharashvili, T., Gogiashvili, J., Lursmanashvili, O. and Devidze, M. 2005. Phase synchronization of slip in laboratory slider system. *Nonlinear Processes in Geophysics*, 12, 1-8.

Chelidze T., Matcharashvili, T. 2005. Complexity of seismic process: implications for earthquake prediction. *Journal of the Georgian Geophysical Society*, 10A, 3- 24.

Chelidze, T., Lursmanashvili, O., Matcharashvili, T. and Devidze, M. 2006. Triggering and synchronization of stick slip: waiting times and frequency-energy distribution. *Tectonophysics*, 424, 139-155.

Chelidze, T., Matcharashvili, T., Javakhishvili, Z., Khutsishvili, T., Mepharidze, E., Zhukova, N. 2006. Influence of periodic variations of Enguri reservoir water level on regional seismic activity. *Journal of the Georgian Geophysical Society*, 11A, 3- 24.

Chelidze T., Matcharashvili, T. 2007. Complexity of seismic process, measuring and applications – A review, *Tectonophysics*, 431, 49-61.

Chelidze, T., Lursmanashvili, O., Varamashvili, N. 2008a. High order synchronization of stick-slip process in the spring-slider model with periodic (tangential) mechanical forcing. *Journal of the Georgian Geophysical Society*, 12A, 16-24.

Chelidze, T., Zhukova, N., Lursmanashvili, O., Matcharashvili, T., Mepharidze, E. 2008b. Stroboscope (synchrogram) technique for revealing high order synchronization of stick-slip process. *Journal of the Georgian Geophysical Society*, 12A, 29-32.

Chelidze, T., Matcharashvili, Lursmanashvili, O., Varamashvili, N. 2008c. Acoustics of stick-slip deformation under external forcing: the model of seismic process synchronization. In: Advanced Topics in Geology and Seismology. D. Triantis, M. Jelenska, F. Vallianatos (Eds). University of Cambridge, *W. S. E. A. S. Press*.pp.:36-43.

Chelidze, T., Lursmanashvili, O., Matcharashvili, T., Varamashvili, N., Zhukova, N., Mepharidze, E. 2009. High order synchronization of stick-slip process: experiments on spring-slider system. *Nonlinear Dynamics*, DOI 10.1007/s11071-009-9536-6.

Chelidze, T., de Rubeis, V., Matcharashvili, T. and Tosi, P. 2010a. Dynamical Changes Induced by Strong Electromagnetic Discharges in Earthquakes' Waiting Time Distribution at the Bishkek Test Area (Central Asia). in: *Geoplanet: Earth and Planetary Sciences*, *Volume* 1, DOI: 10.1007/978-3-642-12300-9; Synchronization and Triggering: from Fracture to Earthquake Processes. Eds.V.de Rubeis, Z. Czechowski and
R. Teisseyre, pp.339-360.

Chelidze, T., Lursmanashvili, O., Matcharashvili, T., Varamashvili, N., Zhukova, N., Mepharidze, E. 2010b. Triggering and Synchronization of Stick-Slip: Experiments on Spring-Slider System. in: *Geoplanet: Earth and Planetary Sciences, Volume* 1, DOI: 10.1007/978-3-642-12300-9; Synchronization and Triggering: from Fracture to Earthquake Processes. Eds.V.de Rubeis, Z. Czechowski and R. Teisseyre, pp.123-164.

Chelidze, T., and Varamashvili, N. 2010c. Models of Stick-Slip Motion: Impact of Periodic Forcing; in: *Geoplanet: Earth and Planetary Sciences, Volume* 1, DOI: 10.1007/978-3-642-12300-9; Synchronization and Triggering: from Fracture to Earthquake Processes. Eds. V.de Rubeis, Z. Czechowski and R. Teisseyre, pp. 23-33.

Chernyak, G. 1978. On the physical nature of seismoelectric effect in rocks. *Izvestia Ac. Sci. U. S. S. R., Physics of Earth*. N2, 108-112 (in Russian).

Costain, J. and Bollinger, J. 2010. Review: Research results in Hydroseismicity from 1987 to 2009. *Bull. Seismol. Soc. Am.* 100, 1841-1858, doi: 10.1785/0120090288.

de Rubeis, V.,Chechovski Z. andTeisseyre R., (Eds). 2010. Synchronization and Triggering: from Fracture to Earthquake Processes. Springer.

Dieterich, J. H. 1979. Modeling of rock friction 1. Experimental results and constitutive equations. *Journal of Geophysical Research*, 84B, 2161-2168.

Ditto, W. L., Rauseo, S. N. and Spano, M. L. 1990. Experimental control of chaos. *Phys.Rev.Lett.*, 65, 3211-3214.

Eckmann, J.-P., Kamphorst, S., Ruelle, D. 1987. Recurrence plots of dynamical systems. *Europhysics Letters*. 4, 973-977.

Evans, D. M., 1966. Denver area earthquakes and the Rocky Mountain Arsenal disposal well. *Mt. Geology*, 3, 23–26.

Goltz, C.1998. Fractal and chaotic properties of earthquakes, Springer, Berlin.

Grassberger, P., Procaccia, I. 1983. Characterization of strange attractors. *Phys. Rev. Letters*, 50, 346-349.

Grasso, J.-R and D. Sornette. 1998. Testing self-organized criticality by induced seismicity. *Jour. Geoph. Res.*, 103, 29 965-29 987.

Gupta, H. K. 1992. *Reservoir-Induced Earthquakes*, 264 pp, New York: Elsevier.

Hainzl, S., T. Kraft, J. Wassermann, H. Igel, and E. Schmedes. 2006. Evidence for rainfall-triggered earthquake activity, *Geophys. Res. Lett.*, 33, L19303, doi:10.1029/2006GL027642.

Hegger, R., Kantz, H. 1999. Practical implementation of nonlinear time series methods: the TISEAN package. *Chaos*, 9, 413-440.

Heki, K. 2003. Snow load and seasonal variation of earthquake occurrence in Japan, Earth. Planet. *Science.lett*. 207, 159-164.

Hill, D. 2010. Surface wave potential for triggering tectonic (nonvolcanic) tremor. *Bull. Seismol. Soc. Am.* 100, 1859-1878, doi: 10.1785/0120090362.

Hill, D., Prejean, S. 2009. Dynamic triggering. In: Earthquake seismology, Volume editor H. Kanamori. *Elsevier*. pp. 257-293.

Jeffrey. J. 1992. Chaos game visualization of sequences. *Computers and Graphics*. 16, 1, 25-33.

Jones, N. 2001. The quake machine. *New Scientists*. June 30, 34-37.

Kanamori, H., and E. E. Brodsky. 2004. The physics of earthquakes. *Rep. Prog. Phys.*, 67, 1429-1496.

Kantz, H., Schreiber T. 1997. Nonlinear time series analysis. Cambridge University Press, Cambridge.

Kartvelishvili, K. Z., Chelidze, T., Goguadze, N. Preliminary results of processing the accelerograms of 2011 great Japan earthquake in Tbilisi: evidence of local dynamic triggering. *Journal of the Georgian Geophysical Society*, Issue (A), Physics of Solid Earth (in press).

Lempel, A., Ziv, J. 1976. On the complexity of finite sequences, *I. E. E. E. Trans. Infor. Theory*, IT-22, 75-81.

Lursmanashvili, O. 2001. Role of exogenous factors in initiation of Caucasus Earthquakes. Journal of the Georgian Geophysical Society, Issue (A), *Physics of Solid Earth*, 6, pp. 22-27.

Lursmanashvili, O., Paatashvili, T., Gheonjian, L. 2010. Detecting quasi-harmonic factors synchronizing relaxation processes: application to seismology. in: *Geoplanet: Earth and Planetary Sciences Volume* 1, DOI: 10.1007/978-3-642-12300-9; Synchronization and Triggering: from Fracture to Earthquake Processes. Eds.V.de Rubeis, Z. Czechowski and R. Teisseyre, pp. 305-322.

Lyubushin A. A. 2009. Synchronization Trends and Rhythms of Multifractal Parameters of the Field of Low-Frequency Microseisms. *Izvestiya, Physics of the Solid Earth*, 45, No. 5, pp. 381–394.

Lyubushin A. A. 2011. Cluster Analysis of Low-Frequency Microseismic Noise. *Izvestiya, Physics of the Solid Earth*, 47, No. 6, pp. 488–495.

Marwan, M. 2003. Encounters with neighborhood, Ph. D Thesis.

Matcharashvili T. and Chelidze, T. 2010. Nonlinear Dynamics as a Tool for Revealing Synchronization and Ordering in Geophysical Time Series: Application to Caucasus Seismicity., in: *Geoplanet: Earth and Planetary Sciences Volume* 1, DOI: 10.1007/978-3-642-12300-9; Synchronization and Triggering: from Fracture to Earthquake Processes. Eds.V.de Rubeis, Z. Czechowski and R. Teisseyre, pp.3-21.

Matcharashvili, T., Chelidze, T., Abashidze, V., Zhukova, N. and Meparidze, E.. 2010. Changes in Dynamics of Seismic Processes Around Enguri High Dam Reservoir Induced by Periodic Variation of Water Level. in: *Geoplanet: Earth and Planetary Sciences, Volume* 1, DOI: 10.1007/978-3-642-12300-9; Synchronization and Triggering: from Fracture to Earthquake Processes. Eds.V.de Rubeis, Z. Czechowski and R. Teisseyre, pp. 273-286.

Matthews, M. and Reasenberg, P. 1988. Statistical methods for investigating quiescence and other temporal seismicity patterns. *Pure and Appl. Geophys.* 126, 357-372.

Meyers, R. (Ed). 2009. Encyclopedia of Complexity and Systems Science. Springer.

Nikolaev, A. V. (Ed.). 1994. Induced Seismicity. Moscow, "*Nauka*", 220 p, (in Russian).

Nikolaev, V. A. 2003. Research of lithospheric stress state on the base of correlation of tidal forces and seismicity. Anakharsys, Moscow (in Russian).

Obara, K. 2003. Time sequence of deep low-frequency tremors in the Southwest Japan Subduction Zone. *Chigaku Zasshi* (J. Geogr.), 112, 837-849.

Packard, N. H., C. J. F. J. and Shaw, R. 1980. Geometry from a time series, *Phys.Rev. Lett.* 45, 712-716.

Peinke, J., Matcharashvili, T., Chelidze, T., Gogiashvili, J., Nawroth, A., Lursmanashvili, O., Javakhishvili, Z. 2006. Influence of periodic variations in water level on regional seismic activity around a large reservoir: Field data and laboratory model. *Phys. Earth Plan. Interiors*. 156, 130–142.

Peng, Z., J. E. Vidale, A. G.Wech, R. M. Nadeau, and K. C. Creager. 2009. Remote triggering of tremor along the San Andreas fault in central California, *J. Geophys. Res.* 114, doi 10.1029/2008JB006049.

Peng, Z., Long, L.and Zhao, P. 2011. The Relevance of High-frequency Analysis Artifacts to Remote Triggering. *hSeismological Research Letters*, 82, 654-660.

Perfettini H., J. Schmittbuhl and J. R. Rice. 2001. Frictional response induced by time-dependent fluctuations of the normal loading. *Journal of Geophysical Research*, 106B, 13455-13472.

Pikovsky, A., Rosenblum, M. G., Kurths. J. 2003. Synchronization: Universal Concept in Nonlinear Science. Cambridge University Press, Cambridge.

Prejean S., Hill, D. 2009. Dynamic triggering of earthquakes. In: *Encyclopedia of Complexity and Systems Science*, R. A. Meyers (Ed.), Springer, pp. 2600-2621.

Quiroga, R., Kraskov, A., Kreuz, T., Grassberger, P. 2002. Performance of different synchronization measures in real data, *Phys. Rev.* E, 65, 041903.

Rapp, P. E., A. A. S. T. and Farwell, L. A. (1993) Filtered noise can mimic low-dimensional chaotic attractors, *Phys. Rev.* E. 47, 2289-2297.

Rapp, P. E., A. A. Z. I. D. and Jumenez-Montero, M. 1994. Phase-randomized surrogates can produce spurious identification of non-random structure, *Phys. Lett.* A. 1992, 27-33.

Rosenblum, M. G., Pikovsky, A., Kurths. J. 1996. Phase synchronization of chaotic oscillators. *Phys. Rev. Lett*., 76, 1804-1808.

Rosenblum, M. G., Pikovsky, A., Kurths. J. 1997. Effect of phase synchronization in driven chaotic oscillators. *I. E. E. E. Trans. C. A. S.-I,* 44. 874-88.

Rosenstein, M. T., C. J. J. and DeLuca, C. J. 1993. A practical method for calculating largest Lyapunov exponents from small data sets, *Physica D* 65, 117-134.

Rubinstein, J. L., Shelly D. R., and Ellsworth W. L. 2010. Non-volcanic tremors. In "*New Frontiers in Integrated Solid Earth Sciences*". S. Cloetingh, J. Negendank,(Eds), Springer, Berlin, doi 10.,1007/1007/978-90-481-2737-5, pp. 287-314.

Ruina, A. 1983. Slip instability and state variable friction laws. *Journal of Geophysical Research*, 88B, 10359-10370.

Scholz, C. H. 1990. The mechanics of earthquakes and faulting. Cambridge Univ. Press. Cambridge.

Scholz, C. H. 1998. Earthquakes and friction laws. *Nature*, 391, 37-42.

Scholz, C. H. 2003. Mechanics of Earthquakes and Faulting. Cambridge University Press, Cambridge.

Schreiber, T. 1993. Extremely simple nonlinear noise-reduction method, *Phys. Rev.* E., 47, 2401-2404.

Schwartz, S. 2009. Episodic Aseismic Slip at plate boundaries. In: *Earthquake Seismology* (Ed. H. Kanamori). Elsevier, 2009, 445-472.

Shapiro, S. A., Kummerow, J., Dinske, C., Asch, G., Rothert, E., Erzinger, J., Kümpel, H.-J. and Kind, R. 2006. Fluid induced seismicity guided by a continental fault: Injection experiment of 2004/2005 at the German Deep Drilling Site (K. T. B.), *Geophys. Res. Lett*., 33, L01309, doi:10.1029/2005GL024659.

Sibson, R. 1994. Crustal stress, faulting and fluid flow. In: Deformation and Fluid Flow *Geological Society, London, Special Publications*, 78, p. 69-84.

Sobolev, G. A. and Ponomarev, A. V. 2003. *Physics of Earthquakes and Precursors*. Moscow, "Nauka", (in Russian).

Sobolev, G., Spetzler, H., Koltsov, A., Chelidze, T. 1993. An Experimental Study of Triggered Stick-slip. *Pageoph.* 140, 79-94.

Sornette, D. 2000. Critical Phenomena in Natural Sciences. Springer.

Sprott, J. 2003. Chaos and Time-Series Analysis. Oxford University Press.

Sprott, J. C., Rowlands, G., 1995.Chaos data analyzer; the professional version. A. I. P., N. Y.

Stich, D., J. Almendros, V. Jiménez, F. Mancilla and E. Carmona. 2011. Ocean noise triggering of rhythmic long period events at Deception Island volcano, *Geophys. Res. Lett*., 38, L22307, doi:10.1029/2011GL049671.

Strogatz, S. 2000. Nonliear Dynamics and Chaos. Westview. Perseus Books Group.

Takens, F. 1981. Detecting strange attractors in turbulence. in: 'Dynamical Systems and Turbulence", D. A Rand and L. S. Young,(eds). Vol.898, *Springer Lecture Notes in Mathematics*, 366-381, Springer- Berliner.

Talwani, P. 1997. On nature of reservoir-induced seismicity. *Pure and Appl. Geophys*, 150, 473-492,

Tamm, I. 1956. Fundamentals of theory of electricity. Moscow (in Russian).

Tarasov, N. T. 1997.Crustal seismicity variation under electric action. *Transactions (Doklady) of the Russian Academy of Sciences*, 353A (3), 445-448,

Tarasov, N., Tarasova, H., Avagimov A. and Zeigarnik, V.1999. The effect of high-power electromagnetic pulses on the seismicity of Central Asia and Kazakhstan. *Volcanology and Seismology* (Moscow), N4-5, 152-160 (in Russian).

Telesca, L. 2010. Analysis of the Cross-Correlation between Seismicity and Water Level in the Koyna Area of India. *Bull. Seismol. Soc. Am.* 100, pp. 2317–2321, doi: 10.1785/0120090392.

Theiler, J., E. S. L. A. G. B. and Farmer, J. 1992. Testing for nonlinearity in time series: the method of surrogate data, *Physica* D 58, 77-94.

Tsallis, C. 1988. Possible generalization of Boltzmann-Gibbs statistics. *J. Stat. Phys.*, 52, 479– 487.

Tsallis, C. 1998. Generalized entropy-based criterion for consistent testing, *Phys. Rev.* E., 58, 1442– 1445.

Varamashvili, N., Chelidze, T. 2004. Sick-slip and electromagnetic field. *Journal of the Georgian Geophysical Society*, 9A, 3-11.

Varamashvili, N., Chelidze, T., Lursmanashvili, O. 2007. Phase synchronization of stick-slip process by periodical mechanical forcing. *Bulletin of Georgian National Academy of Sciences*, 175, 78-83.

Varamashvili, N., Chelidze, T., Lursmanashvili, O. 2008. Phase synchronization of slips by periodical (tangential and normal) mechanical forcing in the spring-slider model. *Acta Geophysica*, 56, 357-371.

Volykhin, A. M., Bragin, V. D., Zubovich, A. P. 1993.Geodynamic Processes in Geophysical Fields, Moscow, Nauka.

Wang C.-J., and Manga, M. 2010. Earthquakes and Water. *Lecture Notes in Earth Sciences.* Springer.

Zbilut, J. P., Webber, C. L. Jr. 1992. Embeddings and delays as derived from quantification of recurrence plots. *Physics Letters* A, 171, 199-203.

Zbilut, J. P., and Webber, C. Jr. 1998. Detecting deterministic signals in exceptionally noisy environments using cross- recurrence quantization, *Phys. Lett.* A.246, 122-128.

Stich, D., J. Almendros, V. Jiménez, F. Mancilla and E. Carmona (2011), Ocean noise triggering of rhythmic long period events at Deception Island volcano, *Geophys. Res. Lett.*, 38, L22307, doi:10.1029/2011GL049671.

In: Earthquakes
Editor: Kostas Konstantinou
ISBN: 978-1-62081-883-1

Chapter 7

IONOSPHERIC EFFECTS OF SEISMOGENIC DISTURBANCES OF THE GLOBAL ELECTRIC CIRCUIT

A. A. Namgaladze and O. V. Zolotov*
Murmansk State Technical University, Russia

ABSTRACT

The seismo-ionosphere coupling phenomena, namely the ionospheric GPS TEC (Total Electron Content) disturbances observed before strong (M>5) earthquakes have been discussed. These precursors have the form of stable local long-living TEC disturbances at the near-epicenter and often at the magnetically conjugated areas. They do not propagate along meridians in contrast with the ionospheric disturbances related with magnetic activity. Basing on a few case studies of relative TEC deviations before the recent strong seismic events (Jan. 12, 2010 Haiti, Jan. 1 and Jan. 2 2011 Argentina and Chile, March 11 2011 Japan earthquake) we discovered 'ban-time' and terminator effects which were not reported in other authors' researches. These new effects are: the sunrise terminator income triggers the TEC anomalies weakness up to their almost full destruction with subsolar point arrival, i.e. at the local time near-noon hours. The sunset terminator passage is often followed by the anomalies renewal.

We propose the possible physical mechanism of the TEC anomalies forming, namely the vertical transport of the F2-layer ionospheric plasma due to the electromagnetic *[E×B]* drift under influence of the electric field of seismic origin. This drift transfers plasma to heights where the ion loss rates are lower (if plasma drifts upwards) or higher (if plasma drifts downwards). As result, the F2-region electron density and TEC increases (or decreases). The consistency of this hypothesis was checked using the first-principle 3D time-dependent Upper Atmosphere Model. Seismogenic electric fields were modeled in two ways: (1) by setting additional electric potentials; (2) by setting vertical electric current flowing between the Earth and the ionosphere at the near-epicenter area as input for the electric potential equation. The model TEC calculation results are close to the

* Murmansk State Technical University, 183010 Murmansk, Russia. e-mail: *Namgaladze@yandex.ru; ZolotovO@gmail.com.*

GPS observed TEC disturbances if the model electric currents needed to create them are of about 10^{-8} A/m^2.

1. Introduction

The weather, climate and space weather (processes related to solar and geomagnetic activity) and their forecasting are extremely important for mankind. Observations and mathematical modeling used together are modern ways for solving the forecasting problems. Modern first principle numerical models are time-dependent, three-dimensional (3D) and global. Until recent years they were being developed for the lower (responsible for the weather and climate; heights < 100 km) and upper (responsible for the space weather; heights > 100 km) atmosphere separately in spite of the lack of any boundary between these two atmospheric regions. Their principal physical difference consists in the amount of the charged particles – ions and electrons, i.e. in their concentration (number density), which is high in the upper atmosphere and low in the lower. Therefore, the electrodynamic processes are very important in the upper atmosphere (thermosphere, ionosphere and magnetosphere).

But they are important in the lower and middle atmosphere (troposphere, stratosphere and mesosphere) as well due to the existence of such lower atmosphere electricity sources as cosmic rays, thunderstorms, soil radioactivity, seismo-dynamic effects etc. The lower atmosphere electric currents are connected with the ionospheric electric currents so that they, closely related to the magnetospheric currents, are creating a common global electric circuit [Harrison et al., 2010; Rycroft et al., 2007]. This circuit consists of currents flowing upwards from thunderstorm current generators, through the ionosphere and down to the Earth's surface as fair-weather currents. The vertical fair-weather current density between the ionosphere and the Earth is about 2-3 pA/m^2 and the total global current is about 1 kA producing a potential difference of about 250-300 kV between the ionosphere and the Earth.

The horizontal potential difference in the ionosphere may reach 100-150 kV (usually 30-50 kV) between the morning and evening sides of the polar cap boundaries at the polar edges of the auroral zones (at about 75 magnetic latitudes) in both northern and southern hemispheres. The cross polar cap potential difference (or drop) is the commonly used characteristic of the high-latitude ionospheric electric fields. Its dependences on solar and geomagnetic activity have been investigated by many authors and used as input in the model electric field calculations.

This potential difference is generated at the magnetopause under the solar wind-magnetosphere interaction and transported into the polar ionosphere along the geomagnetic field lines via so called field-aligned currents (FACs) of the Region-1. They flow down into the ionosphere at the dawn side and up from the ionosphere at the dusk side creating electric fields of several tens or even hundreds of mV/m in the polar caps directed from dawn to dusk. FACs of opposite signs flow at the equatorial edges of the auroral zones (the Region-2 FACs). They shield the inner magnetosphere and the mid-latitude ionosphere from penetrating magnetospheric electric fields. This shielding and the presence of internal atmospheric dynamo processes result in smaller mid-latitude electric fields (of about several mV/m) in comparison with the high-latitude and subauroral ones.

Region-0 FACs are found during intervals of a strictly positive Bz-component of the interplanetary magnetic field (IMF). They may exist in the vicinities of the cusps. The IMF

By component is responsible for the asymmetry of the morning and evening cells of the ionospheric convection (plasma drift trajectories). The intensity of the FACs and the corresponding cross polar cap potential drops (or the electric field strength) are well correlated with the southward IMF Bz-component. It can trigger magnetospheric substorm and storm generation processes.

Another source of ionospheric electric fields is the dynamo-action of the thermospheric winds pushing the ions across the geomagnetic field lines at ionospheric E- and F1 region heights (100-170 km) where electrons are magnetized (electron-neutral collision frequency $\nu_{en} << \Omega_e$ – electron cyclotron frequency) but ions are not (ion-neutral collision frequency $\nu_{in} >> \Omega_i$ ion cyclotron frequency). That is why this height region is referred to as "dynamo-region". Horizontal ionospheric currents of the upper atmosphere flow predominantly in this height range.

In the ionospheric F2-region and above it (heights > 170 km) both electrons and ions are magnetized ($\nu_{en}, \nu_{in} << \Omega_e, \Omega_i$) and electric currents flow first and foremost along the well conducting geomagnetic field lines (FACs); exceptions are plasma drift processes in the magnetosphere like pressure gradient-driven currents.

Due to the high electrical conductivity of the geomagnetic field lines they can be assumed to be electrical equipotentials, so that electric fields along these lines are zero or very small. Electric fields at heights > 170 km are therefore practically always perpendicular to the geomagnetic field lines. These mutually perpendicular electric (***E***) and magnetic (***B***) fields force the ionospheric plasma to move with the so called $\boldsymbol{E}\times\boldsymbol{B}$ plasma drift velocity (equal to $\boldsymbol{E}\times\boldsymbol{B}/\boldsymbol{B}^2$), i.e. in the direction perpendicular to both. In this drift motion, ions and electrons move together with the same velocity, i.e. there is no charge separation and they do not create electric currents.

Global electric field patterns obtained mainly from satellite measurements are presented usually as polar maps of electric potential distributions for different magnetic activity levels and/or different IMF orientations [e.g., Haaland et al., 2006; Förster et al., 2007].

Ionospheric $\boldsymbol{E}\times\boldsymbol{B}$ plasma drifts cause many important peculiarities of the spatial and time variations of the ionospheric plasma, such as the formation of the equatorial anomaly, the main ionospheric trough, the plasmasphere, and ionospheric disturbances related to magnetic storms and substorms as well as strong seismic events.

At high and subauroral latitudes (where the geomagnetic field inclination I is near 90°-60°), the ionospheric plasma drifts affect the electron density mainly due to the divergence/convergence of the horizontal plasma flows. The daytime plasma, having high density, can thus flow across the polar caps to the night side, increasing the density there. Plasma motion will stagnate, where the magnetospheric convection and the co-rotation are oppositely directed with about the same velocity magnitudes. If this occurs outside the sunlit region, the plasma will be lost due to recombination processes to very low density values, forming the so called main ionospheric trough in the dark winter F2-region of the ionosphere at subauroral latitudes.

At middle to low latitudes, the plasma co-rotation with the Earth leads to the formation of the plasmasphere. At these latitudes (where inclination I is small) also the vertical plasma drifts become important. Close to the equator, they cause the so called "fountain effect" forming the F2-region equatorial anomaly. At middle latitudes, they can create increases or decreases of the electron density in the upper ionosphere (heights >200 km – F2-region and plasmasphere) by lifting the plasma up or down across the geomagnetic field lines to heights, where the ion loss rate is lower or higher, respectively. Such vertical movements of the F2-region plasma produce the disturbances of the TEC (Total Electron Content, the amount of electrons in the vertical 1 m^2 crossection column over the ground up to the instrument flight height) – well registered by the GPS (Global Positioning System) observations. Below we shall consider and discuss these effects in the TEC variations observed before strong earthquakes. We associate them with seismogenic sources of the ionospheric electric field disturbances.

Both the ionospheric plasma drifts and the electric currents influence via ion-neutral collisions on the ion and neutral temperatures (Joule heating and thermospheric winds). Thermospheric wind circulation changes the thermospheric gas composition, which in turn influences on the ionospheric ion composition and electron density and so forth. The plasma physical processes within the near-Earth space environment as well as the interactions between the various regions appear to be highly complex. To understand the dynamics of the whole system with all its interrelationships at various levels of complexity is one of the most important reasons to use first principle models for their description. They allow to perform "numerical experiments" and to interpret different measurements on a common basis, which may be obtained at various places far from each other both by ground-based observations and remote or in-situ data gathering with satellites.

This chapter presents and discusses some of most recent measurements of the TEC variations, obtained by the GPS satellites before several strong earthquakes [Jan. 12, 2010 Haiti, Jan. 1 and Jan. 2 2011 Argentina and Chile, March 11 2011 Japan earthquake], and their modelling in the framework of the Upper Atmosphere Model (UAM). The results concern ionospheric electric fields, generated by seismogenic sources, and show their influence on the ionospheric TEC.

2. Observations of the Seismic Impact on the Ionosphere

Historically first data used for searches of the seismo-related impact on ionospheric plasma were obtained using the traditional observation method of HF sounding of the ionosphere. The method is based on the ability of the ionosphere to reflect electromagnetic waves (1-20 MHz). The time between the emission of the signal and its arrival is measured, and then the reflection height is determined. The frequency of the reflected wave is equal to or close to the plasma frequency at the reflection point and it allows one to determine electron density. Several features of the ionosphere plasma parameters variations were revealed using this method and reported as possible precursors:

1) Increase of diffusivity of the E- and F-layers reflections (trails) at the ionogrammes, formation of the sporadic E-layers at the near-epicenter region.
2) Increase (or in some cases decrease) of the critical frequency, altitude of the F2-layer electron concentration maximum.
3) Local nature of the observed disturbances; their lifetime from several hours up to several days before the forthcoming sesmic event.

4) Changes of the sign of the disturbances to the opposite ones before the event realization moment etc. (e.g., [Gokhberg et al., 1982, 1983, 1988; Pulinets, 1998]; see also references in [Pulinets and Boyarchuk, 2004]).

The drawbacks of this method for the seismo-ionosphere coupling mission are:

1) The ability to measure electron density only above the ionosonde and below the electron concentration maximum height;
2) The most ionosondes measure episodically;
3) The network of ionosondes is not dense enough and not uniform. Huge areas without any stations exist over e.g., oceans etc.

Next data sources are satellite observations, also by means of the special satellite projects of the natural hazards monitoring ["COMPASS-1", "COMPASS-2", "Sich-1M", "QuakeSat", "DEMETER"] (see reference list in [Pulinets and Boyarchuk, 2004]). Many authors (e.g., [Gokhberg et al., 1982; Larkina et al., 1989; Fraser-Smith et al., 1990; Hayakawa et al., 2004; Molchanov et al., 2006; Kopytenko et al., 2006; Mullayarov et al., 2007]) reported anomalous disturbances in ULF/VLF band, electric and magnetic fields basing on satellite and ground-based observations. Figure 1 shows seismically disturbed equatorial anomaly observed 1 day prior to the Chile earthquake 12.04.1963 (Alouette-2 data) in comparison to regular quiet-time variations (before and after EQ).

Basing on the satellite observations analysis [Buchachenko et al., 1996] the following latitudinal zones of the ionospheric precursors to earthquakes are discriminated:

1) Large zone: ±30° in latitude relatively the epicenter position; the observed effects: modification of the altitudinal profile of the electron density, changes of the plasma ion composition; special features: seismo-related disturbances of the electron density profile were observed at 03-06 LT and 15-18 LT.
2) Middle zone: ±10° in latitude relatively the epicenter position; the observed effects: generation of electromagnetic emission in ELF (< 30 Hz) band, formation of small-scale plasma density irregularities, quasi-static electric field disturbances; special features: ELF emission, electric fields and small-scale irregularities are conjugated with the epicenter by the geomagnetic field lines and may be observed by satellites at the magnetically conjugated regions as well.
3) Small zone: ≤3 in latitude relatively the epicenter position; the observed effects: generation of the ULF (30 HZ – 3 kHz) band electromagnetic emission, modification of the atmosphere emissions with wavelengths 5577 and 6300Å; special features: the zone of the ULF emission is magnetically conjugated with the earthquake epicenter.

Nowadays there are vast possibilities to investigate the ionosphere modifications of different nature by use of the satellite Global Positioning System (GPS) signals measurements [Liu et al., 2004, 2006; Afraimovich et al., 2004, 2006; Krankowski et al., 2006; Zakharenkova et al., 2007a, 2007b, 2008]. The dense network of GPS receivers (a few thousands all over the world) fulfills simultaneous coverage in global scale with high temporal resolution. The GPS technique provides measurements of the group and phase delays of the signals L_1=1575 MHz and L_2=1228 MHz with a 30-sec interval. The ionospheric delay can be transformed into the content of electrons along the signal path between a GPS satellite and GPS receiver, and then recalculated into its vertical projection. The vertical total electron content (VTEC) is very sensitive to changes of the maximal electron concentration (NmF2) in the F2 layer of the ionosphere.

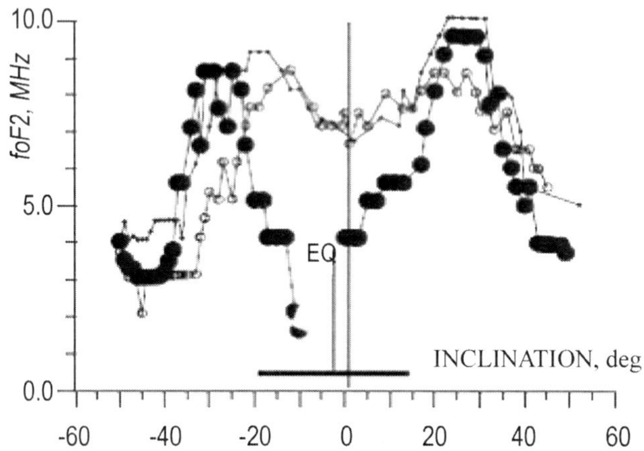

Figure 1. Latitudinal variations of the foF2, MHz (critical frequency of the ionosphere F2-layer): black circles – observed 1 day prior to the Chile earthquake 12.04.1963 according to the Alouette-2 satellite data; two other lines – observed during quiet time (before and after EQ) [Depueva and Ruzhin, 1995]. EQ – the earthquake epicenter position. Inclination – geomagnetic field lines inclination.

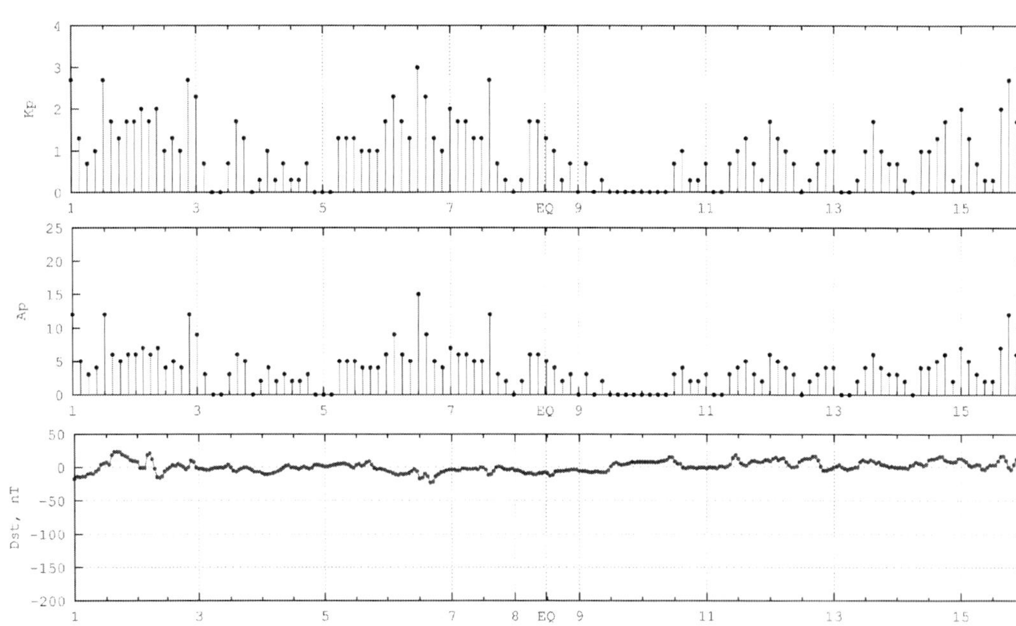

Figure 2. Geomagnetic activity indexes for Jan. 1-15, 2006: (1) Kp-index (top panel); (2) Ap-index (middle panel); (3)Dst-index (bottom panel). The x-axis is the day of the month. EQ – the time of the earthquake.

The extensive studies of the ionospheric earthquake precursors in the GPS TEC measurements carried out in the last years have revealed that for strong mid-latitudinal earthquakes the seismo-ionospheric anomalies very often (usually) look as local TEC increases and they are situated in the immediate vicinity of the earthquake epicenter area. These anomalies do not propagate along the meridians in contrast with the ionospheric disturbances related to magnetic activity. The zone of the anomaly maximum manifestation (TEC enhancement more than 35%) has a spatial scale of some thousands km in longitude and about 1000 km in latitude [Zakharenkova et al., 2007a, 2007b, 2008]. These sizes are in a good agreement with results obtained from the combined analysis of groundbased and satellite ionosonde measurements. It was found out that the spatial scale of the seismo-modified area of the ionosphere at the heights of the F-layer maximum during the strong earthquake preparation time has the diameter of 20-40 in geographical coordinates [e.g., Pulinets, 1998]. The size of the area changes with the earthquake magnitude. Pulinets et al. [2005] analyzed the GPS TEC measurements for the strong Mexico earthquake and found out an anomalous ionospheric modification in the spatial (latitude-longitude) distribution of the TEC deviation. The anomalous enhancement of the TEC was registered 3 days prior to the event and it reached the value of 55% relative to the background conditions. The negative TEC disturbances (TEC decreases) were also observed, for example before the strong Turkey earthquakes of 1999 [Ruzhin et al., 2002] and before several Taiwan and Sumatra earthquakes [Liu et al., 2006; Liu and Chen, 2007].

3. Recent Case Studies of the TEC Disturbances before Strong Earthquakes

3. 1. TEC Disturbances before Kythera Jan. 8, 2006, Mw5.5 Earthquake

The Kythera earthquake took place on Jan. 8, 2006, 11:35 UT (13:35 LT), Southern Greece, (36 20'N, 23 20'E), D 66 km (according USGS), Mw 6.8, URL: http://earthquake.usgs.gov/earthquakes/eqinthenews/2006/ushrak/ . After the main shock event aftershocks were registered during next a few days with maximum magnitude up to Mw 5.5. Geomagnetic situation before the considered earthquake was quiet and changed slowly that allowed us to conclude no disturbances could mask seismo-induced TEC anomalies existed (see Figure 2).

To discriminate spatial characteristics of the ionospheric response on the considered seismic event we calculated TEC deviation maps. Values for the given time were compared with the quiet non-disturbed variation calculated as UT-grouped 7-days running medians. TEC disturbances maps for Jan. 7, 2006 are shown in Figure 3. As one can see, increases of the TEC were observed from 10:00 UT at the near-epicenter area of the forthcoming earthquake. The magnitudes of the anomalies reached ~38-55% relative to the background variation, existed during next 10-12 hours. The TEC increase maximum (~55%) was observed at 18:00 UT-20:00 UT. The disturbed region had local domelike shape with maximum phenomena manifestation near the epicenter. The spatial size of the anomaly (TEC deviation > 35%) was ~4000 km in longitude and ~1500 km in latitude.

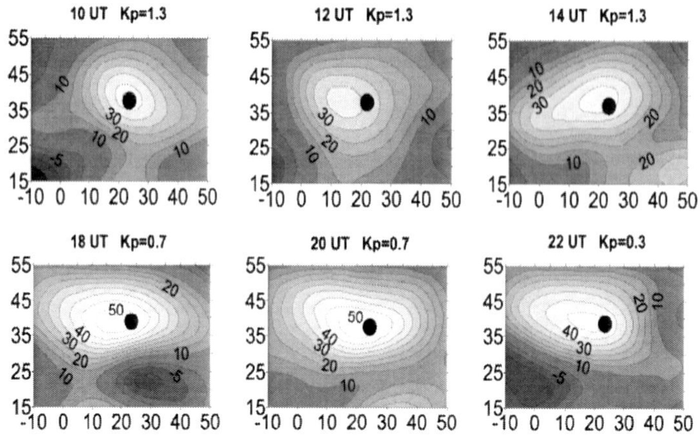

Figure 3. Maps of the TEC disturbances (deviations from quiet background variation, %) for Jan. 7, 2006, Kythera, Southern Greece. Black circle – the earthquake epicenter position. Co-ordinates are geographic latitude vs. longitude.

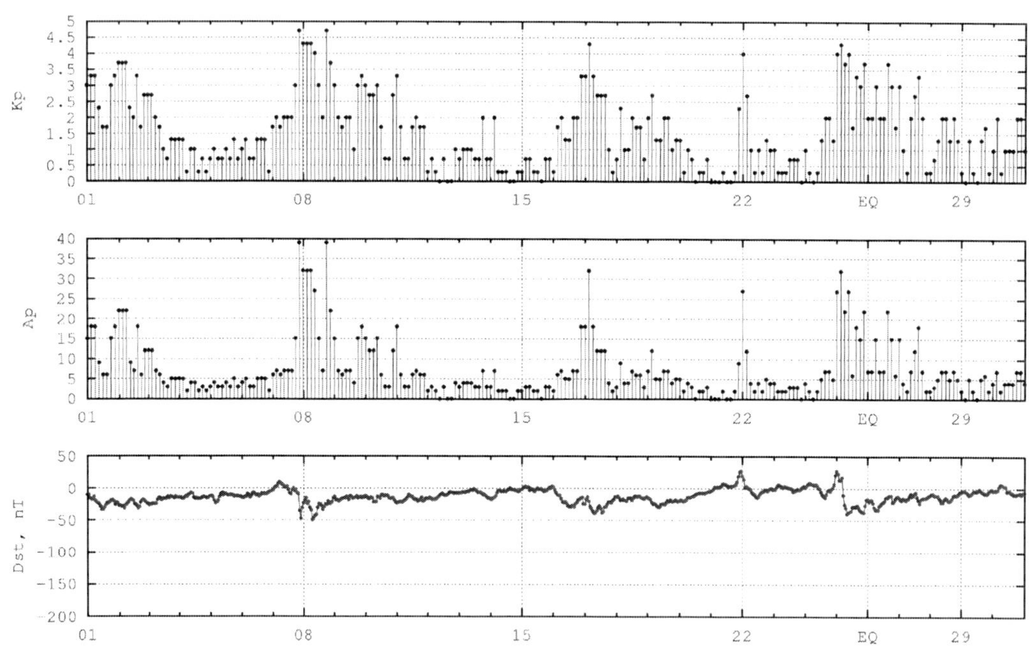

Figure 4. Geomagnetic activity indexes for Sep. 1-31, 2005: (1) Kp-index (top panel); (2) Ap-index (middle panel); (3) Dst-index (bottom panel). The x-axis is the day of the month. EQ – the time of the earthquake.

Therefore, on the 7th of Jan. 2006, a day before the Greece earthquake, the local TEC ionospheric anomaly linked to the epicenter geoposition was detected. As the main shock event and its' aftershocks are within the precursor manifestation area and within a little timeslot, we can not clearly contrast the deposit of each event. We expect that main event, as of greater magnitude and earlier realization time, make major impact on the TEC anomaly formation; aftershocks supply and enlarge the lifetime of the anomaly.

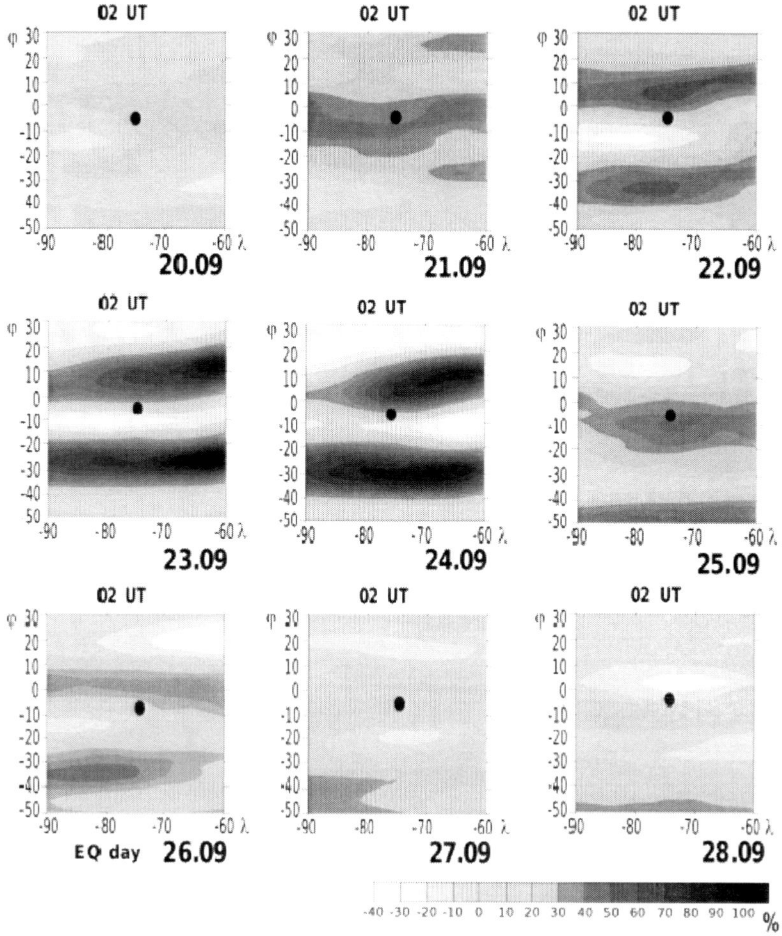

Figure 5. Maps of the TEC disturbances (deviations from quiet background variation, %) for 02 UT of Sep. 20-28, 2005. Black circle – the epicenter position. Co-ordinates are geographic latitude vs longitude.

3. 2. TEC Disturbances before Peru Sep. 25, 2005, Mw 7.5 Earthquake

The Peru earthquake took place on Sep. 26, 2005 at 01:55 UT (Sep. 25, 2005 at 08:55 PM at epicenter), Northern Peru, (5.674 S, 76.409 W), Mw 7.5, URL: http://earthquake.usgs.gov/earthquakes/eqinthenews/2005/usdlad/.

Geomagnetic situation was mainly quiet in the considered period (Sep. 18-24) with episodic moderate disturbances from Sep. 25 (see Figure 4).

In Figure 5 we can see maps of TEC deviations from undisturbed background values for 02UT, September 20-28, 2005. No significant disturbances were observed during Sep. 20, 2005. On Sep. 21 the increased TEC regions were formed above the epicenter area. Then, the 'trough' (inside the anomaly) at the near-epicenter area with camel-case maxima both sides of the geomagnetic equator was formed during Sep. 22. On Sep. 23 the deepening of the 'through' and increase of the magnitudes and spatial area occupied by the anomalous positive structures took place. The destruction of the TEC increased area happened on Sep. 25, 2005 in form of reduction of the anomaly and conjunction of double case maxima into a single structure. During Sep. 27-28, 2005 no significant anomalies were observed.

3. 3. TEC Disturbances before New Zealand Nov. 22, 2004, Mw 7.1 Earthquake

The New Zealand earthquake took place on Nov. 22, 2004, 20:26UT, (46.69 S, 164.78 E), Mw 7.1, URL: http://earthquake.usgs.gov/earthquakes/eqinthenews/2004/usrdbs/.

Geomagnetic activity during Nov. 15-22, 2004 was relatively quiet. On Nov. 21, 2004 TEC reduction in the area, shifted from the epicenter position to north-west, took place. Similar effects were observed at the magnetically conjugated region too, but smaller by magnitude and lifetime. The anomaly distruction happened a day before the seismic event realization.

The considered anomaly looked like decreased TEC region at the near-epicenter area, occupying ~40° in longitude and ~25° in latitude. The amplitude of the modification reached values down to -43% relatively to the undisturbed background variation during one day before the earthquake.

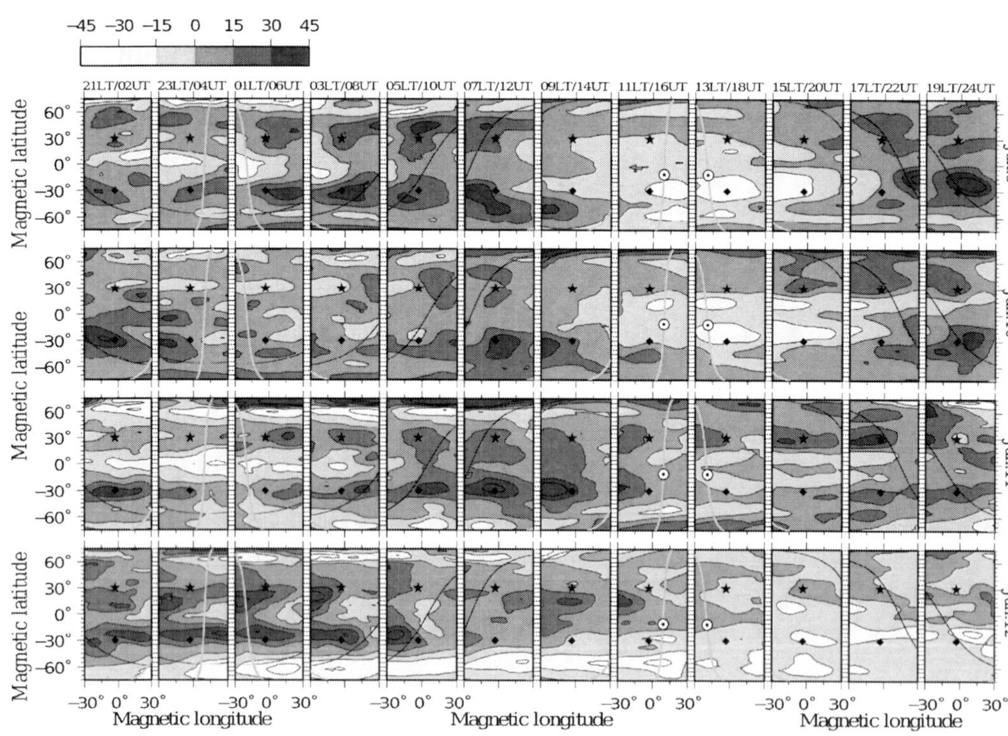

Figure 6. Maps of the TEC disturbances (deviations from quiet background variation, %) for Jan. 9-12, 2010 (from top to bottom) before the Haiti earthquake of Jan. 12, 2010 (21:53UT). Star – the epicenter position. Diamond – the magnetically conjugated point. White circle with a dot – the subsolar point. Black curv – the terminator.

3. 4. TEC Disturbances Prior Haiti Earthquake of January 12, 2010

The Haiti earthquake took place on Jan. 12, 2010, 21:53 UT (16:53 LT), (18.46 N, 72.5 W), Mw 7.0, URL: http://earthquake.usgs.gov/earthquakes/eqinthenews/2010/us2010rja6/.

The geomagnetic situation was relatively quiet for the considered period: the K_p index was in general less then 2 during Jan. 1-12, 2010 reaching distinct maxima values less then 3 for a short period. The A_p index took values of about 5 nT and did not exceed 10 nT for the considered period. The D_{st} index variations were in the range from -15 nT up to 15 nT. Therefore, the observed TEC disturbances could not be explained by the solar and geomagnetic activity. They seem to be caused by processes, originating from the underlying near-Earth atmosphere impact.

To estimate pre-earthquake TEC variations we have built differential (%) TEC maps relative to the quiet background conditions. We used global ionospheric maps (GIM) of the TEC [Dow et al., 2009] provided by the NASA in IONEX format as initial data for the analysis (ftp://cddisa.gsfc.nasa.gov/pub/gps/products/ionex/). The spatial resolution of these TEC data is 5° in longitude and 2.5° in latitude. Corresponding time resolution is 2 hours. In contrast to other ground-based and satellite systems, GNSS-based networks provide almost global coverage and near-permanent (i.e. almost without time gaps) data.

We defined and calculated background TEC values (i.e. undisturbed conditions) as 7-days UT-grouped running observations' medians before the current calculation moment. The resulting TEC difference (%) maps are presented in Figure 6.

As one can see in Figure 6, anomalous TEC enhancements were observed during 06 UT – 14 UT of Jan. 9, 2010, 12 UT – 14 UT of Jan. 10, 2010, 08 UT – 16 UT of Jan. 11, 2010, 04 UT – 10 UT of Jan. 12, 2010 reaching more than 50 % by magnitude relative to the background values [Zolotov et al., 2011a, 2011b]. The amplitudes of the positive anomalies were significantly larger at the magnetically conjugated region. They were strongly fixed at the near-epicenter area. Their lifetime was limited to the night-time hours. The terminator and subsolar point positions (indicated by continuous lines and circle symbols, respectively) have apparently influenced on the behavior of the observed positive structures. Approaching the sunlit ionosphere is correlated with disappearance of the anomalies. Such a behavior of the positive TEC disturbances could be explained by the corresponding disappearance (due to the increased ionospheric conductivity) of the electric field generated by the external electric current. It should be noticed that there exist also negative TEC disturbances around the positive ones but with smaller magnitudes (of about 15%).

3. 5. TEC Disturbances Prior to Jan. 1, 2011, Argentina Mw 7.0 and Jan. 2, 2011, Chile Mw 7.1 Earthquakes

We considered January 1, 2011, 09:56UT/06:56LT, Argentina, Mw 7.0, (26.794 S, 63.079 W), D 576.8 km, URL: http://earthquake.usgs.gov/earthquakes/eqinthenews /2011/usc0000xky/, and January 2, 2011, 20:20 UT/17:20 LT, Chile, Mw 7.1, (38.354 S, 73.275 W), D 25.1 km, URL: http://earthquake.usgs.gov/earthquakes/eqinthenews/2011 /usc0000y49/ earthquakes as one single (or unite) case. The Chile earthquake took place a day after the Argentina earthquake. The positions of both epicenters are close to each other and can be considered as situated within one and the same 'traditional' seismic phenomena manifestation area. Therefore, relaying on the 'traditional' pre-earthquake TEC modification signature (spatial and temporal features), it is impossible to distinguish separately each of the different sources that impact the ionospheric TEC. On the one hand, the depth of the Argentina earthquake epicenter was 576.8 km while the 'traditional' TEC signatures are statistically defined only for earthquakes with less than 80-150 km depth, i.e. ionospheric manifestations of such phenomena have not been investigated at all up to now. On the other hand, the TEC disturbances triggered by this earthquake should, if ever there are any, be masked or merged with Chile's earthquake effects. The TEC difference maps for 30.12.2010-02.01.2011 are presented in Figure 7.

The geomagnetic situation was quiet to moderate for the considered period. During Dec. 22, 2010 – Jan. 3, 2011 (except Dec. 28-30, 2010) the K_p index was less then 3, A_p mostly about 5 nT, and D_{st} was primarily in the range between -20 nT and +10 nT. A moderate magnetic storm happened during Dec. 28-30, 2010: on Dec. 28 the K_p index reached values up to 4, A_p up to 27 nT, and the D_{st} index changed from +13 nT at 10 UT down to -43 nT at 17 UT on 28.12.2010. As the moderate magnetic storm generated in Dec. 28-30, 2010 TEC disturbances of considerable magnitude that are able to mask the seismo-related variations, we excluded the TEC difference maps of this period from the analysis. Therefore, the considered linked to the near-epicenter position long-living TEC disturbances were expected to be due to the going through the underlying atmosphere impact.

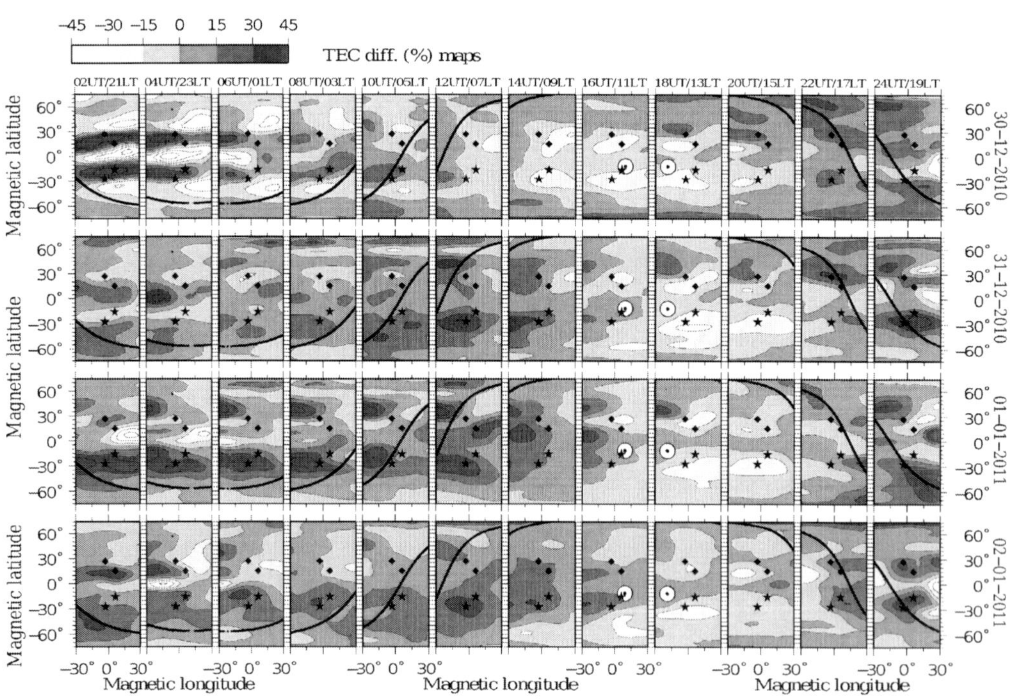

Figure 7. The same as in Figure 6 but for 30.12.2010-02.01.2011 (from top to bottom), 02 UT-24 UT (from left to right) before the Argentina (right star) Jan. 01, 2011, 09:56UT and Chile (left star) Jan. 02, 2011, 20:20 UT earthquakes.

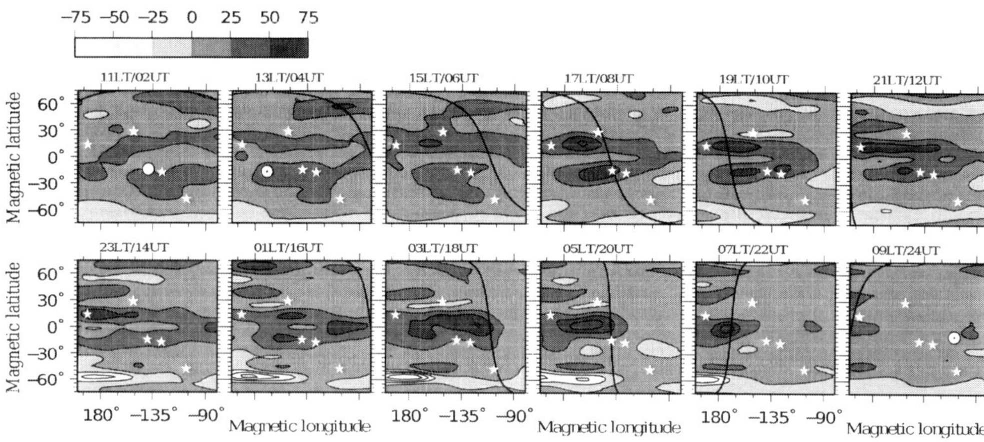

Figure 8. Maps of the TEC disturbances (deviations from quiet background variation, %) for 02 UT – 24 UT, March 08, 2011. Star – the epicenter position. White circle – the subsolar point. Black curve – the terminator line. Local time is for the Japan March 11, 2011 earthquake epicenter position.

In case of the Jan. 1, 2011, Argentina Mw 7.0 and Jan. 2, 2011, Chile Mw 7.1 earthquakes, significant TEC enhancements took place in Jan. 01, 2011, 02UT-14UT at both the Northern and Southern hemispheres. After the passage of the sunrise terminator, positive structures were also observed at the Southern hemisphere during 08-12 UT of Jan. 02. Disturbances at the magnetically conjugated region were insignificant or absent at all. Positive TEC disturbances were also observed after the main shock event realization, possibly as post-effects of the earthquake. The sunlit conditions lead to TEC enhancements decrease up to their full suppression near the subsolar point sector.

3. 6. Case Study of the TEC Disturbances Prior March 11, 2011, M 9.0 Japan Earthquake

While the case of the Haiti earthquake was an example of 'clear' undisturbed by geomagnetic or other seismic events phenomenon manifestation, the Japan earthquake is a case of disturbed ones. The aim of this subchapter is to check the correspondance of the above scheme and to contrast it with real pre-earthquake TEC modifications.

In case of the March 11, 2011, Japan, Mw 9.0, (38.32°N, 142.36 E), D 32 km, URL: http://earthquake.usgs.gov/earthquakes/eqinthenews/2011/usc0001xgp/ earthquake, a sequence of strong seismic events took place close to and associated with the epicenter area and time interval before and after the Mw 9.0 event. The Mw 7.2 March 9, 2011 earthquake (38.424 N, 142.836 E), D 32 km, 02:45UT/11:45LT took place at almost exactly the same epicenter position as the March 11 event. Also on March 9, the second earthquake Mw 6.5, (6.022 S, 149.659 E), 21:24 UT happened. So we have to look at the TEC variations as triggered by a unite source of a spatially distributed cluster of earthquakes. The geomagnetic activity was quiet-to-moderate with episodic periods of moderate-to-strong disturbances.

During March 6-10, 2011 the D_{st} index was within the interval of -22 nT +4 nT without any significant disturbances. On March 10, a moderate magnetic storm took place: D_{st} index varied from -7 nT down to -82 nT on March 11.

As for the previous cases, we defined and calculated the background TEC values (i.e. undisturbed conditions) as 7-days UT-grouped running observations' medians before the current calculation moment. The resulting TEC difference (%) maps are presented in Figure 8.

In case of March 11, 2011, Japan, Mw 9.0, (38.322°N, 142.369 E), D 32 km and associated earthquakes the following TEC disturbances took place: on March 08, 2011 TEC enhancements up to 80% were observed from 06 UT till 20 UT; they firstly were formed at the near-epicenter areas, then their magnitude increased. During 08 UT – 12 UT the anomalies spread out along the parallel at both hemispheres, occupying a larger region. From 12 UT there was a tendency to replenish the Appleton's anomaly. The TEC anomaly comprised an ellipse-like region with the earthquake on its border. At 18 UT-20 UT we see a single increased TEC spot. The passages of the sunrise terminator and later of the subsolar point degraded the anomaly and led to its full decay in the course of the whole day March 09, i.e. no significant TEC disturbances were observed on the day of the Mw 7.2 March 09, 02:45 UT earthquake (happened at nearly the same exact positions of the Mw 9 March 11 and the Mw 6.5, (6.022 S, 149.659 E), 21:24 UT earthquakes). The described above equatorial anomaly modifications are expected to be the TEC response on this equatorial earthquake. We also see some moving positive structures on March 10, but they are likely due to geomagnetic activity driven disturbances.

All the features revealed for the 'clear' case of the Haiti 2010 earthquake were shown to keep alike also for both the activity-disturbed (Chile and Argentina 2011) and multiple-earthquake (Japan March 11, 2011) periods. Those features comprise the terminator and subsolar point related effects: anomalies depression and 'ban' time at the near-noon hours, modification of the shapes of the anomalies under influence of the terminator approach. In case of the Japan 'cluster' earthquakes, the above described features also happened and the conjugated action of equatorial and mid-latitudinal earthquakes could not be extracted or unambiguously separated. The equatorial one makes the main impact in Appleton's anomaly modification. Unfortunately, basing only on the spatial or time interval criteria it is impossible to distinguish the Japan March 9 from the Japan, Mw 9.0, March 11 earthquakes. Also it should be mentioned that the magnitude of the TEC anomaly is larger than in Haiti (Mw 7.0) or Chile (Mw 7.1) and Argentina (Mw 7.0) cases. We expect this is due to the stronger magnitude of the Japan, Mw 9, March 11 earthquake. The warning criteria in this case are applicable only for region- or country-wide areas, but not for smaller ones. We believe that 'traditional' signatures should be extended with the terminator and subsolar point related effects.

Sequential consideration of quiet-to-moderate cases showed that solar and geomagnetic activities can create similar TEC disturbances, but they are not linked to a fixed geoposition. Their lifetime corresponds to the duration of the activity disturbances (characterized by the incices of D_{st}, K_p, etc.) with some time-lag due to the inertness of the ionosphere. They usually originate from the auroral zone and may penetrate down to middle latitudes. Therefore, the applied technique for the TEC disturbances discrimination should be very carefully used at high latitudes (>60°) especially during moderate-to-strong disturbance conditions.

There were no statistical investigations of TEC anomalies related to the deep earthquakes, so that no precursory schemes exist. Nevertheless, the widely accepted spatial scale of TEC precursors did no allow us to separate both (Chile and Argentina) earthquakes from each other. Such combined action, if any, may be expressed as some shift of the anomalously enhanced region.

3. 7. Pre-Earthquake TEC Disturbances Summary

Thus, all recent case studies confirm the known pre-earthquakes TEC signatures reported by e.g., [Liu et al., 2006; Pulinets and Boyarchuk, 2004; Zakharenkova et al., 2007a]:

1) Anomalous strong (30-90% and more deviations relatively to the quiet values) TEC positive or negative modifications (increases or reductions) before the forthcoming seismic event not less than Mw 5 by magnitude linked to the near-epicenter area. Positive ones are more often reported [Zakharenkova, 2007]. The typical zone of the anomaly maximum manifestation extends larger than 1500 km in latitude and 3500-4000 km in longitude. The shapes and dimensions of the disturbed areas are kept rather stable. These anomalies do not propagate along the meridians in contrast with the ionospheric disturbances related to magnetic activity.
2) Local long-living anomaly formation is often reported from several days or hours to couple of weeks before the earthquake release moment. TEC precursors' evolution might be in form of sign change to opposite ones. Before the earthquake release moment or a few hours earlier the anomaly depression up to it's full decay could take place. The lifetime of the seismo-induced deviations of the same sense is from about 4-6 hours up to 12 hours in case of very strong earthquakes.
3) The vertical projection of the epicenter position does not mandatory coincide with the maximum phenomenon's manifestation location.
4) Similar effects are often reported at the magnetically conjugated area.
5) In case of strong low-latitudinal earthquakes there are effects related to the modification of the ionospheric F2-region equatorial anomaly: increase or decrease of the equatorial (Appleton) anomaly with trough deepening or filling [Depueva and Ruzhin, 1995; Depueva et al., 2007].This traditional pre-earthquake TEC modification scheme, reported by many authors, should be extended with the following two new ones:
6) "Ban-time" effect at the near-noon hours [Namgaladze et al., 2011], i.e. anomaly reduction with subsolar point income took place, and
7) Sun-rise terminator income should trigger anomaly depletion while sun set terminator income trigger the anomaly recovery. It should be noticed that many ULF precursors also dominate at night-time (see, e.g. Hobara et al. [2004]).

We extended the traditional signature with items (6)-(7), derived from our recent pre-earthquake TEC variations analysis and also not presented in other authors' publications. We will discuss below possible physical mechanisms of generation of that combined signature (disturbances) in frame of Global Electric Circuit (GEC).

4. Pre-Earthquake Anomalous TEC Disturbances: Physical Interpretation and Modelling

4. 1. Physical Interpretation

Seismogenic electric fields generated near the tectonic faults are important elements of the Global Electric Circuit (GEC) [e.g., Harrison et al., 2010; Pulinets and Boyarchuk, 2004, and reference there in; Pulinets and Ouzounov, 2010; Sorokin and Chmyrev, 1999; Sorokin et al., 2007; Freund, 2011]. We shall consider below their ionospheric effects mainly in the Total Electron Content (TEC) variations as possible ionospheric precursors of earthquakes.

Many authors [e.g., Pulinets, 1998; Sorokin and Chmyrev, 1999; Sorokin et al., 2005a, 2005b, 2006, 2007] strongly rely on the hypothesis of a seismogenic electric field related to the vertical turbulent transportation of injected aerosols and radioactive particles (radon isotopes). The increase of the atmospheric radioactivity level during the earthquake formation leads to the enlargement of the ionization and electric conductivity of the near-ground atmosphere. The joint action of these processes leads to an intensification of an electric field in the ionosphere up to the value of units-tens mV/m [Chmyrev et al., 1989]. Ionosphere response on such electric fields were investigated mainly using 1D models and for lower ionosphere [Hegai et al., 1997; Kim et al., 2002].

Freund [2011] proposed another mechanism of the near-ground atmosphere layer ionization basing on so called “positive holes”: most crustal rocks contain dormant electronic charge carriers in the form of peroxy defects; when rocks are stressed, peroxy links break, releasing electronic charge carriers, known as positive holes. The positive holes are highly mobile and can flow out of the stressed subvolume. F. Freund expects this mechanism to be significantly more efficient than the above-named radon-related ones.

TEC is formed mainly by the F2-layer plasma density. NmF2 and TEC disturbances related with magnetic activity are created by the neutral atmosphere (neutral gas composition, internal gravity waves and winds) and electric field variations. But it is impossible to localize neutral atmosphere disturbances at a limited area, they will propagate away from the source. Therefore, there are strong arguments in favor of the idea of an electric field of seismic origin as the main cause for the TEC anomalies observed before earthquakes:

1) The geomagnetically conjugate ionospheric precursors [Pulinets et al., 2003];
2) Effects related to the ionospheric F2-region equatorial (Appleton) anomaly controlled by the zonal electric field [Depueva and Ruzhin, 1995; Pulinets and Legen’ka, 2002];
3) “Ban-time” at the near-noon hours [Namgaladze et al., 2011].

It was proposed [Namgaladze et al., 2007] that the main reason for the appearance of local TEC anomalies in the form of an increased (decreased) total electron content of the ionosphere, observed on the base of GPS (Global Positioning System) signals measurements, is the vertical drift of F2-region ionospheric plasma upward (downward) under the influence of a zonal electric field of seismogenic origin directed to the east (west). At middle latitudes the vertical upward component of the electromagnetic drift, created by an eastward electric field, leads to an increase of the F2 region electron concentration maximum (NmF2) and TEC due to the plasma transport up to regions with lower concentration of the neutral molecules O_2 and N_2 and, consequently, with lower loss rate of the dominating ions O^+ in the ion-molecular reactions [Brunelli and Namgaladze, 1988]. An electric field of the opposite direction (westward) creates the opposite – negative – effect in NmF2 and TEC. In the low latitude regions (near the geomagnetic equator) an increased eastward electric field leads to a deepening of the Appleton anomaly minimum ("trough" over the magnetic equator in the latitudinal distribution of electron concentration) due to the intensification of the fountain-effect.

4. 2. Modelling

To check this hypothesis of the zonal electric field as the most probable cause for the observed TEC disturbances before earthquakes, model calculations were carried out with the UAM - the global numerical model of the Earth's upper atmosphere [Namgaladze et al., 1988, 1991, 1998a, 1998b]. The aims to be solved were:

1) To check in principle the ability to generate in the ionosphere additional electric potentials of seismic origin and create the required zonal electric field at the near-epicenter area;
2) To calculate this electric field and the corresponding TEC disturbances;
3) To estimate the efficiency of the proposed mechanism and to compare simulated TEC deviations with those, which are usually treated as precursors at low- and mid-latitudinal sectors.

The Upper Atmosphere Model is a global, three-dimensional, time-dependent, numerical model simulating the thermosphere, ionosphere, plasmasphere and inner magnetosphere of the Earth as a single system. It was initially developed at the Kaliningrad observatory (now West Department) of IZMIRAN [Namgaladze et al., 1988, 1991] and then extended at the Polar Geophysical Institute of the Russian Academy of Sciences and at the Murmansk State Technical University [Namgaladze et al., 1998a, 1998b]. The model includes the equations of the continuity, momentum and heat balance for neutral and charged particles and the electric potential equation and calculates the concentrations, velocity vectors and temperatures of basic neutral (O_2, N_2, O) and charged (NO^+, O_2^+, O^+, H^+ and e) components of the atmosphere at the altitude range from 60 km to $15R_e$.

The following continuity equations are solved in the model:

$$\partial n_\alpha/\partial t + \nabla (n_\alpha \mathbf{V}_\alpha) = Q_\alpha - L_\alpha, \; \alpha = n, i, e, \qquad (1)$$

where the subscripts n, i and e refer to neutral components, ions and electrons, respectively; n_α, Q_α, L_α are the concentration, the production and loss rates of α-gas, respectively; $\mathbf{V}_\alpha$ is the velocity vector of α-gas.

Momentum equations:

$$\rho_\alpha[dV_\alpha/dt + \Omega\times(\Omega\times r) + 2\Omega\times V_\alpha] = F_\alpha, \; \alpha = n, i, e, \qquad (2)$$

where Ω is the Earth's angular velocity vector; r is the radius-vector from the Earth's centre; ρ_α is the mass density of α-gas; F_α is the force acting on the α-gas volume unit.

Heat balance equations:

$$\rho_\alpha c_{v\alpha} dT_\alpha/dt + p_\alpha \nabla \mathbf{V}_\alpha = \nabla(\lambda_\alpha \nabla T_\alpha) + P_{Q\alpha} - P_{L\alpha} + P_{T\alpha}, , \alpha = n, i, e, \quad (3)$$

where $c_{v\alpha}$ is the specific heat at constant volume of α-gas; λ_α is the thermal conductivity coefficient of α-gas; T_α is the α-gas temperature; p_α is the α-gas pressure; $P_{Q\alpha}$, $P_{L\alpha}$ are the rates of heating and heat loss of α-gas; $P_{T\alpha}$ – the rate of the elastic and inelastic exchange of heat between α-gas and other gases.

The equation for the potential φ of the electric field $\mathbf{E} = -\nabla\varphi$ is solved numerically taking into account the dynamo-action of the thermospheric winds:

$$\nabla [\hat{\sigma}(\nabla\varphi - \mathbf{V}\times\mathbf{B}) - \mathbf{j}_m - \mathbf{j}_s] = 0, \quad (4)$$

where $\hat{\sigma}$ is the ionospheric conductivity tensor; **V** is the neutral wind velocity vector; **B** is the magnetic field; $\mathbf{j}_m$ is the magnetospheric current density, $\mathbf{j}_s$ – the density of the seismogenic electric current flowing between the Earth and ionosphere.

The equations (1-3) take into account the photodissosiation of molecular oxygen, photoionization by the direct and scattered solar radiation and corpuscular ionization of neutral components, ion-molecular reactions, dissosiative recombination of molecular ions, charge exchange reactions and transfer processes (ambipolar diffusion, ion drag, ion-ion friction and electromagnetic drift) for atomic ions, neutral gas pressure gradients, molecular and turbulent diffusions, neutral-ion and viscous frictions, heating by UV and EUV solar radiation, Joule heating and heating by precipitating energetic particles, heat conductivity, heating of neutral gas due to chemical reactions and heat losses due to radiation.

The system of equations (1-4) is completed by initial and boundary conditions. At the upper boundary (h = 520 km) of the model neutral atmosphere we assume that:

$$\partial \mathbf{V}/\partial r = \partial T/\partial r = 0, \quad (5)$$

and all neutral components are in the diffusion equilibrium there. At the lower boundary (h = 80 km) the wind velocity is taken according to the geostrophical approximation and the neutral temperature and densities are taken from the NRLMSISE-00 empirical thermospheric model [Picone et al., 2002]. The neutral atmosphere and lower ionosphere (h = 80 – 175 km) parameters and electric field potential are calculated in the spherical geomagnetic coordinate system.

The upper ionosphere (h > 175 km), plasmasphere and magnetosphere parameters are calculated in the dipole geomagnetic coordinate system. The boundary conditions for the model ionosphere are given near the bases of the field lines in the northern and southern hemispheres at height of 175 km. The atomic ion densities at this boundary are obtained from photochemical equilibrium conditions. The values of the ion and electron temperatures at this boundary are calculated from the equations of heat balance. We assume that geomagnetic field lines with $L \geq 15$ (L – parameter of McIlwain) are open and ion densities and heat fluxes are set equal to zero at the altitude of $15R_E$, where R_E is the Earth radius.

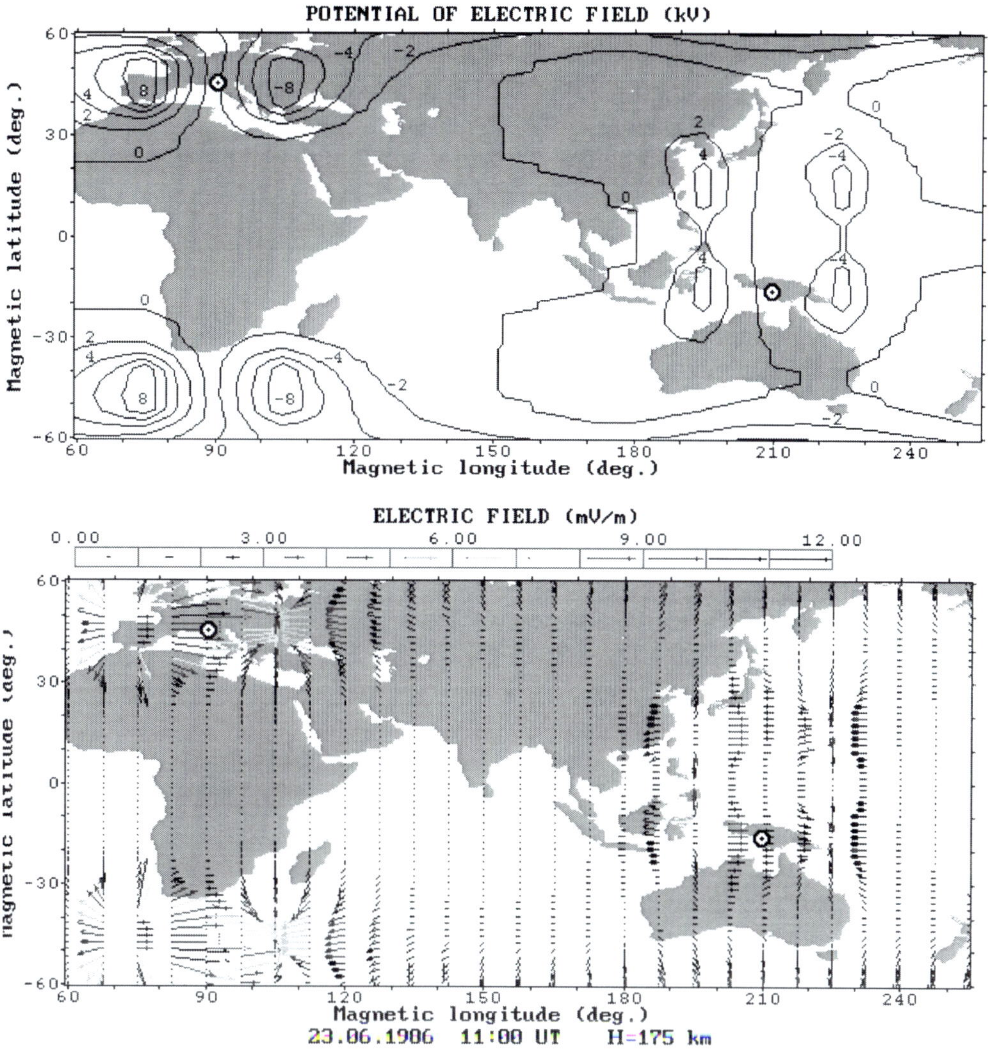

Figure 9. Model calculated electric potential (top panel) and horizontal electric field vectors (bottom panel) generated by the joint action of the background and additional (seismogenic) sources for two earthquake epicenters indicated by open circles with point. NB: The electric field of seismic origin appears at the near-epicenter region and at the magnetically conjugated area 'simultaneously' due to the very high ('infinite') conductivity of the dipole geomagnetic field lines. That electric field acts at both hemispheres simultaneously - at the epicenter and at the magnetically conjugated area. The ionospheric effects of such field may be non-symmetrical due to the differences in background conditions like illumination, etc.

Initial conditions are the stationary solution of the modelling equation system. To obtain this solution we integrate the (1-4) equations until the results of integration do not differ under continuating of integration.

The detailed expressions for all coefficients and terms of the equations (1-4) can be found in [Namgaladze et al., 1988; Brunelli and Namgaladze, 1988].

The UAM takes into account the non-coincidence of the geodetic and geomagnetic axes and gives the opportunity to model upper atmosphere parameters variations with the coincided axes. Besides the fully self-consistent version the model can be configured alternatively using the empirical models, for example, the model of neutral composition and temperatures NRLMSISE-00 and the empirical horizontal wind model HWM-93 [Hedin et al., 1996] as the corresponding model parts.

Firstly, the upper atmosphere state, presumably foregone a strong earthquake, was modeled [Namgaladze et al., 2009a, 2009b] by means of switching-on additional sources of electric field in the UAM electric potential equation which was solved numerically jointly with all other UAM equations (continuity, momentum and heat balance) for neutral and ionized gases. These sources were switched on and maintained permanently for 24 h in the form of additional positive and negative potentials with values of 2 and 5 kV (in case of the low-latitudinal sources) and of 10 kV (in case of the mid-latitudinal source) on the western and eastern boundaries of near-epicenter areas, respectively.

We investigated 2 near-epicenter areas [Namgaladze et al., 2009b] with sizes of 10 in latitude and 30 in longitude; the epicenters were situated at these points with the following magnetic coordinates: (1) (45N, 90); (2) (15S, 210). These sizes approximately correspond to the horizontal sizes of those regions of increased TEC values, which were found, e.g., in [Pulinets et al., 2003; Zakharenkova et al., 2007b]. The first region is a typical mid-latitude ionosphere, the second region is a near-equatorial ionosphere, for which the electric field effects are more essential than at middle latitudes. For the simulations we chose a magnetically quiet day of June solstice at high solar activity.

The calculated electric potential pattern and horizontal electric field vectors for the sources of 5 and 10 kV per mesh point are shown in Figure 9 for the quiet and seismo-disturbed conditions. It illustrates the regions, where an eastward electric field above the prospective epicenters of the future earthquakes appeared, when imposing (switching on) additional potentials. The appearance of similar magneto-conjugated regions in the opposite hemisphere is visible. The symmetry of the potential and electric field relative to the geomagnetic equator is due to the ideal electric conductivity of the plasma along the geomagnetic field lines and, accordingly, by their electric equipotentiality. The amplitudes of the additional eastward electric fields are about 2-4 mV/m in case of the low-latitude source and 4-10 mV/m in case of the mid-latitude source. They exceed the background quiet fields (of about 0.2 and 1 mV/m, correspondingly), but they are noticeably smaller than the quiet high-latitude electric fields of magnetospheric origin (15-25 mV/m) obtained in the model calculations.

The calculated ionospheric effects in TEC and foF2 (F2-layer critical frequency) created by the additional electric field sources are shown in Figures 10 and 11. As we can see in these Figures, the action of the near-equatorial source intensifies the equatorial anomaly of the F2-layer in the near-epicentral area of the ionosphere by deepening the minimum of f F2 over the magnetic equator and displacing the anomaly crests from the equator to the middle latitudes. This behavior agrees very well with the Alouette-1 and 2 observations [Depueva and Ruzhin, 1995; Depueva et al., 2007; Ruzhin and Depueva, 1996].

The action of the mid-latitudinal source leads to an increase of foF2 and TEC in the near-epicentral and in the magnetically conjugated areas. The amplitude of the enhancements and its spatial sizes are in a good agreement with the corresponding characteristics of precursors in the TEC observations. Although the electric field magnitudes in case of the mid-latitudinal source in our calculations are 2-5 times larger than in the low-latitudinal case, the evoked effects are weaker than in the near-equatorial regions.

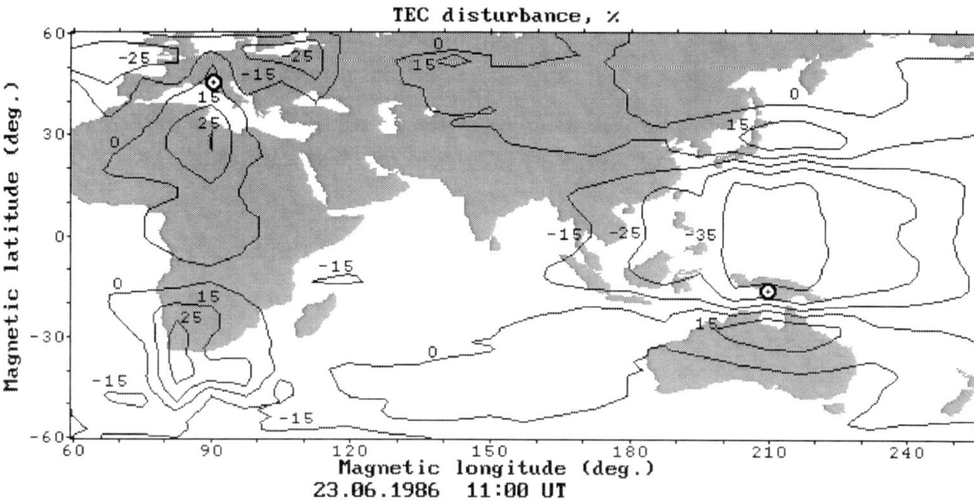

Figure 10. Model calculated relative TEC disturbances (%) created by the additional (seismogenic) sources for two earthquake epicenters indicated by the markers.

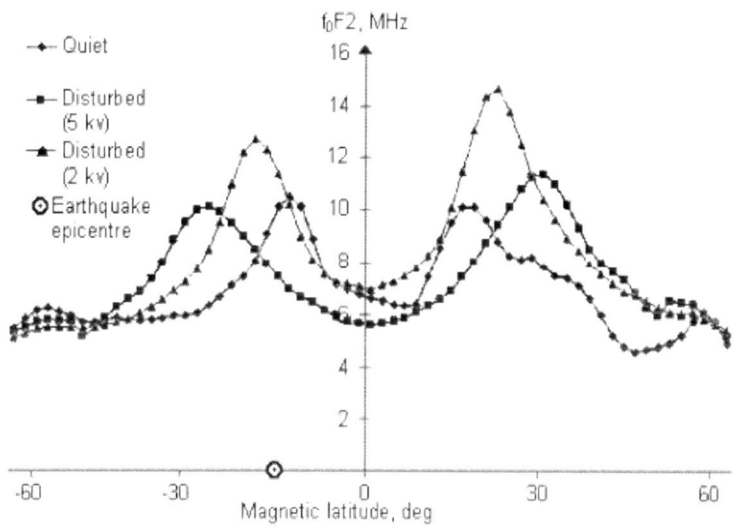

Figure 11. Model calculated latitudinal variations of the F2-layer critical frequencies along the magnetic meridian of the equatorial earthquake epicenter located at 15 magnetic latitude for quiet and disturbed conditions.

Presented in [Namgaladze et al., 2009a, 2009b] simulations in the first approximation proved that NmF2 and TEC disturbances observed before earthquakes can be generated by the vertical drift of the F2-region ionospheric plasma under the influence of a zonal electric field of seismogenic origin. In case of TEC enhancements in mid-latitudes and deepening of the equatorial anomaly trough, this field is directed to the east and induces an electromagnetic drift of the plasma across the geomagnetic field with velocity directed straight upwards over the magnetic equator and upwards and polewards at mid-latitudes. The pattern of spatial distribution of the seismogenic origin electric field potential has been proposed. For the existence of an eastward electric field in the near-epicentral area it is necessary to dispose positive electric charges on the western boundary of this area and negative charges on the eastern boundary.

However, we did not discuss the mechanism of the generation of the seismogenic electric field and we did not explain, why the positive charges should accumulate on the western boundary of the near-epicentral region, and negative charges on its eastern boundary or vice versa. We showed only what kinds of electric field were able to produce TEC and NmF2 disturbances observed before strong earthquakes.

The next step was to reproduce TEC variations before concrete earthquake events instead of synthetic model cases, and to investigate other types of electric field sources (dipole-like, positive and negative) and their spatial configurations. Numerical simulations were carried out for

1) The Peru September 26, 2005 earthquake. and
2) The Kythira seismic event on January 8, 2006, Greece, at mid-latitudes [Zakharenkova et al., 2008].

It was shown that in case of the Kythira earthquake a simulated dipole-like source agreed better with the experimental GPS TEC data than in cases of monopole (positive) additional sources. Variations of the TEC disturbances for the low-latitude region (Peru earthquake) had the form and the structure of the equatorial anomaly ("trough"). The increase of the eastward electric field leads to the deepening of the equatorial anomaly minimum ("trough" over the magnetic equator in the latitudinal distribution of electron density) due to an intensification of the fountain-effect. The dipole-like and single sign (positive) additional electric field sources generate similar disturbances in ionospheric TEC (in contrary to the mid-latitudinal cases). A noticeable difference turns up in the magnitude of these variations: the dipole-like sources generate a deeper equatorial trough. The amplitudes of the peaks (the "camel case" maxima) remain nearly at the same level or are slightly larger. That behavior agrees well (at least qualitatively) with the observed Peru pre-earthquake TEC variations.

Thus, both cases reveal a good agreement with the observations, the dipole-like and monopole sources have less differences in their manifestation at low latitudes.

It should be noticed that the applied "manually fixing" technique has two drawbacks:

1) It is hard to explain the nature of the "manually set" electric potential sources (which should be charges) on the borders of the near-epicenter area, especially their accumulation;
2) The approaching sunrise terminator and its corresponding changes of the ionospheric conductivity due to the well-conducting sunlit ionosphere should depress the additional electric potential and the corresponding disturbances up to full disappearance down to zero.

A more physical approach was realized by Sorokin et al. [2005a, 2006, 2007] by setting vertical electric currents flowing between the Earth and ionosphere at the considered region. This 'electric current technique' was used for the case of the Haiti Jan. 12 earthquake.

4. 3. TEC Disturbances Prior Haiti Earthquake of Jan. 12, 2010: UAM Modelling

In case of the 'vertical external electric current technique' the upper atmosphere state, presumably foregone a strong earthquake, is modeled by means of switching-on additional external electric current sources (but not additional electric potentials as in the 'electric potential technique', described in the previous case) at the lower boundary (80 km) in the UAM electric potential equation which was solved numerically jointly with all other UAM equations (continuity, momentum and heat balance) for neutral and ionized gases.

To estimate an upper limit for the magnitudes of the vertical electric current applied, we looked through the publications available. Sorokin et al. [2005a, 2006, 2007] calculated the ionospheric electric field related to external electric current variations in the lower atmosphere. This current is formed presumably due to the convective upward transport of charged aerosols and their gravitational sedimentation in the lower atmosphere. This effect is related to the occurrence of ionization sources due to seismic-related emanation of radon and other radioactive elements into the lower atmosphere. Freund [2011] proposed other sources of the near-ground atmosphere layer ionization – so called "positive holes" – which he expected to be significantly more efficient then the named radon-related ones. According to Sorokin et al. [2005a, 2005b, 2007], an external current density of about 10^{-6} A/m^2 within an area of about 200 km in radius (approximately 130 000 km^2) is required to create an electric field of several mV/m in the ionosphere.

To build model difference maps (of the electric potential, zonal and meridional electric field and TEC) we firstly performed a regular calculation without any additional electric current sources (set as lower boundary condition) to use the results as quiet background values. Then an external electric current flowing between the lower atmosphere and the ionosphere over the Haiti earthquake (Jan. 12, 2010, 21:53 UT, Mw 7.0) epicenter area has been used as model input for the calculations of the ionospheric electric field and the corresponding TEC variations. Several spatial configurations and magnitudes of these currents have been taken into consideration:

1) "Point" current (equivalent to one cell) sources of different signs and magnitudes, given in a single node of the numerical grid.
2) "Line" sources.

According to UAM simulations point current sources with magnitudes of about 10^{-5} A/m^2 and 10^{-6} A/m^2 given in a single grid node (corresponding to one grid cell of 2 × 5, and averaged vertical electric current densities of 5×10^{-6} A/m^2 and 5×10^{-7} A/m^2, respectively) induced very intense vertical drift motions and extremely strong and unrealistic TEC disturbances. Point sources of 10^{-9} A/m^2 and 5×10^{-9} A/m^2 triggered TEC disturbances not exceeding 15-30% in magnitude.

The "line" kind sources have been simulated as vertical external currents with a magnitude of 4×10^{-8} A/m^2 directed from the ionosphere to the Earth set centric at 9 numerical grid nodes with 5° longitudinal steps along the magnetic parallel of the earthquake epicenter. It corresponds to an external electric current density of 2×10^{-8} A/m^2 set on the region of approximately ~200 km × ~4000 km (2° along the meridian and 40° along the parallel). The generated TEC disturbances have reached ~20-50% by magnitude depending on the current's spatial distribution and lifetime. Simulation results are presented in Figure 12 and 13.

As one can see, the additional electric potential generated by the external current reduces down to zero when approaching the terminator in both (modelled and observed) cases. Corresponding TEC disturbances disappear later than the electric potential with a time-lag of about 4-6 hours. The simulations underestimate the observed magnitude of anomalous TEC increase at the northern and overestimate it at the southern hemisphere. Both (modelled and observed) cases show stronger TEC magnitudes at the magnetically conjugated region in comparison with the near-earthquake epicenter area.

The calculated latitude-altitude variations of the electron density for 08 UT, on Jan. 11, 2010 along the -20°, 0°, and +20° longitudes are presented in Figure 14. As it follows out of this Figure, the effects in comparison with non-disturbed conditions (i.e. without seismic electric current sources) are: an electron density 'trough' filling at the geomagnetic equator and formation of inhomogeneities. At -20° longitude a mid-latitude maximum stratification took place with a trough formation in the latitude range of ~30 -40°; at +20° longitude we see an electron concentration increase at 300 km height and a depletion of the magnetic tubes higher up. Main changes took place at ~200 – 600 km altitude range.

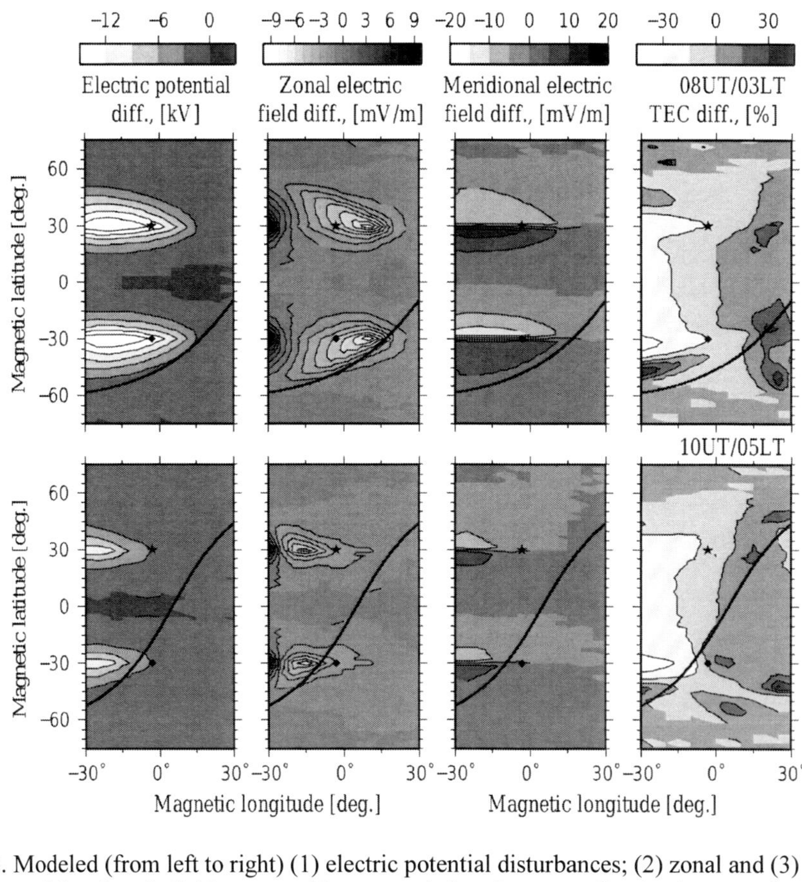

Figure 12. Modeled (from left to right) (1) electric potential disturbances; (2) zonal and (3) meridional components of the electric field generated by external electric current flowing between the Earth and the ionosphere and (4) regional map of the TEC deviations relative to the non-disturbed conditions for the 08UT/03LT (top panel) and 10UT/05LT (bottom panel). Star – the earthquake epicenter position (Haiti). Diamond – the magnetically conjugated point. Black curve – the terminator.

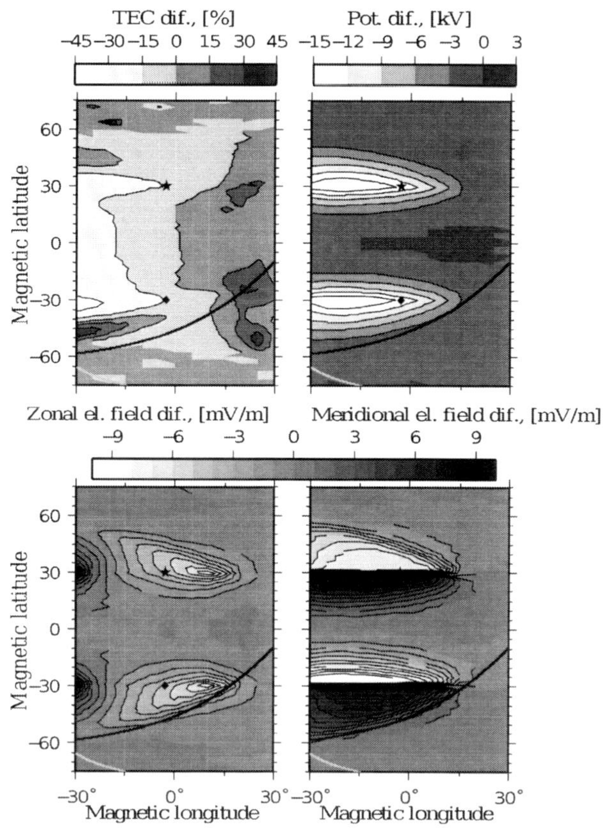

Figure 13. Modeled deviations from quiet background conditions: (1) TEC (top left); (2) electric potential (top right); (3) zonal and (4) meridional electric field for 08 UT (03 LT) of Jan. 11, 2010. Star – the earthquake epicenter position (Haiti). Diamond – the magnetically conjugated point. lg[$^{Ne, m^{-3}}$].

Summarizing the numerical simulation results with external electric currents flowing between the lower atmosphere and ionosphere due to seismo-induced conductivity changes of the underlying atmosphere column, one can state that they reproduce the main features of the TEC variations. In both the GPS observations and model cases we have:

1) Areas of increased TEC existed and are localized in the near-epicenter area.
2) Magnetic conjugation took place.
3) TEC disturbances at the magnetically conjugated area were stronger by magnitude than in the near-epicenter region.
4) The amplitude of TEC enhancements reached 50% and more.
5) Negative structures (areas of the TEC reduction) existed westward and equatorward relative to the positive ones.
6) Approaching the sunrise terminator with the well-conducting sunlit ionosphere triggered a shift of the anomaly to the night-side and causes the subsequent suppression of the TEC variations both at the epicenter and in the magnetically conjugated regions.
7) The anomalies did not exist during the near-noon hours. Such behavior and the lifetime agreed with [Akhoondzadeh and Saradjian, 2011] results. The obtained morphological features also agreed with [Pulinets and Tsybulya, 2010].

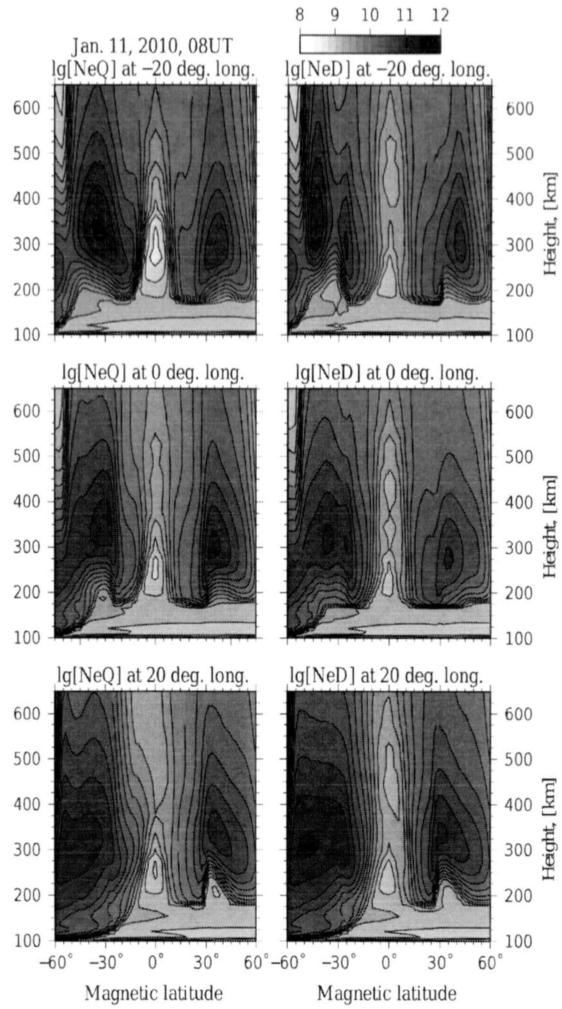

Figure 14. The latitude-altitude variations of the electron density along the geomagnetic meridians -20° (upper panel); 0° (middle panel); +20° (lower panel) calculated for Jan. 11, 2010, 08UT (03LT). At the left column – the quiet variations, at the right – the variations disturbed by the seismogenic electric field.

We should notice that a few discrepancies between the observations and the model results persist:

1) The simulations underestimate the amplitudes of the TEC anomaly at the northern hemisphere and overestimate them at the southern one in comparison with the observations.
2) Negative structures are stronger than the observed ones by magnitude in the model case.
3) The increased TEC area occupies a smaller region in the model case.
4) The isoline patterns and the magnitudes differ from the observed ones but not drastically. The discrepancies between the simulation results and the GPS observations could be due to the rather simplified assumptions about the external electric field sources, their density distribution and acting regime.

Night domination effects, the existence of a "ban-time" and the terminator-driven TEC anomaly suppression should be due to the changes of the conductivity of the ionosphere. It is related to the crossing of the sunrise terminator, where the well-conducting sunlit ionosphere leads to a depression of the electric potential disturbances and to a reduction of the electric field, generated by the externally driven electric current.

CONCLUSION

We have considered some recent results concerning the TEC anomalies observed before strong earthquakes, the modeling of the seismogenic ionospheric electric fields and their effects on TEC variations. Presented case studies of individual earthquakes confirm the traditional pre-earthquake TEC variations' signatures treated as precursors:

1) Anomalous strong (30-90% and more) TEC positive or negative deviations relative to the quiet background conditions before the forthcoming seismic event not less than Mw5 by magnitude linked to the near-epicenter area. These anomalies do not propagate along the meridians in contrast with the ionospheric disturbances related to magnetic activity.
2) The typical zone of the anomaly maximum manifestation extends larger than 1500 km in latitude and 3500-4000 km in longitude.
3) Anomaly living time is from several hours or days to couple of weeks before the earthquake release moment.
4) Analogous effects at the magnetically conjugated area are often reported.
5) In case of strong low-latitudinal earthquakes there are effects related to the modification of the ionospheric F2-region equatorial anomaly. From the analysis of the relative TEC deviations before a few strong recent seismic events (Jan. 12, 2010 Haiti; Jan. 1 and 2 2011 Argentina and Chile; March 11, 2011 Japan) these pre-earthquake TEC signatures are extended with terminator coming and 'ban'-time effects not presented earlier.

TEC is formed mainly by the F2-layer plasma density. NmF2 and TEC disturbances related to magnetic activity are created by the neutral atmosphere disturbances (neutral gas composition, internal gravity waves and winds), mainly horizontal motions and electric field variations. But it is impossible to localize neutral atmosphere disturbances at a limited area, they will propagate away from the source. We think the results reported by [Klimenko et al., 2011a, 2011b] to be doubtful: the physical nature of fixing AGW in a local region is unclear. The model technique used to set AGW as input needs more discussions: authors set oscillations in the given nodes, and those oscillations are for some unnamed reason do not propagate horizontally. Thus, it is not AGW but the manually set oscillations at the given grid nodes with the fixed amplitude.

We propose [Namgaladze et al., 2007] that the main reason for the appearance of local TEC anomalies in the form of an increased (decreased) total electron content of the ionosphere, observed on the base of GPS (Global Positioning System) signals measurements, is the vertical drift of F2-region ionospheric plasma upward (downward) under the influence of a zonal component of seismogenic electric field directed to the east (west). At middle latitudes the vertical upward component of the electromagnetic drift, created by an eastward electric field, leads to an increase of the F2 region electron concentration maximum (NmF2) and TEC due to the plasma transport up to regions with lower concentration of the neutral molecules O_2 and N_2 and, consequently, with lower loss rate of the dominating ions O^+ in the ion-molecular reactions [Brunelli and Namgaladze, 1998]. An electric field of the opposite direction (westward) creates the opposite – negative – effect in NmF2 and TEC. In the low latitude regions (near the geomagnetic equator) an increased eastward electric field leads to a deepening of the Appleton anomaly minimum ("trough" over the magnetic equator in the latitudinal distribution of electron concentration) due to the intensification of the fountain-effect. Therefore, we expect the local vertical F2-region plasma drift as the principle reason for the observed disturbances.

The reported night-time ULF anomalies also support this mechanism as magnetic tubes with heightened electron concentration correspond to the better ULF waves' propagation conditions.

The global numerical model of the Earth's upper atmosphere (UAM – Upper Atmosphere Model) was used for the presented investigations. Ionospheric electric field patterns were calculated, taking into account both the dynamo-action of neutral winds and the external current drivers at the lower and upper boundaries of the ionospheric dynamo-region. These currents transfer electric charges from the magnetosphere to the dynamo-region's upper boundary and from the near Earth's lower atmosphere to the lower boundary.

Electric fields generated by the interaction of the solar wind and its embedded interplanetary magnetic field with the geomagnetic field as well as geomagnetic field-aligned currents have been studied already many years, their existence is out of doubt. Very high electrical conductivity along the geomagnetic field lines ensures reliable electrical coupling between the magnetosphere and the ionosphere. Another situation takes place with the seismogenic electric fields due to very low (in comparison with the ionized layers above ~80 km) conductivity of the neutral air between the Earth and the ionosphere. Nevertheless it is not zero and varies significantly from place to place, for example, from "good weather" regions to thunderstorm areas. The average vertical electric current density between the Earth and the ionosphere is about 2-3 pA/m^2 and the corresponding average electric potential difference amounts to about 250-300 kV. In thunderstorm areas, the vertical electric current density increases by several orders in magnitude. The same takes place presumably near tectonic fault regions before earthquakes due to the motion of (or accumulated stresses between) tectonic plates.

Some of the recent results presented above on TEC variations before earthquakes, which were obtained using GPS data, give strong evidences in favor of the hypothesis on seismogenic electric fields as the main cause for these variations. The persistence near epicenter of the TEC anomalies not propagating along meridians, their geomagnetically conjugate occurrence, and their tendency to avoid sunlit time periods constitute these evidences. The presented above UAM calculations show that a vertical electric current of about 10^{-8} A/m^2 at an area of about 1000 km $\times$ 4000 km can create the electric fields of several mV/m in the nighttime ionosphere which in turn can produce the TEC variations of up to 50%, which turns out to be very similar to the observed ones. Such current requires an increase of the concentration of charged particles of about 10^4 times in comparison with the quiet conditions (if velocity does not change significantly). The obtained estimations of the external electric current of seismic origin are close to the upper limits of the measured vertical electric currents above the thunderstorm regions and tropical forests [Making and Ogawa, 1984; Davydenko et al., 2009; Le Mouël et al., 2010] but do not exceed them.

The electric coupling between the tectonic faults and the ionosphere requires the presence of emerging electric charges (or ionization sources) near the faults. Two mechanisms were considered in the last years: the ionization by radioactive radon emission and the "positive hole" mechanism by Freund. Possibly, they are not competing but both acting. The detailed physics and chemistry of the lower (near the Earth surface) atmosphere in presence of these ionization sources should be developed in the future.

ACKNOWLEDGMENTS

The authors acknowledge:

1) NOAA National Geophysical Data Center / NGDC and the World Data Center for Geomagnetism, Kyoto for the A_p, K_p and D_{st} indexes' data.
2) US Geological Survey and National Earthquake Information Center for the earthquakes' data, URL: http://earthquake.usgs.gov/.
3) IGS community and Crustal Dynamics Data Information System (CDDIS) for the TEC Global Ionosphere Maps [Noll et al., 2009; Noll, 2010].
4) University of Hawaii for Generic Mapping Tools [Wessel and Smith, 1998]. The authors are grateful to Irina E. Zakharenkova [West Department of Pushkov Institute of Earth Magnetism, Ionosphere and Radiowaves Propagation of the Russian Academy of Science, Kaliningrad, Russia] and Boris E. Prokhorov [Helmholtz Centre Potsdam, GFZ German Research Centre for Geosciences; University Potsdam, Applied Mathematics, Interdisciplinary Center for Dynamics of Complex Systems (DYCOS)] for the help in TEC relative deviations (%) maps processing. We thank Dr. K.I. Konstantinou (Associate Professor, Dept of Earth Sciences, National Central University, Taiwan) for the useful comments that helped us to improve the manuscript.

REFERENCES

Afraimovich, E. L., Astafyeva, E. I., Oinats, A. V., Vasukevich, Yu.V., Zhivetiev, I. V.: Global electron content as a new index of solar activity. Comparison with I. R. I. modeling results. *I. R. I. News*. 13. N. 1. (2006).

Afraimovich, E. L., Astafieva, E. I., Gokhberg, M. B., Lapshin, V. M., Permyakova, V. E., Steblov, G. M., Shalimov, S. L.: Variations of the total electron content in the ionosphere from GPS data recorded during the Hector Mine earthquake of October 16, 1999, California. *Russian Journal of Earth Sci*. 6. N. 5. 339-354 (2004).

Akhoondzadeh, M., Saradjian, M. R.: T. E. C. variations analysis concerning Haiti (January 12, 2010) and Samoa (September 29, 2009) earthquakes. *Adv. Space Res*. 47. N. 1. 94-104 (2011).

Brunelli, B. E., Namgaladze, A. A.: Fizika ionosfery. *Nauka*, Moscow. 527 p. (1988). (in Russian).

Buchachenko, A. L., Oraevskii, V. N., Pokhotelov, O. A., Sorokin V. M., Strakhov, V. N., Chmyrev, V. M.: Ionospheric precursors to earthquakes. *Phys. Usp.* 39 959–965 (1996) doi: 10.1070/PU1996v039n09ABEH001550.

Chmyrev, V. M., Isaev, N. V., Bilichenko, S. V., Stanev, G. A.: Observation by space-borne detectors of electric fields and hydromagnetic waves in the ionosphere over on earthquake center. *Phys. Earth Planet. Inter.* 57. 110-114 (1989).

Davydenko, S. S., Thomas, C. M., Maribeth, S. Modeling the electric structures of two thunderstorms and their contributions to the global circuit. *Atmos. Res.* 91(2-4). 165-177. (2009).

Depueva, A. Kh., Ruzhin, Yu.Ya.: Seismoionospheric fountain-effect as analogue of active space experiment. *Adv. Space Res.* 15. N 12. 151-154 (1995).

Depueva, A. Kh., Mikhailov, A. V., Devi, M., Barbara, A. K.: Spatial and time variations in critical frequencies of the ionospheric F region above the zone of equatorial earthquake preparation. *Geomagnetism and Aeronomy*. 47. N. 1. 129-133 (2007). doi:10.1134/S0016793207010197.

Dow, J. M., Neilan, R. E., Rizos, C.: The International G. N. S. S. Service in a changing landscape of Global Navigation Satellite Systems. *J. of Geodesy*. 191-198 (2009). doi:10.1007/s00190-008-0300-3.

Förster, M., Paschmann, G., Haaland, S. E., Quinn, J. M., Torbert, R. B., McIlwain, C. E., Vaith, H., Puhl-Quinn, P. A., Kletzing, C. A.: High-latitude plasma convection from Cluster E. D. I.: Variances and solar wind correlations. *Ann. Geophys.* 25. N. 7. 1691-1707 (2007).

Freund F.: Pre-earthquake signals: Underlying physical processes. *J. Asian Earth Sci.* (2011). doi:10.1016/j.jseaes.2010.03.009.

Fraser-Smith, A. C., Bernardy, A., McGill, P. R., Ladd, M. E., Helliwell, R. A., Villard Jr., O. G.: Low frequency magnetic field measurements near the epicenter of the Loma-Prieta earthquake. *Geophys. Res. Lett.* 17. N. 9. 1465–1468. (1990). doi: 10.1029/GL017i009p01465.

Gokhberg, M. B., Morgounov, V. A., Yoshino, T., Tomizawa, I.: Experimental measurement of electro-magnetic emissions possibly related to earthquakes in Japan. *J. Geophys. Res.* 87. 7824- 7828. (1982). doi: 10.1029/JB087iB09p07824.

Gokhberg, M. B., Pilipenko, V. A., Pokhotelov, O. A.:On the seismic precursors within the ionosphere. Izvestiya Academy of Sciences U. S. S. R., *Series Physics of the Earth*. 10. 17–21 (1983) (in Russian).

Gokhberg, M. B., Kustov, A. V., Liperovsky, V. A., Liperovskaya, R. Kh., Kharin, E. P., Shalimov, S. L.: About disturbances in F-region of ionosphere before strong earthquakes. Izvestiya Academy of Sciences U. S. S. R., *Series Physics of the Earth*. 4. 12–20 (1988) (in Russian).

Hayakawa, M., Molchanov, O. A., NASDA/UEC team: Summary report of NASDA's earthquake remote sensing frontier project. *Phys. Chem. Earth.* 29. 617-625. (2004). doi:10.1016/j.pce.2003.08.062.

Haaland, S. E., Paschmann, G., Sonnerup, B. U. Ö.: Comment on "A new interpretation of Weimer et al.'s solar wind propagation delay technique" by Bargatze et al. *J. Geophys. Res.* 111. A06102 (2006). doi:10.1029/2005JA011376.

Harrison, R. G., Aplin, K. L., Rycroft, M. J.: Atmospheric electricity coupling between earthquake regions and the ionosphere. *J. Atmos. Solar-Terr. Phys.* 72. N. 5-6. 376-381 (2010).

Hobara, Y., Koons, H. C., Roeder, J. L., Yumoto, K., Hayakawa, M.: Characteristics of U. L. F. magnetic anomaly before earthquakes. *Phys. Chem. Earth.* 29 (2004). 437–444. doi: 10.1016/j.pce.2003.12.005.

Hegai, V. V., Kim, V. P., Nikiforova, L. I.: A possible generation mechanism of acoustic-gravity waves in the ionosphere before strong earthquakes. *J. Earthquake Predict. Res.* 6. 584-589. (1997).

Hedin, A. E., Biondi, M. A., Burnside, R. G., Hernandez, G., Johnson, R. M., Killeen, T. L., Mazaudier, C., Meriwether, J. W., Salah, J. E., Sica, R. J., Smith, R. W., Spencer, N. W., Wickwar, V. B., Virdi, T. S.: Revised Global Model of Thermospheric Winds Using Satellite and Ground-Based Observations. *J. Geophys. Res.* 96. 7657–7688 (1996).

Kim, V. P., Pulinets, S. A., Hegai, V. V.: Theoretical model of possible disturbances in the nighttime mid-latitude ionospheric D-region over an area of strong earthquake preparation. Radiophys. *Quantum Electronics.* 45. 262-268. (2002).

Kopytenko, Yu. A., Ismaguilov, V. S., Hattori, K., Hayakawa M.: Determination of hearth position of a forthcoming strong E. Q. using gradients and phase velocities of U. L. F. geomagnetic disturbances. *Phys. Chem. Earth.* 31. 292-298. (2006). doi:10.1016/j.pce.2006.02.004.

Krankowski, A., Zakharenkova, I. E., Shagimuratov, I. I.: Response of the ionosphere to the Baltic Sea earthquake of 21 September 2004. *Acta Geophysica.* 54. N. 1. 90-101 (2006).

Klimenko, M. V., Klimenko, V. V., Zakharenkova, I. E., Pulinets, S. A., Zhao, B., Tsidilina, M. N.: Formation mechanism of great positive T. E. C. disturbances prior to Wenchuan earthquake on May 12, 2008. *Adv. Space. Res.* 48. 488-499. (2011a). doi:10.1016/j.asr.2011.03.040.

Klimenko, M. V., Klimenko, V. V., Karpov, I. V., Zakharenkova, I. E.: Simulation of Seismo-Ionospheric Effects Initiated by Internal Gravity Waves. *Russian J. Phys. Chem.* B. 5. N. 3. 393–401. (2011b). doi: 10.1134/S1990793111030109.

Larkina, V. I., Migulin, V. V., Molchanov, O. A., Kharkov, I. P., Inchin, A. S., Schvetcova, V. B.: Some statistical results on very low frequency radiowave emissions in the upper ionosphere over earthquake zones. *Physics of the Earth and Planetary Interiors.* 57. 100—109. (1989). doi:10.1016/0031-9201(89)90219-7.

Le Mouël, J.-L., Gibert, D., Poirier, J.-P. On transient electric potential variations in a standing tree and atmospheric electricity. *Comptes Rendus Geosciences.* 342(2). 95-99. (2010).

Liu, J. Y., Chuo, Y. J., Shan, S. J., Tsai, Y. B., Pulinets, S. A., Yu, S. B.: Pre-earthquake ionospheric anomalies monitored by G. P. S. *T. E. C. Annales Geophys.* 22. 1585-1593 (2004).

Liu, J. Y., Chen, Y. I., Chuo, Y. J., Chen, C. S.: A statistical investigation of preearthquake ionospheric anomaly. *J. Geophys. Res.* 111. A05304 (2006). doi:10.1029/2005JA011333.

Liu, J. Y., Chen, Y. I.: Precursors in ionospheric G. P. S. T. E. C. of the 26 December 2004 M9.3 Sumatra earthquake. XXIV I. U. G. G. General Assembly. Perugia, Italy. Session JSS010. (2007).

Rycroft, M. J., Odzimek, A., Arnold, N. F., Füllekrug, M., Kułak, A., Neubert, T.: New model simulations of the global atmospheric electric circuit driven by thunderstorms and electrified shower clouds: The roles of lightning and sprites. *J. Atm. Solar-Terr.Phys.* 69. 2485-2509. (2007). doi:10.1016/j.jastp.2007.09.004.

Making, M., Ogawa, T. Responses of atmospheric electric field and air-earth current to variations of conductivity profiles. *J. Atmos. Terr. Phys.* 46(5). 431-445. (1984).

Molchanov, O., Rozhnoi, A., Solovieva, M., Akentieva, O., Berthelier, J. J., Parrot, M., Lefeuvre, F., Biagi, P. F., Castellana, L., Hayakawa, M.: Global diagnostics of the ionospheric perturbations related to the seismic activity using the V. L. F. radio signals collected on the D. E. M. E. T. E. R. satellite. *Nat. Haz. Earth System Sci.* 6. 745–753. (2006).

Mullayarov, V. A., Karimov, R. R., Kozlov, V. I.: Variations in thunderstorm V. L. F. emissions propagationg over the epicenters of earthquakes. *J. Atm. Solar-Terr. Phys.* 69. 1513-1523. (2007).

Namgaladze, A. A., Korenkov, Yu. N., Klimenko, V. V., Karpov, I. V., Bessarab, F. S., Surotkin, V. A., Glushchenko, T. A., Naumova, N. M.: Global model of the thermosphere-ionosphere-protonosphere system. *Pure and Applied Geophysics.* 127. N. 2/3. 219-254 (1988). doi:10.1007/BF00879812.

Namgaladze, A. A., Korenkov, Yu. N., Klimenko, V. V., Karpov, I. V., Surotkin, V. A., Naumova, N. M.: Numerical modeling of the thermosphere-ionosphere-protonosphere system. *J. Atmos. Terr. Phys.* 53. N. 11/12. 1113-1124 (1991). doi:10.1016/0021-9169(91)90060-K.

Namgaladze, A. A., Martynenko, O. V., Namgaladze, A. N.: Global model of the upper atmosphere with variable latitudinal integration step. *Int. J. of Geomagnetism and Aeronomy*. 1. N. 1. 53–58 (1998a).

Namgaladze, A. A., Martynenko, O. V., Volkov, M. A., Namgaladze, A. N., Yurik, R. Yu.: High-latitude version of the global numeric model of the Earth's upper atmosphere. *Proc. of the M. S. T. U.* 1. N. 2. 23-84 (1998b). URL: *http://goo.gl/8x9f2*.

Namgaladze, A. A., Shagimuratov, I. I., Zakharenkova, I. E., Zolotov, O. V., Martynenko, O. V.: Possible mechanism of the TEC enhancements observed before earthquakes. In: Proc. XXIV IUGG General Assembly. Perugia, Italy. (2007).

Namgaladze, A. A., Klimenko, M. V., Klimenko, V. V., Zakharenkova, I. E.: Physical mechanism and mathematical modeling of earthquake ionospheric precursors registered in total electron content. *Geomagnetism and Aeronomy.* 49. N. 2. 252-262 (2009a). doi:10.1134/S0016793209020169.

Namgaladze, A. A., Zolotov, O. V., Zakharenkova, I. E., Shagimuratov, I. I., Martynenko, O. V.: Ionospheric total electron content variations observed before earthquakes: Possible physical mechanism and modeling. *Proc. of the M. S. T. U.* 12. N. 2. 308-315 (2009b). URL: *http://goo.gl/A8cLx.*

Namgaladze, A. A., Zolotov, O. V., Prokhorov, B. E.: The TEC signatures as strong seismic event precursors. In: Proc. XXX General Assembly of URSI Istanbul, Turkey. (2011).

Noll, C., Bock, Y., Habrich, H., Moore, A.: Development of data infrastructure to support scientific analysis for the International G. N. S. S. Service. *J. of Geodesy.* 309–325 (2009).

Noll, C.: The Crustal Dynamics Data Information System: A resource to support scientific analysis using space geodesy. *Adv. Space Res.* 45. N. 12. 1421-1440 (2010).

Picone, J. M., Hedin, A. E., Drob, D. P., Aikin, A. C.: NRLMSISE-00 empirical model of the atmosphere: Statistical comparisons and scientific issues. *J. Geophys. Res. A* 107. 1468. (2002). doi: 10.1029/2002JA009430.

Pulinets S. A.: Seismic activity as a source of the ionospheric variability. *Adv. Space Res.* 22. N. 6. 903-906 (1998).

Pulinets, S. A., Legen'ka, A. D.: Dynamics of equatorial ionosphere during the preparation of strong earthquakes. *Geomagnetism and Aeronomy*. 42. N 2. 239-244 (2002). (in Russian).

Pulinets, S. A., Legen'ka, A. D., Gaivoronskaya, T. V., Depuev, V. Kh.: Main phenomenological features of ionospheric precursors of strong earthquakes. *J. Atmos. Solar-Terr. Physics*. 65. 1337-1347 (2003).

Pulinets, S. A., Boyarchuk, K.: *Ionospheric Precursors of Earthquakes*. Springer, Berlin, Germany. 315 p. (2004).

Pulinets, S. A., Leyva Contreras, A., Bisiacchi-Giraldi, G., Ciraolo, L.: Total electron content variations in the ionosphere before the Colima, Mexico, earthquake of 21 January 2003. *Geofisica Internacional*. 44. N. 4. 369-377. (2005).

Pulinets, S., Ouzounov, D.: Lithosphere–Atmosphere–Ionosphere Coupling (L. A. I. C.) model – An unified concept for earthquake precursors validation. *J. Asian Earth Sci.* (2010). doi:10.1016/j.jseaes.2010.03.005 (in Press).

Pulinets, S. A., Tsybulya, K. G. Unique variations of the total electron content in the preparation period of Haitian earthquake (M7.9) on January 12, 2010. *Geomagnetism and Aeronomy*. 50. N. 5. 686-689 (2010). doi: 10.1134/S0016793210050166.

Ruzhin, Yu. Ya., Depueva, A. Kh.: Seismoprecursors in space as plasma and wave anomalies. *J. Atmosph. Electr.* 16. N. 3. 251-288 (1996).

Ruzhin, Yu. Ya., Oraevsky, V. N., Shagimuratov, I. I., Sinelnikov, V. M.: Ionospheric precursors of earthquakes revealed from GPS data and their connection with "sea-land" boundary. Proceed. *16th Wroclaw E. M. C. Symposium*. 723-726. (2002).

Sorokin, V. M., Chmyrev, V. M.: Modification of the ionosphere by seismic related electric field. In: Hayakawa, M. (Ed.) Atmospheric and ionospheric electromagnetic phenomena associated with earthquakes. Terra Scientific Publishing Company (T. E. R. R. A. P. U. B.), Tokyo. 805–818 (1999).

Sorokin, V. M., Chmyrev, V. M., Yaschenko, A. K.: Theoretical model of DC electric field formation in the ionosphere stimulated by seismic activity. *J. Atmos. Solar-Terr. Phys.* 67. 1259-1268 (2005a).

Sorokin, V. M., Yaschenko, A. K., Chmyrev, V. M., and Hayakawa, M.: D. C. electric field amplification in the mid-latitude ionosphere over seismically active faults. *Nat. Haz. Earth Sysem Sci.* 5. 661–666 (2005b).

Sorokin, V. M., Yaschenko, A. K., Hayakawa, M.: Formation mechanism of the lower ionosphere disturbances by the atmosphere electric current over a seismic region. *J. Atmos. Solr-Terr. Phys.* 68. 1260-1268 (2006).

Sorokin, V. M., Yaschenko, A. K., Hayakawa, M.: A perturbation of D. C. electric field caused by light ion adhesion to aerosols during the growth in seismic-related atmospheric radioactivity. *Nat. Haz. Earth System Sci.* 7. 155–163 (2007).

Wessel, P., Smith, W. H. F.: New, improved version of Generic Mapping Tools released. *E. O. S. trans.* 79. 579 (1998).

Zakharenkova, I. E., Shagimuratov, I. I., Krankowski, A., Lagovsky, A. F.: Precursory phenomena observed in the total electron content measurements before great Hokkaido earthquake of September 25, 2003 (M=8.3). *Studia Geophysica et Geodaetica*. 51. N. 2. 267-278. (2007a).

Zakharenkova, I. E., Shagimuratov, I. I., Krankowski, A.: Features of the ionosphere behavior before Kythira 2006 earthquake. *Acta Geophysica*. 55. N. 4. 524-534 (2007b).

Zakharenkova, I. E., Application of the G. P. S.-signals measurements for ionospheric earthquakes precursors' detection, PhD thesis, Immanuel Kant State University, Kaliningrad, Russia. (2007). (in Russian).

Zakharenkova, I. E., Zolotov, O. V., Namgaladze, A. A., Shagimuratov, I. I., Martynenko, O. V.: Preseismic Modification of the Ionosphere for Greece 2006 Earthquake: G. P. S. T. E. C. Measurements and Modeling Results. In: Proc. International Conference "Fundamental Space Research". Sunny Beach, Bulgaria. pp. 371-374 (2008).

Zolotov, O. V., Prokhorov, B. E., Namgaladze, A. A., Martynenko, O. V.: Variatsii polnogo electronnogo soderjanija ionosferi v period podgotovki zemletrjasenii. *Khimicheskaja Fizika*. 30. N. 5. 84-87. (2011a). (in Russian).

Zolotov, O. V., Prokhorov, B. E., Namgaladze, A. A., Martynenko, O. V.: Variations of the ionosphere Total Electron Content before earthquakes. *Chem. Phys*. 5. N. 3. 435-438. (2011b). doi: 10.1134/S1990793111030146.

In: Earthquakes
Editor: Kostas Konstantinou

ISBN: 978-1-62081-883-1

Chapter 8

Development of Multi-Component Borehole Instrument for Continuous Monitoring of Crustal Deformation

Zuxi Ouyang
Institute of Crustal Dynamics, China Earthquake Administration, Beijing, China

Abstract

Deep borehole observation can eliminate effectively various Earth's surface interferences, providing considerable advantage over other geophysical measurements made at the ground, and improve greatly our capability to detect very small deformations and vibrations in the surrounding strata. This new technology will promote consequently our understanding of presently active geological processes and its relation to earthquakes and tectonics.

A new RZB multi-component borehole instrument is an integrated packages consisting of six channels for the strainmeters (4 horizontal components, 1 vertical component, and 1 for system reference), two channels for the tiltmeters, three channels for broadband strain seismometers, and one channel for the ground temperature. Furthermore, another four channels provide the assistant observation including well temperature, well water level, atmospheric pressure and electronic compass.

The objective of setting four strainmeters in the horizontal plain is to measure changes in four diameters apart from each other at 45° of the cylinder, which allows us to record shear strain and areal strain changes. Each measurement unit for strain, tilt, and broadband strain seismometer as well as ground temperature consists of transducers and a circuit board distributing data highway and power to individual gauges assembled in a stainless-steel pressure cylinder with length of 1m and outside diameter of 102mm. A cable, connecting the sensors up hole, provides simultaneously a digital data connection via an RS-485 link and an analogue signal transmission line for seismometers. All data are stored in digital format in the acquisition system at the ground with 1-Hz sampling rate for strain, tilt and temperature, 200-Hz sampling rate for broadband strain seismometer. Flexible combination of diversified sensors and their duplicate parts enables us to make good project that can adapt to differing circumstances.

Because of its principle of operation, multi-component borehole instrument is permanently cemented in the bottom of an open hole to assure good coupling with the surrounding rock in order to get meaningful information about changes in the Crustal stress state. These new borehole instruments have been installed at stations in Beijing and Fujian, Sichuan, Yunnan provinces. The instruments can run at high sensitivity to provide strain resolution of 0.1 nanostrains with basic dynamic range of ±500 microstrain and tilt resolution of 1 nanorad with basic dynamic range of ±500 microrad. Up to now these instruments have observed good tidal signals, strain-steps associated with large earthquakes, and strain changes of geophysical interests.

1. INTRODUCTION

It is well known that the stress state of the Earth's Crust is very important aspect, because continuous increase in the intensity of the stress state may result in an occurrence of an earthquake [Silver, 1998; Dragert, 2001]. The seismologists all over the world therefore paid much attention to the problem of continuous monitoring of the changes in the Crustal strain filed in order to study earthquake prediction.

In the last half century borehole observation techniques to monitor changes in the Earth's strain field as a function of time has experienced three develop steps in China. In the 1960s magneto elastic stress transducers were introduce into China and used in measurements of Crustal stress state, including in engineering geology application and earthquake prediction study [Hast, 1958]. Some measurement data and valuable experiences were accumulated. The measurement resolution of magneto-elastic stress transducers was about 0.1kgf/cm^2, equivalent to strain changes of ~10-7 in rocks. A mechanical clamping system was employed to maintain sensor's contact with the surrounding rock. It was also realized later on that the measurement resolution of the transducer used was low and the contact condition by mechanical installation method did not reach the level of stability required for good quality data. The contemporary research into continuous borehole observation instrument of Crustal deformation abroad was the Sacks-Everson dilatometer as a representative example [Sacks et al., 1971]. The dilatometer has high strain sensitivity, as 10^{-10}~10^{-11} , to detect volume changes in the surrounding rock, but the observed data present neither changes of principle strains nor the information about the shear strains in the strain state, thus some deficiency restricted its application in practice. This is the first phase.

To 1980s Z.OUYANG and Q.WANG developed successively "RZB-1 Capacitance borehole strainmeter" [Ouyang, 1977; Ouyang et al., 1988], "ZX-79 Vibration string borehole strainmeter" [Wang, 1985]. M. Gladwin produced GTSM instrument, also entitled "Tensor strainmeter" [Gladwin, 1984]. Based on the Sacks-Everson dilatometer, S. Sakata invented a borehole three-component strainmeter [Sakata, 1981]. These newly developed borehole instruments, representing the developmental direction of borehole observation technology then, were characterized by high strain sensitivity, capability of measuring changes of all horizontal strain components, in-situ calibration of the sensitivity factor and downhole installation with cement. This is the phase two of the technology development.

United States brought forward a grand EarthScope Project in the turn of the century [Earth Scope Project Plan, 2001]. The project is made up of following 4 parts: US Array, PBO, SAFOD and InSAR. The geodetic component of EarthScope, PBO-A Plate Boundary Observatory-consists of two elements, which are GPS network of 975 receivers and borehole strainmeter network of ~200 instruments, as well as 5 Long-Base Laser Strainmeters. Up to

July 2010, 87 GTSM Tensor Strainmeters have been installed in boreholes at a depth between 150-250m, according to a unitive standard. Seismometer, pore pressure transducer, borehole tiltmeter and GPS receiver were installed successively in the same borehole at some sites. A multifunctional geophysical observation station is ultimately setting up based on application of multifold borehole instruments [PBO whitepaper, 1999; Silver et al., 1998]. Unlike the technology employed in USA, a new type of borehole instruments "The multi-component observation system of Crustal activity" was developed in Japan near the end of last century. H. Ishii developed the new multi-component borehole instrument in which manifold sensors for monitoring of strain, tilt, acceleration, geomagnetism and ground temperature, as well as associated measurement electronic units were integrated in an instrument assembly, and carried out applications at dozens of boreholes. These instruments are installed in boreholes at a depth between 150m~800 m, the deepest one among them up to now being 1200m [Ishii et al. 2001, 2003].

Based on previous experiences in developing RZB-1 borehole strainmeter, a new multi- component borehole observation system for deep borehole observatories has been developed in China in recent years. This new device, namely RZB-3, is an integrated borehole instrument including of strain, tilt, strain-seismometer, ground temperature observation units and an electronic compass. The downhole measuring device resides near the bottom of the hole, in which the installation depth is 150~300m to isolate the instrument from most of the human and environmental induced disturbances [Ouyang, 1977; Ouyang et al., 1988; 2009; 2010]. Some new RZB-3 instruments were installed at a few sites of tectonic interests in Beijing and Fujian, Sichuan, Yunnan provinces. The instruments have observed strain and tilt tidal signals, strain-steps associated with large earthquakes, and strain seismic waves.

Table 1. Types of Multi-component Observation

Instrument	Sensors/Component Integrated	Resolution / Dynamic Range	Integration Type	Reference
PBO Borehole Observatory.	Strain / 3-horizontal, 1-redundant Tllt / 2 Seismometer / 3 Pore pressure / 1 GPS	10^{-10}/ 1×10^{-3}	Integration In Borehole.	2004
Multi-component Borehole Instrument.	Strain / 3-horizontal 2-inclined,1-vetical Tilt / 2 Temperature / 1 Accelerometer / 3 Magnetic field / 4	10^{-9}	Integration In Instrument.	2001
RZB-3 Multi-component Borehole Observation System.	Strain / 4-horizontal 1-vertical,1-refferance Tilt / 2 Temperature / 1 Strain Seismometer / 3 Electronic Compass / 1	10^{-10}/$\pm0.5\times10^{-3}$ 10^{-9}rad/$\pm0.5\times10^{-3}$rad 10^{-3} ℃	Integration In Instrument.	2008

The technology development of borehole observation techniques in the third phase may be characterized as integration of manifold borehole geophysical observation devices and observation in deep boreholes. There are two ways to achieve multi component wide band observation of Crustal deformation in a deep borehole (Table 1). Obviously, integration in an instrument has advantages of system reliability and easy installation. At present the content of downhole multi-component observation system mainly includes strainmeter, tiltmeter, seismometer, strain seismometer, magnetometer, geothermometer, electronic compass and pore pressure gage, and the downhole instrument was to be installed at depth about 150-600m.

In this chapter we describe the basic types of deep borehole multi-component observatory developed recently at United States, Japan and China, especially the introduction to the construction, characteristics and primary application of Chinese RZB borehole instruments. We start in §2.1 with a discussion of the system outline, system integration technique and its reliability are discussed in §2.2. Data and primary results are examined in §3.1-§3.4, in each case, the discussion will be quite general, focusing on the features of the new borehole multi-component instrument.

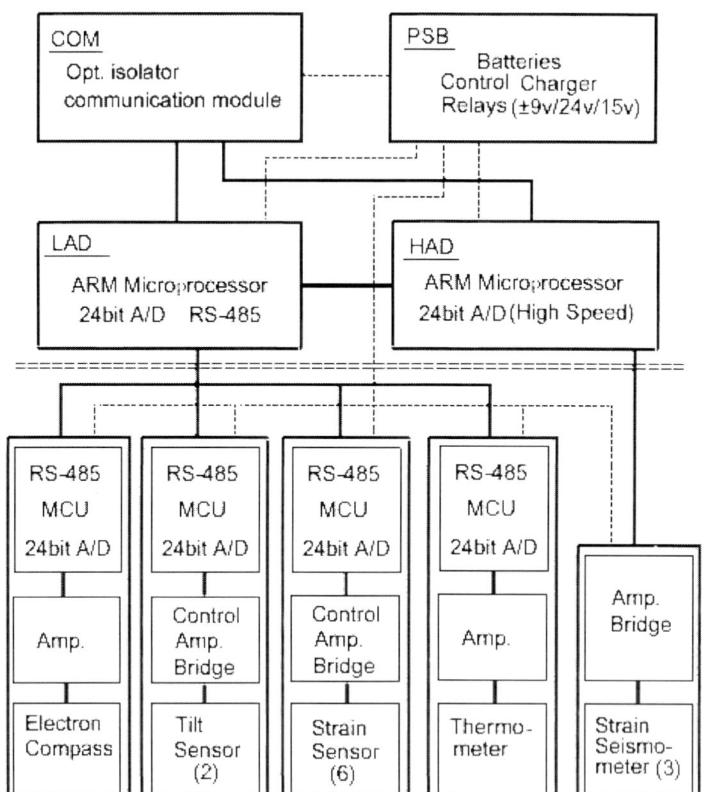

Figure 1. The schematic diagram of new RZB-3 system. The system consists of downhole packages and ground electronic box. These two parts are connected through the RS-485 Standard Data bus. The dashed horizontal line separates downhole components from those on the ground. Thin lines= power distribution; and thick lines=analog and digital signal lines; Amp=Amplifier; A/D= analog-to-digital converter; MCU=microprocessor circuit unit; LAD=low-sampling rate A/D; HAD=high sampling rate A/D; COM=communication unit; PSB=power supply box.

2. RZB-3 Multi-Component Borehole Observation System

2. 1. System Outline

New RZB-3 Multi-component borehole observation system consists of a downhole measuring instrument and a surface electronic box (Figure 1). The downhole integrated instrument is composed of a strain package (6 strain sensors: 4 for horizontal, 1 for vertical and 1 for system reference), a tilt package (2 tilt sensors), a strain seismometer package (3 strain sensors), a thermometer, and an electronic compass, housed in a compositive cylinder with outer diameter of 102 mm and length of 1500mm ~ 5200mm, depending on how many observation devices are incorporated in the instrument. The downhole instrument is coupled with the surrounding rock using special cement and scaled for deployment in 130-146mm borehole. The installation depth for the system is about 150-300m.

The RZB-3 strainmeter package resembles the RZB-1 strainmeter, using the same principle and mechanical design. The capacitance displacement transducer with sensitivity better than 0.01 nm becomes preferred sensors for application in designs of strainmeters, tiltmeters and strain seismometers. Differential capacitance transducers were adopted to measure plate displacement caused by deformation or tilt of the borehole. Figure 2(A) shows the schematic representation of the principle of operation of the strain sensors. The center capacitive plate, supported on a base, with a pair of closely spaced capacitor plates, mounted on an opposite pillar, constitutes a differential capacitance bridge. Four such strain sensors, oriented 45 degrees apart, are mounted on a steel cylinder to overlap each other. As the cylinder deforms, the central plate moves between two outer plates resulting in that two gaps vary differentially. It causes output changes of the bridge proportional to the variation in diameter of the cylinder. Because the strain sensor has a fairly broad frequency band response, so it can also be used to record strain seismic waves [Jones, 1970; Ouyang, 1977; Ouyang et al., 1988; 2009].

A)

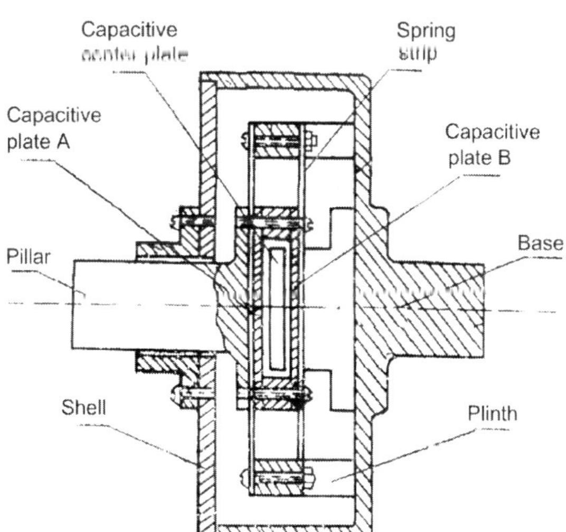

Figure 2. (Continued on next page)

B)

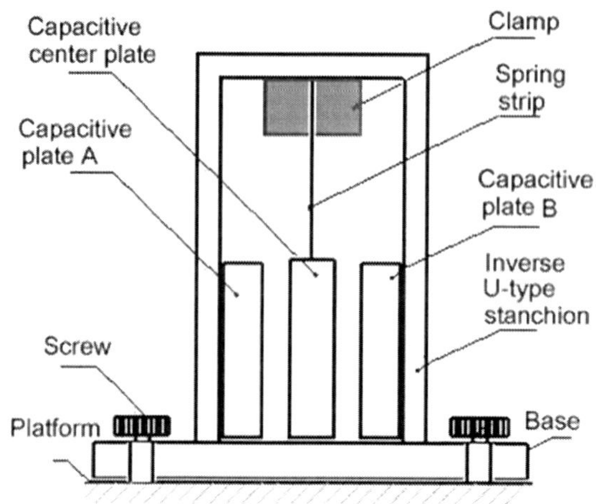

A) Schematic diagram for the strain sensor. (B) Schematic diagram for the tilt sensor.

Figure 2. Schematic diagram for transducers.

For borehole tiltmeter, the traditional vertical pendulum configuration was used in design of tilt sensors. A suspended mass, hanging on an inverse U-type stanchion as the central plate, with two outer metal plates, mounted on the stanchion, forms a three plate capacitance displacement transducer. As the ground inclines, the suspended mass swings freely forth and back under the influence of gravity, the mass position is varied and detected by the differential capacitance bridge. A schematic diagram of the principle of tilt sensor's operation is depicted in Figure 2(B). The borehole tiltmeter employs two orthogonal tilt sensors to provide the complete tilt vector data. We also need a framework to keep the sensors on scale in case of large tilt changes during the instrument's operation. Thus two sensors are mounted on a platform that can be leveled by means of electric motors. The leveling operation will take place at initial system installation and, subsequently, whenever the tilt sensors move away from the zero position by >80% of the basic dynamic range.

Design of the electronic measurement and control circuits for downhole manifold components, such as strain and tilt unit, was unitive. There are two circuit boards as shown in Figure 1. The major roles of the MCU circuit board are to acquire signals from each sensor, convert analog signals to digital data in 24bit and send out the converted digital data to the LAD via RS-485 data bus. It also distributes the power to individual sensor modules. Although strain-seismic waves are also detected by capacitance displacement sensors, its outputs will be amplified and send out in analogue form other than strain and tilt signals.

The surface electronic box comprises four main sections (Figure 1): LAD board (low-sampling rate A/D converter), HAD board (high-sampling rate A/D converter), COM board (communication unit) and PSB board (power supply box). A multi-core cable, including a twisted-pair RS485 serial link, provides connection for digital and analogue signal transmission and power supply between the downhole devices and the ground electronic box. Besides sending the data sampling and other control signal downhole, the LAD collects data from MCU and the time reference signals from an external time reference through the serial link and synchronizes the internal RTC. The data collected by RS-485 data bus are handled by an ARM microprocessor and processed to give lower sampling-rate data. All data are stored in digital format in the storage memories on LAD and HAD boards. 1 Hz sampling rate is the normal data rate for strain, tilt and ground temperature; 200-Hz sampling rate for strain seismic waves. In fact, there are another three channels provided on LAD circuit board as the access of assistant observation including well temperature, well water level and atmospheric pressure.

Each station site must have a reliable IP connection for forwarding data to the data acquisition computer located in each datacenter of provinces or autonomous regions. The COM board is optically isolated from the communication module. The whole system is powered by 12V rechargeable batteries with charging equipment such as solar panels or mains supply charger, and the downhole devices are powered by ±9V through another voltage regulator.

The main specifications of RZB-3 system are listed as follows:

- Arbitrary combination of manifold sensors.
- 4 strain sensor configuration.
- Strain package resolution: 0.1nano strain; basic dynamic range: ±500 microstrain.
- Tilt package resolution: 1nanorad; basic dynamic range: ±500microrad.
- Geothermometer: 0.0005C.
- Strain seismograph: 1 ~ 200 (sps).
- Standard connection: RS-485.
- Outer diameters: ≤ 102mm.

Figure 3. Photograph showing RZB-3D instrument installed at Zhangzhou station, Fujian Province just before the downhole device was sent to the 253m deep borehole on December 3, 2008. The instrument integrated packages including Electronic compass, Borehole strainmeter, Reserved borehole strainmeter, Borehole strain seismometer, Borehole tiltmeter and Geothemometer (From above to bottom).

2. 2. The Reliability and System Integration

The RS-485 is among the most successful standards for industrial applications, and allows for bidirectional, multi-point, party-line communications, with data rates to 10M bits/sec (10 Mbps) and line lengths to 1200 meters. Differential transmission provides good noise immunity. In our new system, all downhole integrated units are connected together by RS485 data bus, forming a node, and also linked with ground electronic box. The distributed control system composed of an ARM microprocessor on LAD board and several downhole MCU systems is finally implemented.

Having sacrificed retrieve for repair in the event of instrument malfunction, reliability therefore becomes the most important affair. In order to promote the system's reliability, we adopted mainly two approaches. Firstly, we developed digital capacitance measuring circuits based on charge balancing principle. The converted data from the following digital filter is scaled, and the output result can be read through the serial interface. Then, we designed a digital capacitance strain and tilt transducer using the circuits. These new digital transducers are accurate, robust and small sized. Another important characteristic of the new RZB-3 system is the terseness and reliability as a result of using developed digital transducers. Secondly, the new system has necessary backup downhole measurement modules to guarantee higher reliability, thanks to arbitrary combination of manifold sensors of the RZB-3 system at the expense of a rather small amount of cost without any modification to other packages of the instrument.

As an example, Figure 3 shows a new RZB-3D instrument installed at Zhangzhou in a borehole at a depth of 253m. The system is an integrated instrument consisting of a strain package (4 horizontal sensors and 1 reference), a reserved strain package (4 horizontal sensors and 1 reference), a tilt package (2 tilt sensors), a borehole strain seismometer package (3 sensors), a geothermometer and an electronic compass. This multi-component borehole instrument is about 4.63 meters in length and 100kg in weight. The reason for including a redundant borehole strain package is the requirement for high reliability.

2. 3. Installation of the Instrument

The strainmeters must be bonded well to the surrounding rock to record meaningful data about changes in the strain state. This is accomplished by cementing the borehole multi component observation instrument in the bottom of an open hole in intact, thick rock. Although factors impacting on successful strain monitoring are many, but following three aspects are essential to build a good crustal strain observation station. First of all, site selection is a very difficult task given the various criteria that need to be met to ensure high-quality measurements, specially in the geologically complex and townified areas. In any case it is strictly necessary to study local geology and associated fault processes in detail. Next, Borehole drilling is a very critical process. The key issues in drilling are associated with the type of the drilling rig, the techniques used and demand for core recovery. However, the core recovery higher than 90% is a quality requisite for the segment to install the downhole instrument.

Finally, after completion of the borehole drilling and casing, the first installation step is to check the quality of the borehole from viewing the core samples and all geophysical logging data in the field. When it is determined that the borehole is clear and acceptable, it is time to start the installation process. The relative orientation of each sensor is already fixed and known, their absolute azimuth can be subsequently determined at installation time by electronic compass. The final step is grouting. In the PBO design for installation, a quantity of grout was lowered into the hole with a container first. After withdrawing that container, the strainmeter was lowered into the hole and sunk through the grout to be completely immersed.

Having considered that the deep borehole installation needs more time, the existing procedure needs to be modified. We send the borehole observation instrument and a pipe simultaneously into the well first, than pump the cement into the borehole via the conduit. In all previous cases for installation of RZB type borehole strainmeter, we use above-mentioned methods, only one time out of ~40 in which we found that the strainmeter did not record long-term compressing strain from grout curing and borehole relaxation. We generally use simple criteria to estimate the installation quality. The measure is the degree to which the strainmeter is undergoing net compression, can be estimated by computing the areal strain from the raw strain data. The compression area strain, strainmeter suffered within one or two weeks, is larger than 1×10^{-5}, the installation could be considered eligible.

3. Data and Primary Results

RZB-1 borehole strainmeter is the simplest instrument of RZB family. From the 1980s up to the present, we have set up about 40 borehole strainmeters, including the experimental network of 12 stations in west China, set up in the later 1980s, in particular. Most of these instruments have been working for over 15 years. In recent years, 7 new RZB-3 borehole multi component instruments were installed in Beijing, and Fujian, Yunnan and Sichuan provinces.

a)

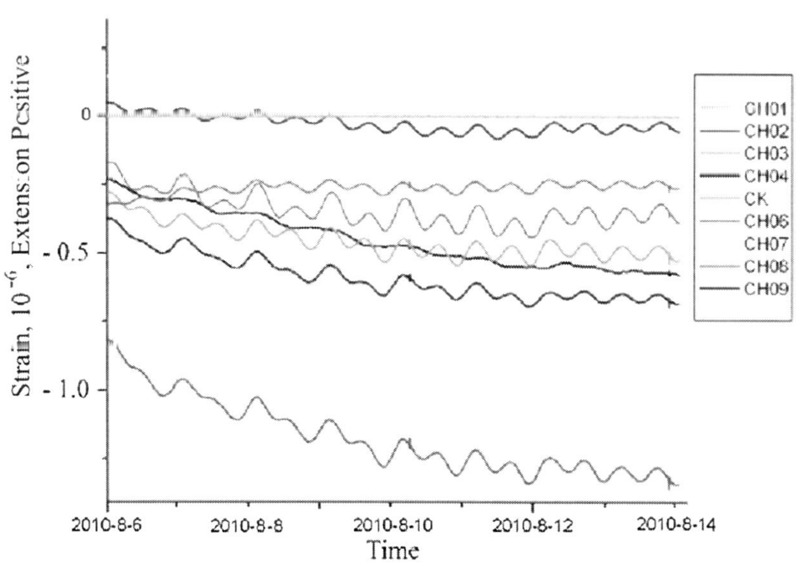

Figure 4. (Continued on next page)

b)

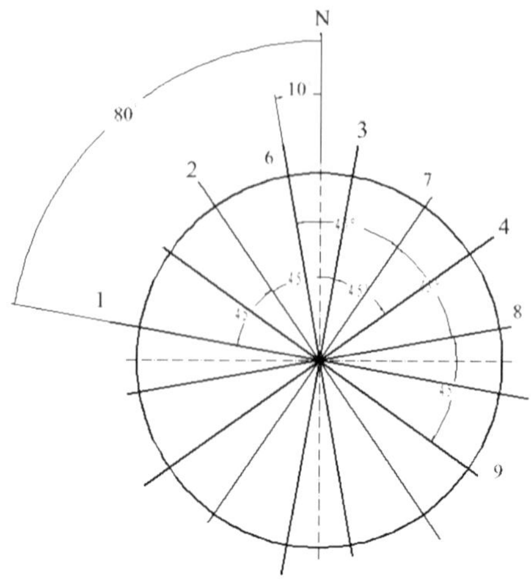

Figure 4. Borehole strains recorded at the Baishan, Beijing. (A) Raw borehole strain data of 8 components and CK channel for the system reference observed for the time period between August 6, 2010 and August 14, 2010, (B) The orientation of 8 strain transducers for the Baishan station.

3. 1. Noise Level of the System

An experimental station was set up at the Baishan, Beijing on September 28, 2008. A RZB-3B borehole strainmeter was installed in a well at depth of 215m. The instrument consisted of a 4 component strainmeter, another 4 component strainmeter as backup, a reference sensor and an electronic compass. Figure 4 shows ~9 days of raw strain data from Baishan.

The strainmeter runs at almost maximum sensitivity to provide sub-nanostrain resolution at a normal sampling rate of 1 sample/second (sps) on each sensor. Figure 4(B) shows the orientations of 8 sensors, and the earth tides can be seen clearly on 8 channels (Figure 4(A)). The instrument has a reference sensor which has the same construct as other strain sensors but is isolated from borehole deformation. The reference sensor is specially used to monitor the noise and electric interferences of the system. The LAD board provides various data with lower sampling rate according to the report format criterion. Figure 5 shows a typical example from Baishan station, we found that data of 1 sample/minute (spm) have much higher signal-to-noise ratio and the noise level of the reference sensor is improved to better than 0.1 nanostrain demonstrating that the low noise level is one of the new RZB system characteristics.

3. 2. Long-Term Strain Observation

Figure 6 is the Long term data observed at periods from October 2008 to May 2009, following the installation of RZB-3 instrument at the Baishan station. It can be seen that the strain data becomes more stable in nearly 3-6 months later after installation. The system recorded the whole installation process including cement solidification and stress adjustment in borehole surroundings without any manual operation to extend the basic measurement range which is ±500 microstrain. The total dynamic range of new RZB-3 system can be extended to $\pm 1\times 10^{-3}$. The large dynamic range is another important characteristic of new RZB-3 borehole observation system. Based on differential three capacitive plate design, this newly developed digital capacitive transducer achieves not only higher resolution but also larger dynamic range.

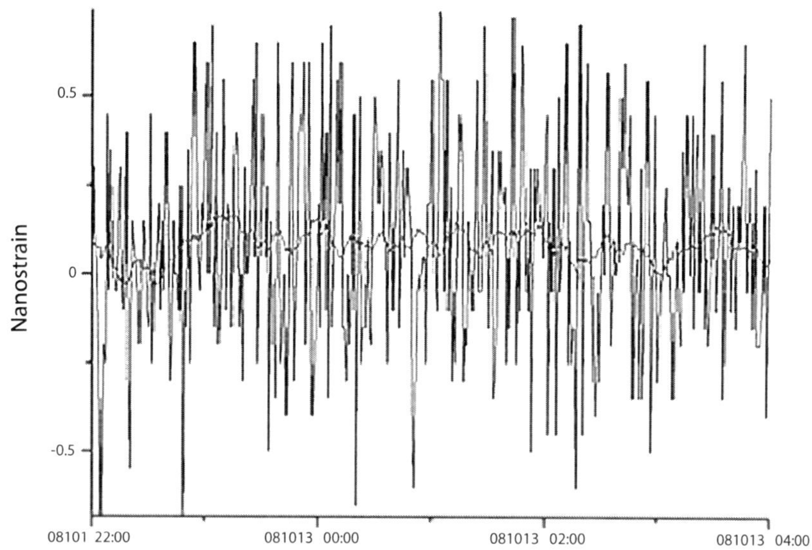

Figure 5. The system reference data from the Baishan station for the time period of 6 hours on 15 days after the installation. The red line is the filtered 1 sample/minute data.

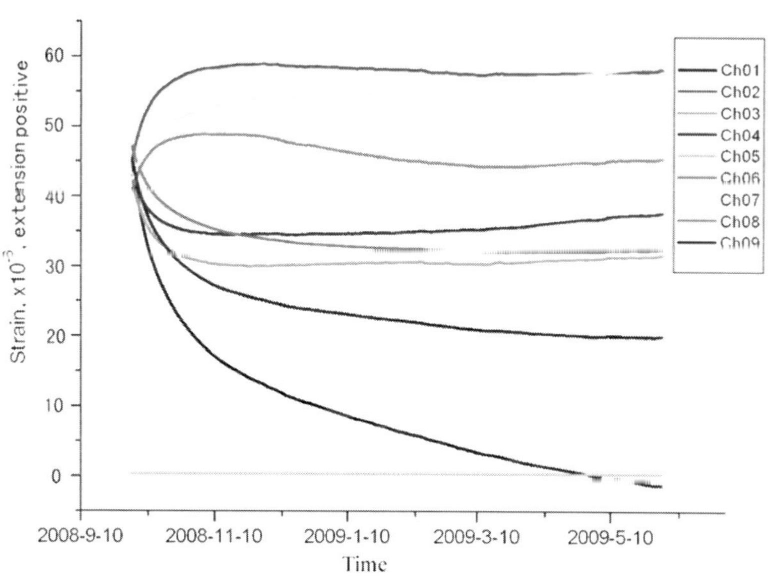

Figure 6. The raw data of multi-component borehole strains observed at Baishan station between October 4, 2008 and June 2, 2009. The instrument integrated two borehole strain packages, so that therefore there are 8 strain sensors, CH1~CH4 and CH6~ch9, on different azimuth shown in Figure 4 (B). CH5 is the transducer for system reference which is isolated from borehole deformation.

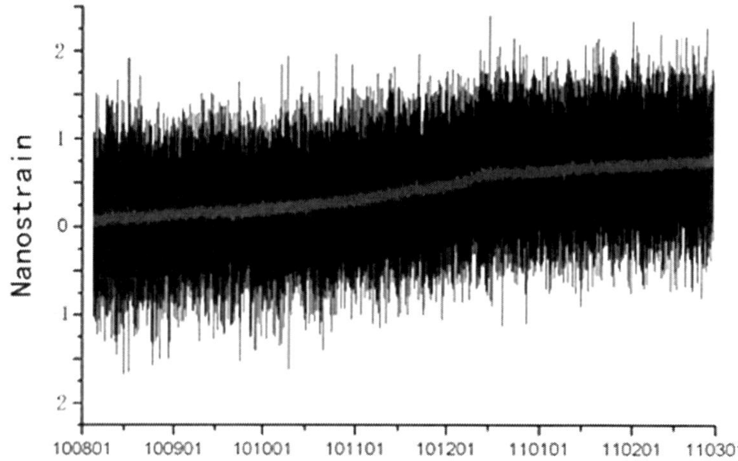

Figure 7. The system reference data observed at Baishan station between August 3, 2010 and February 28, 2011. The red line is filtered 1 sample/minute output.

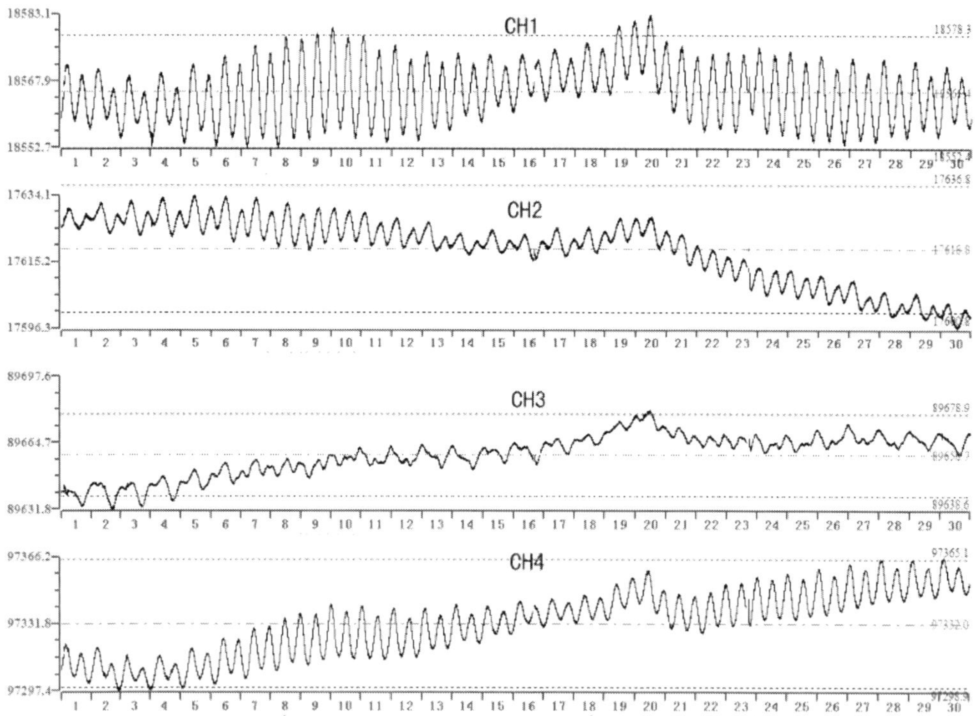

Figure 8. Borehole strains recorded at the Zhangzhou, Fujian Province in September, 2010. Amplitude scale is given in counts of the 24-bit A/D converter.

Although there were one or two strain components in any strainmeter package being extending, the areal strain is contracting at all the time. The contractive trend presumably caused mainly by curing of the grout (over a period of 6-12 months) and relaxation of the borehole.

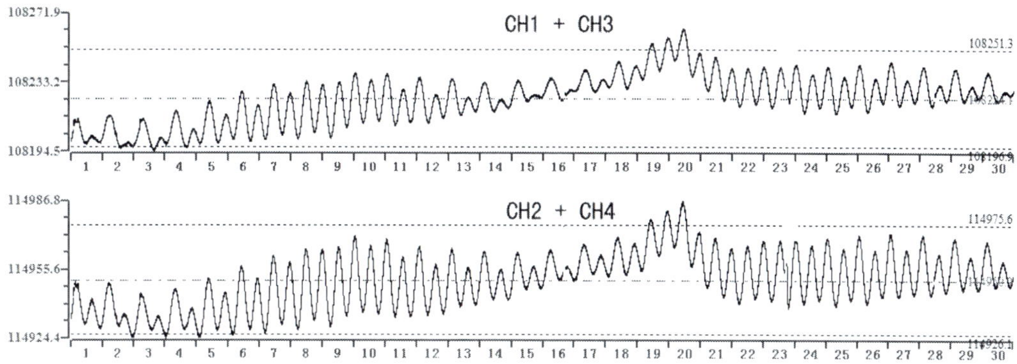

Figure 9. Areal strains recorded at the Zhangzhou, Fujian RZB-3 Station in September, 2010. CH1+CH3 and CH2+CH4 are two results of the area strain, separately calculated from two groups of orthogonal strain sensors, see text for explanation. Amplitude scale is given in counts of the 24-bit A/D converter.

Table 2. M2 areal strain tide

Time (Year-month)	Tide Factor	Error	Relative Error	Phase Lag (degree)
2010-01	0.3563	0.0038	0.0106	3.858
2010-02	0.3464	0.0053	0.0153	3.8058
2010-03	0.3609	0.0158	0.0437	-0.5845
2010-04	0.3434	0.0041	0.012	3.0237
2010-05	0.3396	0.0038	0.0113	3.8836
2010-06	0.3378	0.0032	0.0094	3.6369
2010-07	0.3359	0.0021	0.0063	3.1003
2010-09	0.332	0.0058	0.0176	3.8351

Obviously, the strain data becomes more stable in nearly two years later after installation. Figure 7 shows the recorded data of system reference sensor for the period between August, 2010 and March, 2011, the system drift becomes lower than 0.2×10^{-9} / month showing that the instrument itself is stable and the drift of measurement circuits of the system is very low.

3. 3. Multi-Component Observation and Tide Analysis

Another experimental station was set up at Zhangzhou, Fujian Province on December 7, 2008. A RZB-3D Borehole Multi Component Observation System was installed in a borehole at depth of 252m (Figure 3). Figure 8 is the data of 4 borehole strain components, CH1-CH4, recorded in September, 2010.

Borehole deformation theory indicated that when you have two strain sensors, placed orthogonally, the strain sum combination of the two sensors equals to the areal strain in the surrounding rock. There are 4 linear strain sensors in a RZB borehole strainmeter package, so we have two groups of orthogonal sensors. The areal strain obtained from one group should be the same as from another [Jaeger, 1976; Ouyang et al., 1988; Qiu et al., 2005]. Area strains calculated from above 4 strain components were presented in Figure 9.

Tide analyses were applied to evaluate the data quality. We filtered the raw data to hourly samples and used 30 day window to calculate the M2 tide constituents by using the EIS2000 program. The results were presented in an East-North coordinate system [Jiang et al. 2000]. The phase of the M2 areal strain tide agrees with calculated phases for most instruments, 4 of 7 RZB-3 strainmeters, that we checked. We found that for most RZB-3 strainmeters the M2 tide constituents of areal strain reconciled with calculated tides well and remained consistent over a long time. For example, In table 2 listed are the results of M2 areal strain tide analyses for Zhangzhou station in period from January to September, 2010. M2 phase of observed area strain is within about 5°with areal strain phase calculated by EIS2000. On the other hand, the phase of the shear components is moderately agreed with calculated one.

Figure 10 is the data of two tilt components recorded at Zhangzhou in June 2010. Tidal tilts are well recorded by this instrument. We found that EW tilt tidal amplitude was distinctly bigger than other stations. Having considered Zhangzhou station is located 28 km to the coast, large amplitude is presumably caused by local ocean tidal loading. The reasons for the drift on NS tilt channel are unknown for the time being.

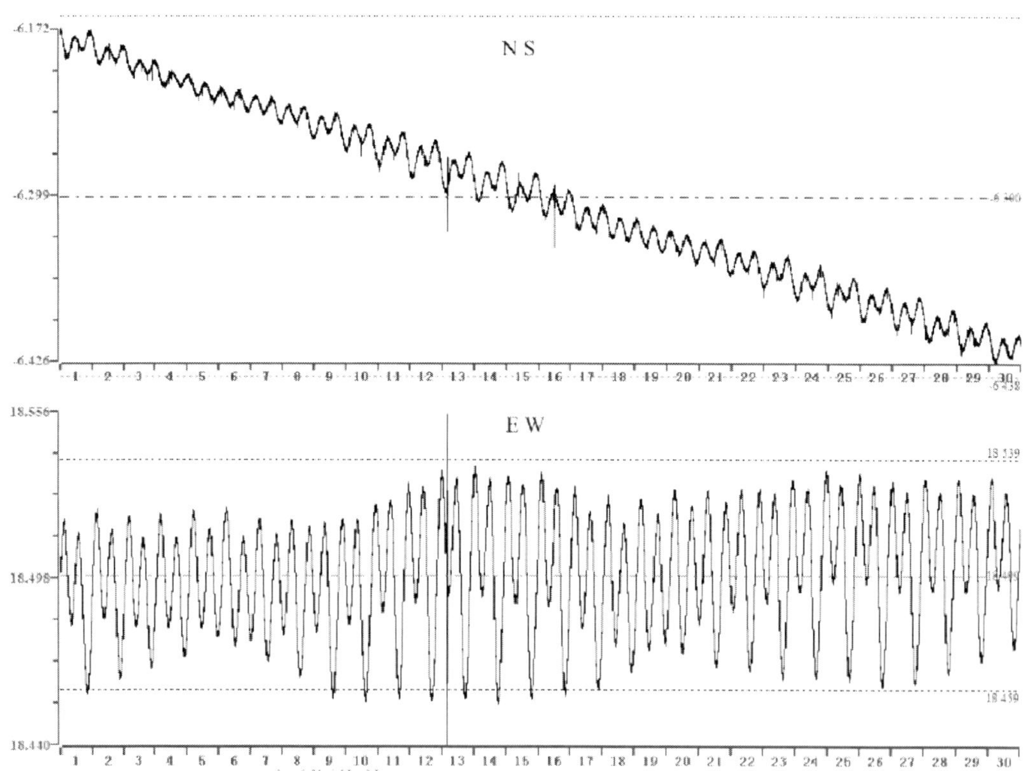

Figure 10. Borehole tilt recorded at the Zhangzhou, Fujian Province in Jun, 2010. Amplitude scale is given in counts of the 24-bit A/D converter. An earthquake was recorded on 29 April 1999, but our recording rate at that time was inadequate to capture the waveform.

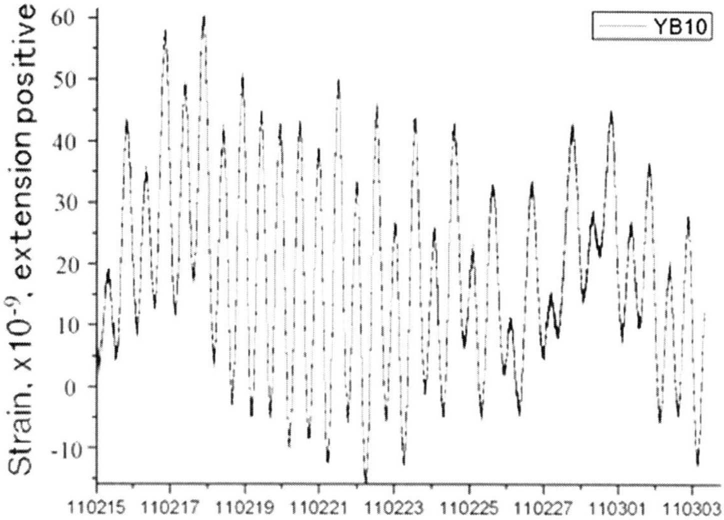

Figure 11. Vertical strains recorded at the Zhangzhou (Longwen Middle School), Fujian RZB-3 Station in February, 2011.

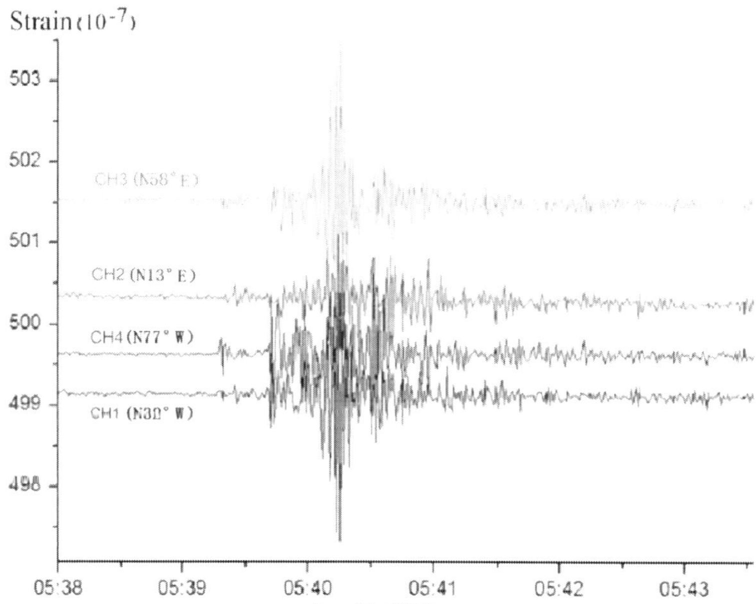

Figure 12. The strain seismic waves of Suining earthquake recorded at Jiangyou station, Sichuan. The data rate is 10 sample/second.

Tide analyses indicated that amplitude relative error of M2 tilt tide remained basically about 1%, but in most cases the M2 phases recorded by RZB tiltmeter consistently differed by about 45° from the phases predicted by the EIS2000. These phase errors are considered likely due to the measurement error of the tiltmeter direction.

Figure 11 shows ~17 days of vertical strain data recorded at Zhangzhou (Longwen Middle School), FUJIAN in Feb, 2011. Zhangzhou LMS station (Longwen Middle School) is part of the experimental network of multi-component observation system, the LMS station is set up on Jan 2, 2011 and situated at a distance of ~7km from Zhangzhou station, Fujian. Vertical Tidal strains are well recorded with the borehole tiltmeter package of RZB-3 multi-component observation system. Tide analyses indicated that amplitude relative error of M2 vertical strain tide is about 1.2%.

Figure 12 is the strain seismic waves recorded by new RZB multicomponent observation system at the Jiangyou station, west of Sichuan province. This station was set up on November 3, 2009. Strain seismic wave series is taken from a data file of the Suining earthquake with magnitude 5.1 occurred on 31 January 2010, 05:39:48, about 110km away. By comparing of strain seismic waves from Jiangyou with general digital records of the Suining earthquake observed at adjacent seismic station, we found that the shape and onset time of main seismic phases were extreme similar between these two wave series.

3. 4 Uniformity Inspection of Four Strain Components

As discussed above, there are 4 linear strain sensors in a RZB borehole strainmeter package, so we have two groups of orthogonal sensors.

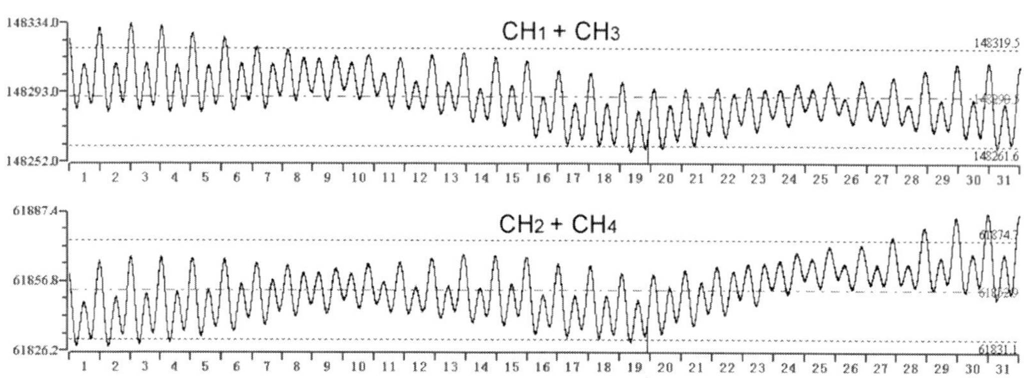

Figure 13. Two time serials of the areal strain observed at Zhangzhou station during December, 2009.

Table 3. Results of correlation analysis

Time (year - month)	Correlation
2009 - 1	0.9998
2009 - 2	0.9994
2009 - 3	0.9984
2009 - 4	0.9687
2009 - 5	0.8937
2009 - 6	0.9816
2009 - 7	0.9910
2009 - 8	0.9953
2009 - 9	0.9993
2009 - 10	0.9991
2009 - 11	0.9974
2009 - 12	0.9968

The area strain results from two combinations out of four sensors should be uniformity. Figure 13 is an example showing good uniformity of two time serials of the areal strain observed at Zhangzhou station. We have calculated the correlation between two data sets taking one month as time window to evaluate the data quality. Correlation factors between area strains (CH1 + CH3) and (CH2 + CH4) are listed in Table 3 for the year 2009. We found that the strong correlation exists between the two area strains showing good quality of areal strain measurement obtained with new RZB-3 multi-component borehole instrument.

CONCLUSION

In the turn of the century, the borehole strain project has progressed from a small instrumentation research effort to a larger-scale monitoring program. But there are a lot of problem with regard to Crustal deformation measurement that is still open. Where is a right place to set up a borehole crustal deformation observatory? What depth is appropriate to observe good data reflected tectonic activity? How to install a borehole instrument successfully?

Needless to say, the measurement of Crustal strains is not an easy task but the observed data will be more reliable and abundant that administer to the understanding of the nature of Crustal strain field and search for quality assurance construction procedures of observation networks of Crustal strain field.

Newly developed RZB-3 borehole multi-component observation system is characterized by high ratio of signal to noise, large dynamic range, arbitrary combination of manifold transducers and backup modules, compact construction and low cost. The objective of doing these work is to improve measurement capabilities of the borehole instrument in different aspects, such as in data self-checkout by using 4 strain component configuration, in promotion of reliability with application of digital transducers, as well as in order to find more independent evidences of crustal deformation pertinent to earthquakes and volcano eruptions through deep borehole multi component observations.

REFERENCES

Dragert, H., K. Wang, and T. S. James, A Silent Slip Event on the Deeper Cascadia Subduction Interface, *Science*, 292, 1525-1528, 2001.

Earth Scope Project Plan - A New View into Earth, 2001, *http://www.unavco.net/project.*

Gladwin, M. T., 1984, High precision multi-component borehole deformation monitoring, *Rev. Sci. Instrum.*, *55*: 2011–2016.

Hart, R. H. G. Gladwin, M. T. Gwyther, R. L. Agnew D. C. and Wyatt, F. K. 1996, Tidal calibration of borehole strainmeters: removing the effects of local inhomogeneity, *J. Geophys. Res.*, 101: 25, 553–25, 571.

Hast, N., 1958, The measurement of rock pressure in mines [M], Stockholm.

Ishii H., 2001, Development of new multi-component borehole instrument. *Report of Tono Research Institute of Earthquake Science,* 6: 5–10 (in Japanese).

Ishii, H. T. Yamauchi, Y. Asai, M. Okubo, S. Matsumoto, and Aoki, H. 2003, Continuous multi-component monitoring of crustal activities by a newly developed instrument installed in a 1200m deep borehole –The deepest multiple observation in the world consisting of stress, strain, tilt, seismic waves, geomagnetism, temperature. Paper presented at the XXII General Assembly of I. U. G. G., Sapporo, Japan.

Jaeger, J. C. and Cook, N. G. W., 1976, *Fundamentals of Rock Mechanics,* [M], Chapman and Hall, (London), 2nd ed., 1976.

Jiang J. Li S. L. Zhang Y. B. et al., 2000, *Information manage and software system E. I. S. 2000 for seismic precursors*, Beijing: Seismic Press, (In Chinese).

Jiang J. X. Yin G. H. Liu L. B., 1998, Borehole strain steps to the pending earthquakes in Xinjiang, China, E. O. S., 1998, Vol. 79(45), supplement.

Jones, R. V., 1970, The Pursuit of Measurement, *P. I. E. E.*, V. 117, No. 6.

Linde, A. T. Gladwin, M. T. Johnston, M. J. S. Gwyther, R. L. and Bilham, R. G., 1996, A slow earthquake sequence on the San Andreas fault, *Nature*, 383, 65–68.

Ouyang, Z. X., 1977, RDB-1 Borehole Capacitance Strainmeter, [A], *The proceeding of 1st National Assembly on Crustal Stress*, 337-348, (in Chinese).

Ouyang, Z. X. and Zhang, Z. R., 1988, A coupling method of borehole strainmeters with the rock well, Zhao G. G. ed. *Crustal Structure and Crustal stress*, V.2: 1–10. Beijing: Seismic Press, (in Chinese).

Ouyang, Z. X. Li, B. Y. Jia, W. J. et al., 1988, A borehole crustal strain monitoring system, Zhao, G.G. ed. *Crustal structure and Crustal stress*,V. 2: 11–20. Beijing: Seismic Press, (in Chinese).

Ouyang, Z. X. Zhang, J. Chen, Z. et al. 2009 New progress in multi-component observation of Crustal deformation in deep boreholes. *Recent Developments in World Seismology*, 2009(11): 1~13 (in Chinese).

Ouyang, Z. X., 2010, A new type of multi-component borehole instrument for continuous monitoring of crustal deformation, *Rock Stress and Earthquakes*, Xie (ed.), 809-814, Taylor and Francis Group, London,2010.

P. B. O. white paper, 1999, *http: //www.unavco.ucar.edu/community/publications/PBO whitepaper.pdf.*

P. B. O. Strainmeter Products, 2011, *http://pboweb.unavco.org/stations/.*

Qiu, Z. H. Shi, Y. L. Ouyang, Z. X. 2005, Absolute calibration of 4 component borehole strainmeters, *Earthquake*, 25(3): 27–34, (in Chinese).

Qiu, Z. H. Shi, Y. L. and Ouyang Z. X. 2005, Relative in-situ calibration of 4 component borehole strainmeters [J]. *Journal of Geodesy and Geodynamics*, (1): 118-122, (in Chinese).

Roeloffs, E., (2010): Evidence for aseismic deformation rate changes prior to earthquakes, Annual Review of Earth and Planetary Sciences, Vol. 34: 591-627.

Silver, P., Y, Bock, D. C. Agnew, et al., 1998, A Plate Boundary Observatory, *I. R. I. S. newsletter*, 16, 3 and 7-9, 1998.

The Plate Boundary Observatory (P. B. O.) Component of the Earth Scope M. R. E. Proposal, 2002, *http://pboweb.unavco.org.*

Sacks, I. S., Suyehiro, S., Evertson, D. W., Yamagishi, Y. 1971, Sacks-EvertsonStrainmeter, Its Installation in Japan and some preliminany resslts concerning Strainmeter, Its Installation in Japan and some preliminary results concerningstrain steps. *Papers in*

Meteorology and Geophysics, 22: 195-208. strain steps, *Meteorology and Geophysics*, 22: 195-208.

Sakata S., 1981, On the concept of some newly invented borehole three component strainmeters, [J] (In Japanese with English Abstract), *Rep. Nat. Res. Center Disaster Pre.*, 25: 95-126.

In: Earthquakes
Editor: Kostas Konstantinou

ISBN: 978-1-62081-883-1

Chapter 9

NEW MODEL FOR AVOIDING COLLAPSES OF REINFORCED CONCRETE STRUCTURES DURING STRONG EARTHQUAKES

Alexandros-Dimitrios G. Tsonos*
Aristotle University of Thessaloniki, School of Engineering,
Department of Civil Engineering, Thessaloniki-Macedonia-Greece

ABSTRACT

In this study a new well established model is presented and commented, that saves human lives because it helps avoiding collapses in the reinforced concrete constructions during strong earthquakes. Using this model the initial formation of plastic hinges in the beams and the final concentration of the damages only in the beams is accurately assured. The model also assures that the columns and the beam-column joints can remain intact. This model can be applied for the design of modern R/C constructions, as well as for the design of strengthening schemes of old R/C constructions by the use of reinforced concrete jackets. The model can also predict the type of earthquake damages of old structures but also the earthquake damages of the modern structures. Further by the use of this model a new technique can be applied which helps avoiding the serious disadvantages related to the currently used FRPs for the strengthening of old R/C structures. In this direction a mixed type strengthening system is proposed which includes strengthening of the beam – column joints with a conventional reinforced concrete jacket and the other parts of the structures with FRPs. This mixed type system is calculated by the use of the new proposed model.

* Aristotle University of Thessaloniki, School of Engineering, Department of Civil Engineering, Thessaloniki-Macedonia-Greece, GR-541 24, Tel.: +30 2310/995667, e-mail: *tsonosa@civil.auth.gr*.

1. Introduction

Damage incurred by earthquakes over the years has indicated that many reinforced concrete (R/C) buildings, designed and constructed during the 1960s and 1970s, were found to have serious structural deficiencies today, especially in their columns and beam – column (b/c) joints. These deficiencies are mainly due to lack of a capacity design approach and/or poor detailing of the reinforcement. The beam – column region of these buildings has often been found to contain no shear reinforcement. As a result, lateral strength and ductility of these structures were minimal and hence some of them collapsed [Paulay and Park 1984, Karayannis et al. 1998, Park 2000]. The studies by Tsonos 2006, 2007 provide structural engineers with useful information about the safety of the new reinforced concrete frame structures incorporating full seismic detail in current building codes. Sometimes, this could be jeopardized during strong earthquakes by premature joint shear failures. The joints could at times remain the weak link of the structures even today. One of the most popular pre - and post – earthquake retrofitting methods for columns and beam – column joints is the use of reinforced concrete jacketing. The wrapping of reinforced concrete members with fiber – reinforced polymer FRP sheets including carbon (C), glass (G) or aramid (A) fibers, bonded together in a matrix made of epoxy, vinylester or polyester, has been used extensively throughout the world in numerous retrofit applications in deficient old reinforced concrete buildings. These are recognized as alternative strengthening systems to conventional methods, such as reinforced concrete jacketing [Ilki 2006, Mosalam 2006]. However, it is worldwide recognized that the beam – column joints of old reinforced concrete structures cannot be effectively retrofitted with FRPs [Tsonos 2008, Karayannis and Sirkelis 2008].

Investigations into recent earthquake damage in Greece [Korinth, 1981; Kalamata, 1986; Aegion, 1995; Athens 1999] have shown that, in many cases, damaged areas of reinforced concrete buildings were localised in beam-column connections. Furthermore, considering the commonly accepted idea that failure of joints may quickly lead to general failure [Park and Paulay, 1975], the important issue of preserving the beam-column joints intact after strong earthquakes has arisen. A new model published in the studies by Tsonos [1999, 2006, 2007] assured the initial formation of plastic hinges in the beams but also the final concentration of damages only in these elements. This model also assures that the columns and the beam – column joints can remain intact during strong earthquake damages of the modern structures. This model can be applied for the design of modern R/C constructions, as well as for the design of strengthening schemes of old but also for modern structures by the use of reinforced concrete jackets [Tsonos 1999, 2001a, 2001b, 2002, 2008, 2010] offering in all cases higher level of safety. The model can also predict the type of earthquake damages of old structures but also the earthquake damages of the modern buildings.

By the use of the model a new technique is proposed which helps avoiding the serious disadvantages related to the currently used FRPs for the strengthening of old R/C structures. For this purpose a mixed type of strengthening system is proposed in this study, which consists of strengthening the beam – column joints with conventional reinforced concrete jacket and the other parts of the structures with FRPs. This mixed type system was also calculated by the use of the new proposed model.

This model has been used by Park and Mosalam for the evaluation and establishment of their work [see also Report by Park and Mosalam 2009]. Further Bakir [2003] considers that the prevailing models for the analysis of joints are the model by Paulay and Park and the model by Tsonos. Recently, Eligehausen et al. [2011] and Karayannis et al. [2011] use this model for verification purposes of their own approaches whereas this model (also called as Tsonos' model by these authors) is considered by one of them as a leading one in the field.

Moreover, it is mentioned that recently (September 2011) the author of this study was awarded a prize for this model by the Rector of the Aristotle University of Thessaloniki.

In the following paragraphs this new model is presented and commented. Applications of the model are also demonstrated.

2. A Summary of the Proposed Model

A new model published in the studies by Tsonos [1996, 1997, 1999, 2006, 2007] predicts the beam-column joint ultimate shear strength.

Figure 1(a) shows a reinforced concrete exterior beam-column joint for a moment resisting frame and Figure 1(b) shows the internal forces around this joint. The shear forces acting in the joint core are resisted:

(i) Partly by a diagonal compression strut that acts between diagonally opposite corners of the joint core [see Figure 1(c)], and
(ii) Partly by a truss mechanism formed by horizontal and vertical reinforcement and concrete compression struts. The horizontal and vertical reinforcement is normally provided by horizontal hoops in the joint core around the longitudinal column bars and by longitudinal column bars between the corner bars in the side faces of the column. Both mechanisms depend on the core concrete strength. Thus, the ultimate concrete strength of the joint core under compression/tension controls the ultimate strength of the connection. After failure of the concrete, strength in the joint is limited by gradual crushing along the cross - diagonal cracks and especially along the potential failure planes (Figure 1(a)).

For instance, consider the section I - I in the middle of the joint height (Figure 1(a)). In this section, the flexural moment is almost zero. The forces acting in the concrete are shown in Figure 1(d).

Each force acting in the joint core is analysed into two components along the X and Y axes (Figure1(d)). T_i are the tension forces acting on the longitudinal column bars between the corner bars in the side faces of the column. Their resultant is ΣT_i. An equal and opposing compression force ($-\Sigma T_i$) must act in the joint core to balance the vertical tensile forces generated in the reinforcement. This compression force was generated by the resultant of the vertical components of the truss mechanism's diagonal compression forces D_1, D_2 ...D_ν. Thus $D_{1y} + D_{2y} + \ldots + D_{\nu y} = \Sigma T_i = T_1 + T_2 + T_3 + T_4$. The column axial load is resisted by the compression strut mechanism. The summation of vertical forces equals the vertical joint shear force V_{jv}:

D_{cy} + $(T_1+...+T_4) = D_{cy} + D_{sy} = V_{jv}$ (2a)

↓ ↓

compression strut truss model.

The summation of horizontal forces equals the horizontal joint shear force V_{jh}:

$D_{cx} + (D_{1x}+ D_{2x}...+D_{vx}) = D_{cx} + D_{sx} = V_{jh}$ (2b)

The vertical normal compressive stress σ and the shear stress τ uniformly distributed over section I - I are given by equations (3) and (4) below.

$$\sigma = \frac{D_{cy} + D_{sy}}{h_c^{'} x b_c^{'}} = \frac{V_{jv}}{h_c^{'} x b_c^{'}} \tag{3}$$

$$\tau = \frac{V_{jh}}{h_c^{'} x b_c^{'}} \tag{4}$$

where h'_c and b'_c are the length and the width of the joint core respectively.

It is now necessary to establish a relationship between the average normal compressive stress σ and the average shear stress τ (Figure 1 (e)). From Eqs. (3) and (4):

$$\sigma = \frac{V_{jv}}{V_{jh}} \cdot \tau \tag{5}$$

It has been shown that:

$$\frac{V_{jv}}{V_{jh}} = \frac{h_b}{h_c} = \alpha \text{ (Eurocode 8, Park and Paulay 1975, Paulay and Park 1984)} \tag{6}$$

where α is the joint aspect ratio.

The principle (σ_I = maximum, σ_{II} = minimum) stresses are calculated:

$$\sigma_{I,II} = \frac{\sigma}{2} \pm \frac{\sigma}{2} \sqrt{1 + \frac{4\tau^2}{\sigma^2}} \tag{7}$$

Equation 8 (Figure 1(e), Tsonos 2007) was adopted for the representation of the concrete biaxial strength curve by a 5th degree parabola.

$$-10\frac{\sigma_1}{f_c} + \left[\frac{\sigma_{II}}{f_c}\right]^5 = 1 \tag{8}$$

where f_c is the increased joint concrete compressive strength due to confinement by joint hoop reinforcement, which is given by the model of Scott et al 1982 according to the equation

$$f_c = \mathrm{K} \cdot f_c^{'} \tag{9a}$$

Also, $f_c^{'}$ is the concrete compressive strength and K is a parameter of the model [Scott et al. 1982] expressed as

$$K = 1 + \frac{\rho_s \cdot f_{yh}}{f_c^{'}} \tag{9b}$$

where ρ_s is the volume ratio of transverse reinforcement and f_{yh} is its yield strength.

Substituting Eq. (5), (6) and (7) into Eq. (8) and using $\tau = \gamma\sqrt{f_c}$ gives the following expression:

$$\left[\frac{\alpha\gamma}{2\sqrt{f_c}}\left(1+\sqrt{1+\frac{4}{\alpha^2}}\right)\right]^5 + \frac{5\alpha\gamma}{\sqrt{f_c}}\left(\sqrt{1+\frac{4}{\alpha^2}}-1\right) = 1 \tag{10}$$

Assume here that $$x = \frac{\alpha\gamma}{2\sqrt{f_c}} \tag{11}$$

And

$$\psi = \frac{\alpha\gamma}{2\sqrt{f_c}}\sqrt{1+\frac{4}{\alpha^2}} \tag{12}$$

Then expression (10) is transformed into:

$$(x+\psi)^5 + 10\psi - 10x = 1 \tag{13}$$

The solution of the system of Eqs. (11) to (13) gives the beam-column joint ultimate strength

$$\tau_{ult} = \gamma_{ult}\sqrt{f_c} \text{ (MPa).}$$

This system is solved each time for a given value of the joint aspect ratio using standard mathematical analysis. The joint ultimate strength τ_{ult} depends on the increased joint concrete compressive strength due to confining f_c and on the joint aspect ratio α.

For simplicity reasons the presentation of the above methodology was chosen for exterior beam – column joints, while it is the same for interior ones. For the development of the above formulation it has been assumed that bond conditions of the beam and column bars anchoring or passing through the joint region are generally favourable.

The validity of the formulation was checked using test data for more than 120 exterior and interior beam – column subassemblages that were tested in the Structural Engineering Laboratory at the Aristotle University of Thessaloniki, as well as data from similar experiments carried out in the United States, Japan and New Zealand [Tsonos 2007]. The applied loading procedure used in these subassemblages was a typical one commonly used in this field of research. According to this method the subassemblages were subjected to many cycles of loading applied by slowly imposed earthquake-type loads or displacements. The vast majority of the experimental research studies in this direction adopt similar test implementation. The strain rate of the load applied corresponded to static conditions. Nevertheless during a real seismic excitation, the strain rate $\dot{\varepsilon}$ is higher than the rate corresponding to static conditions. Soroushian and Sim [1986] showed that an increase in $\dot{\varepsilon}$ with respect to static conditions leads to a moderate increase in the strength of concrete:

$$f_{c,dyn} = [1,48 + 0,160 \cdot \log \dot{\varepsilon} + 0,0127 (\log \dot{\varepsilon})^2] \cdot f_{c,stat} \tag{14}$$

Scott et al. [1982] tested column specimens with various amounts of hoop reinforcement under strain rates ranging from $0,33 \cdot 10^{-5}\,\sec^{-1}$ (static loading), to $0,0167\,\sec^{-1}$ (seismic loading). Their test results conformed with the results obtained from expression (14). Using the aforementioned expression it is estimated that for a strain rate of $\dot{\varepsilon} = 0,0167\,\sec^{-1}$ concrete strengths increase by about 20% (compared with the static one). An expression similar to (14) can be found in the C.E.B. Code [1990]. Thus the strengths exhibited by the subassemblages used in this study during the tests are somewhat lower than the strengths they would exhibit if subjected to load histories similar to actual seismic events.

3. The Contribution of the Proposed Model to the Human Safety

1. The Model Assures the Formation of Plastic Hinges in the Beams and the Integrity of the Columns and the Beam – Column Joints during Strong Earthquakes

Using this model we can predict with a high degree of precision the joint ultimate strength $\tau_{ult} = \gamma_{ult} \sqrt{f_c}$.

The improved retention of strength in the tested beam – column subassemblages as the values of the ratio $\tau_{cal} / \tau_{ult} = \gamma_{cal} / \gamma_{ult}$ decrease was demonstrated. For $\tau_{cal} / \tau_{ult} \leq 0.50$ the beam – column joints of the subassemblages behaved excellently during the tests and remained intact at the conclusion of the tests [Tsonos 1997, 1999, 2006, 2007].

$\tau_{cal} = \gamma_{cal} \sqrt{f_c}$ is the acting calculated joint shear stress.

τ_{cal} is calculated from the horizontal joint shear force assuming that the top reinforcement of the beam yields (Figure 1(a)). In this case the horizontal joint shear force is expressed as:

$$V_{jhcol} = 1.25 A_{s1} \cdot f_y - V_{col} \tag{15}$$

where A_{s1} is the top longitudinal beam reinforcement (Figure 1(a)), f_y is the yield stress of this reinforcement and V_{col} is the column shear force (Figure1(a)). Development of inelastic relations at the faces of joints of reinforced concrete frames is associated with strains in the flexural reinforcement well in excess of the yield strain. Consequently, joint shear force generated by the flexural reinforcement is calculated for a stress of $1.25 f_y$ in the reinforcement (ACI – ASCE 352 – 02).

It is clear that a significant innovation is introduced in the design of earthquake – resistant reinforced concrete structures which involves also an important contribution for the safety of these structures and also for the people who live inside them (Figure 3(a)).

Thus in the old reinforced concrete structures the “strong columns and weak beams” rule resulted by the flexural strength ratio, which must be higher than 1.0, was completely unknown. Thus often in the old buildings exist much stronger beams than the columns framing at the joints.

However in the modern reinforced concrete structures the “strong columns – weak beams” method does not secure the safe formation of plastic hinges only in their beams and consequently the formation of the safe elastoplastic mechanism with only a beam failure mechanism. In Figure 2 are shown failure modes of four subassemblages designed according to modern codes which were imposed to strong seismic loading. However two of them exhibited premature joint shear failure from the early stages of the seismic loading [Tsonos 2006, 2007]. This occurs because sometimes the modern codes do not include any earthquake – resistant design of beam – columns joints for shear strength (new Greek Codes) or the methods which adopt (Eurocodes 2 and 8) do not secure the safe and exclusive formation of plastic hinges in the beams of the structures [Tsonos 2006, 2007]. Thus even today the modern codes do not secure the safety of new structures. In some cases, safety could be jeopardized during strong earthquakes by premature joint shear failures (Figure 2, Figure 3(b)). The joints could at times remain the weak link even for structures designed in accordance with current model building codes.

This serious problem can be solved by the use of the proposed in this study new model and new method. Thus using this model and taking into consideration in the joint regions that $\tau_{cal} / \tau_{ult} \leq 0.50$ and of course using the rule “strong columns – weak beams” it can be secured that the exclusive formation of plastic hinges in the beams of a reinforced concrete structure but also the damage concentration only in the beams. The columns and the beam – column joints would remain intact during strong earthquakes. The safety of the structures and the lives of the people inside them are significantly secured (Figure 3(b)).

This innovative method significantly improves the design approach of the modern codes.

The question arises: How a model that gives the ultimate strength of a reinforced concrete beam-column joint and predicts the actual value of the joint shear stress can also be used for the prediction of the actual values of shear forces and moments developed in the columns and in the beams, framing at the joint, during strong earthquakes. The answer can be found in Paulay and Priestley 1992, who clearly demonstrated that shear forces acting in beam – column joints are significantly higher than those acting in their adjacent columns. Thus, joints fail earlier than columns during a strong earthquake motion.

Consequently, a model that predicts the actual value of the joint's shear stress could also generally predict the actual values of shear forces and moments resisted by the structure in the beams and columns, in the vicinity of the joint region during strong earthquakes.

It is worth mentioning at this point that predicting the actual values of connections shear stresses during an earthquake also involves predicting the actual values of the structures flexural strength ratio M_R ratio with the same accuracy.

2. The Model can be used to Predict the Failure Mode of the Structures in the Vicinity to the Joint Regions

Therefore, when the calculated joint shear stress τ_{cal} is greater or equal to the joint ultimate strength $(\tau_{cal} = \gamma_{cal}\sqrt{f_c} \geq \tau_{ult} = \gamma_{ult}\sqrt{f_c})$, then the predicted actual value of connection shear stress will be near τ_{ult} $(\tau_{ult} = \gamma_{ult}\sqrt{f_c})$. This is because the connection fails earlier than the adjacent beam(s). When the calculated joint shear stress τ_{cal} is lower than the connection ultimate strength $(\tau_{cal} = \gamma_{cal}\sqrt{f_c} < \tau_{ult} = \gamma_{ult}\sqrt{f_c})$, then the predicted actual value of connection shear stress will be near τ_{cal} because the connection permits its adjacent beam(s) to yield.

$(\tau_{ult} = \gamma_{ult}\sqrt{f_c})$ is calculated from the solution for the system of Eqs.(11) to (13).

From the above analysis it is clear that the model gives the failure mode of the structures (beam failure, joint failure, column failure). However, for $(\tau_{cal}/\tau_{ult} \leq 0.50)$ a pure beam failure mode is secured.

This methodology is very useful for the old reinforced concrete structures in order to determine their failure mode during strong future earthquakes after finding their reinforcing details and their concrete compressive strengths e.g. to find out the crucial beam – column joint or column failures in case it would happen which would involve dangerous collapses for these structures.

After the determination of the predicted failure modes, the retrofitting schemes of the old structures with the use of the model would be easier and safer designed. So the strengthened buildings would behave in the future strong earthquakes by a safe beam failure mode [Tsonos 1999, 2001a, 2001b, 2002].

The proposed model is also very useful for the new structures designed according to modern codes in order to determine what mode of failure would exactly exhibit and how dangerous would this be for the inhabitants living inside during strong earthquakes [Tsonos 2006, 2007].

Crucial questions about the seismic behavior of new structures would be raised:

1) Whether they exhibit an exclusive beam failure mode as they were theoretically designed to behave according to the modern codes, or
2) Whether the beam – column joints of these modern structures are potentially as in the old structures their weak links and would demonstrate during strong earthquakes a premature joint shear failure, which could lead to dangerous collapses.

With the use of the new model the potential inadequacy of the modern structures could be detected and using again this model a retrofitting safe scheme for these modern structures could be introduced.

4. An Innovative Solution Based on the Proposed Model Is Introduced for Curing the Disadvantages of the FRP Strengthening Method

For the first time an innovative solution is proposed which ensures a satisfactory and perhaps perfect seismic performance of existing old reinforced concrete buildings strengthened by using composite materials FRPs. The weak point in using such materials in repairing and strengthening reinforced concrete structures is the area of the beam-column joints.

Indeed, all the strengthened subassemblages in the beam-column region with composite materials FRPs described in the literature demonstrated in the best cases a mixed type failure during seismic type loading.

Thus, whereas during the first cycles of loading their beams yielded, yet during the following cycles a large part of damage of these strengthened subassemblages was concentrated in their joint regions.

Of course this failure mode is highly dangerous for the people who live in old buildings retrofitted in post – earthquake or pre-earthquake cases. The representative failure mode of the subassemblage of Figure 4(b) clearly demonstrates this critical situation [Tsonos 2009].

The whole strengthened beam-column joint region of this specimen not only failed but also was removed (i.e. leaving a hole in this location) during the last cycles of loading. This exactly is the reason why the Greek Code of the Repair and Strengthening of Reinforced Concrete Buildings (2010) does not allow the use of composite materials for the strengthening of reinforced concrete beam – column joints. The innovative solution presented in this study for curing the above inadequacy consists of strengthening the joint regions of old structures with a local reinforced concrete jacket designed according to the proposed model while retrofitting the columns can be achieved by using composite materials FRPs.

In order to investigate the effectiveness of the proposed solution a new beam-column subassemblage identical with the above subassemblage retrofitted by FRPs, was strengthened by the proposed innovative mixed type technique. The seismic performance of this specimen was optimal. The damage was concentrated only in the critical region of the beam, while the columns and the joint region were intact at the conclusion of the test (Figure 4(a), Tsonos 2009).

Conclusion

1) The design assumptions of modern codes do not soundly help avoiding a premature joint shear failure and do not secure the development of the optimal failure mechanism with plastic hinges in the beams near their adjacent column according to the requisite "strong column/ weak beam". The proposed model and method secure the exclusive formation of plastic hinges in the beams and the concentration of the damage only in the critical regions of the beams. According to the proposed model the columns and the beam – column joints remain intact during strong earthquakes.
2) The proposed model can be used both for the old structures and the modern structures in order to determine their failure mode during future earthquakes. The model can also be used in order to choose and propose the best retrofitting scheme for both the old and modern reinforced concrete structures, if this scheme is inevitable.
3) An innovative solution is presented in this study for the first time. The mixed type technique, by using local reinforced concrete jacketing in the joint region designed according to the proposed model while the columns were upgraded by composite materials, eliminated the disadvantages of the application of composite materials FRPs for the strengthening of old buildings structures, due to the ineffective strengthening of the beam – column joints by FRPs.

References

ACI-ASCE Committee 352, "Recommendations for design of Beam-Column Connections in Monolithic Reinforced Concrete Structures (ACI 352R-02)", Reported by Joint ACI-ASCE Committee 352, American Concrete Institute, Farmington Hills, Mich., 2002, 37 pp.

Bakir, P. G., "Seismic resistance and mechanical behaviour of exterior beam–column joints with crossed inclined bars", Structural Engineering and Mechanics, *An International Journal*, 2003, Vol. 16, No 4, pp. 493–517.

C. E. B. (1993), C. E. B. – F. I. P. Modes Code 1990, Bulletin d' Information C. E. B., 213/214, Lausanne.

El-Nabawy A., El-Din Fahmy Taher S., Khalil A.-H., El-Metwally S. E., "Behavior of reinforced high-strength concrete beam–column joint. Part II: Numerical simulation", Structural Concrete, *F. I. B.*, 2004, Vol.5, No 3, pp.101–112.

EN 1992-1-1: 2004, Eurocode 2: Design of Concrete Structures - Part 1-1: *General Rules and Rules for Buildings*, pp. 224.

EN 1998-1: 2004, Eurocode 8: Design of Structures for Earthquake Resistance -Part 1: *General Rules, Seismic Actions and Rules for Buildings, Stage* 51, pp. 229.

Greek Code of the Repair and Strengthening of Reinforced Concrete Buildings, *Earthquake Planning and Protection Organization* (E. P. P. O.), September 2010, pp. 230. (in Greek).

Ilki, A., "FRP Strengthening of RC Columns (Shear, Confinement and Lap slices)", *Retrofitting of Concrete Structures by Externally Bonded FRPs with emphasis on Seismic Applications*, F. I. B. (C. E. B.-F. I. P.) Bulletin 35, *Technical Report*, 2006, pp. 123-142.

Karayannis, C., Chalioris, C. and Sideris, K. "Effectiveness of R/C Beam-Column Connection Repair using Epoxy Resin Injections", *Journal of Earthquake Engineering*, Vol. 2, No. 2, 1998, pp. 217-240.

Karayannis C., Sirkelis G. "Effectiveness of R/C Beam-Column Connection Repair using Epoxy Resin Injections", *Earthquake Engineering and Structural Dynamics*, Vol. 37, 2008, pp. 769-790.

Karayannis C. G., Favvata M. J. and Kakaletsis D., "Seismic Behavior of Infilled and Pilotis RC Frame Structures with Beam-Column Joint Degradation Effect", *Engineering Structures*, Vol. 33, 2011, pp. 2821-2831.

Mosalam, K., "Seismic Retrofitting of R. C. Beam-Column Joints using FRP", Retrofitting of Concrete Structures by Externally Bonded FRPs with emphasis on Seismic Applications, F. I. B. (C. E. B.-F. I. P.) Bulletin 35, *Technical Report*, 2006, pp. 143-166.

New Greek Earthquake Resistant Code (E. R. C.-1995), Athens, 1995, pp. 145, (in Greek).

New Greek Code for the Design of Reinforced Concrete Structures (C. D. C. S.-1995), Athens, 1995, pp. 167, (in Greek).

Park, R. and Paulay, T., "Reinforced Concrete Structures", John Wiley Publications, New York, 1975, pp. 769.

Park, R., "A Summary of Results of Simulated Seismic Load Tests on Reinforced Concrete Beam – Column Joints, Beams and Columns with Substandard Reinforcing Details", *Journal of Earthquake Engineering*, Vol. 6, No.2, 2000, pp. 147-174.

Park S. and Mosalam K. M., "Shear Strength Models of Exterior Beam-Column Joints without Transverse Reinforcement", Pacific Earthquake Engineering Research Center, University of California , Berkeley, PEER NEES, 2009/106, November 2009, pp. 87.

Paulay, T., and Park, R., "Joints of Reinforced Concrete Frames designed for Earthquake Resistance", *Research Report* 84-9, Department of Civil Engineering, University of Canterbury, Christchurch, New Zealand, 1984, pp. 71.

Paulay, T. and Priestley, M. J. N., "Seismic design of reinforced concrete and masonry buildings", 1992, *John Wiley and Sons*, pp. 767.

Scott, B. D., Park, R., and Priestley, M. J. N., "Stress-Strain Behavior of Concrete Confined by Overlapping Hoops at Low and High Strain Rates", *ACI Journal*, Proceedings V. 79, No. 1, Jan.-Feb. 1982, pp. 13-27.

Sharma, A., Eligehausen, R., Reddy, G., R., "A new Model to Simulate Joint Shear Behavior of poorly detailed Beam-Column Connections in R. C. Structures under Seismic Loads, Part I: Exterior Joints", *Engineering Structures*, No 33, 2011, pp.1034-1051.

Soroushian, P., and Sim., J., "Axial Behavior of Reinforced Concrete Columns under Dynamic Loads", *ACI Journal*, Proceeding V.83, No. 6, Nov.-Dec. 1986, pp.1018-1025.

Tsonos, A. G., "Towards a New Approach in the Design of R/C Beam-Column Joints", Technika Chronika, *Scientific Journal of the Technical Chamber of Greece*, V. 16, No. 1-2, 1996, pp. 69-82.

Tsonos, A. G., "Shear Strength of Ductile Reinforced Concrete Beam-to-Column Connections for Seismic Resistant Structures", *Journal of European Association for Earthquake Engineering*, No. 2, 1997, pp. 54-64.

Tsonos, A. G., "Lateral Load Response of Strengthened Reinforced Concrete Beam-to-Column Joints", *ACI Structural Journal Proceedings*, V. 96, No. 1, 1999, pp. 46-56.

Tsonos, A. G., "Seismic Retrofit of R/C Beam-to-Column Joints using Local Three-Sided Jackets", *Journal of European Earthquake Engineering*, No. 1, 2001a, pp. 48-64.

Tsonos, A. G., "Seismic Rehabilitation of Reinforced Concrete Joints by the Removal and Replacement Technique", *Journal of European Earthquake Engineering*, No. 3, 2001b, pp. 29-43.

Tsonos, A. G., "Seismic Repair of Exterior R/C Beam-to-Column Joints using Two-Sided and Three-Sided Jackets", Structural Engineering and Mechanics, *An International Journal*, V. 13, No. 1, 2002, pp. 17-34.

Tsonos, A. G., "Cyclic Load Behaviour of Reinforced Concrete Beam-Column Subassemblages designed according to Modern Codes", *Journal of European Earthquake Engineering*, No 3, 2006, pp. 3-21.

Tsonos, A. G., "Cyclic Load Behaviour of Reinforced Concrete Beam-Column Subassemblages of Modern Structures", *ACI Structural Journal*, Vol. 194, No. 4, 2007, pp. 468-478.

Tsonos, A. G., "Effectiveness of CFRP - Jackets and RC - Jackets in Post-Earthquake and Pre-Earthquake Retrofitting of Beam-Column Subassemblages", *Engineering Structures*, No 30, 2008, pp. 777-793.

Tsonos, A. G., "Ultra-high performance fiber reinforced concrete: an innovative solution for strengthening old R/C structures and for improving the FRP strengthening method", Keynote Speaker, *Materials Characterisation Computational Methods and Experiments IV*, 2009, The New Forest, U. K., pp. 273-284.

Tsonos, A. G., "Performance Enhancement of R/C Building Columns and Beam-Column Joints through Shotcrete Jacketing", *Engineering Structures*, No 32, 2010, pp.726-740.

Reviewed by Professor Chris G. Karayannis.

In: Earthquakes
Editor: Kostas Konstantinou
ISBN: 978-1-62081-883-1

Chapter 10

INCREASING COMMUNITY POTENTIAL TO MANAGE EARTHQUAKE IMPACTS: THE ROLE OF SOCIAL AND CULTURAL FACTORS

***Douglas Paton*[1*] *and Li-ju Jang*[2 #]**
[1]School of Psychology, University of Tasmania, Launceston, Australia
[2]School of Medical Sociology and Social Work, Chung Shan Medical University, Taiwan

ABSTRACT

Recognition of the growing risk from earthquakes faced by communities in many countries, and particularly those situated on the Ring of Fire, has stimulated interest in promoting people's capacity to co-exist with the hazardous aspects of their environment. From a community perspective, one important contribution to this goal involves encouraging the adoption of preparedness measures that increase the ability of people and communities to cope with and adapt to the environmental impact of earthquakes. However, despite the attention and financial resources devoted to encourage people to do so, the fact that levels of preparedness in communities at risk from earthquakes are generally low makes it important to determine why this is the case and to identify ways of encouraging sustained community preparedness. This chapter contributes to understanding the factors that influence household and community preparedness for earthquakes by identifying how people interpret their risk and make decisions about how to manage the risk pose by living in seismically active areas.

This chapter discusses how people's decisions about preparing for earthquakes derives less from the objective use of information and more from how community members interpret earthquake risk and information about it. It describes how interpretive processes at the person (outcome expectancy), community (community participation, collective efficacy), societal (empowerment, trust) and cultural levels of analysis interact to predict levels of earthquake preparedness. The chapter accommodates the fact that earthquakes occur in many countries and acknowledges that the practical utility

* Correspondence author: Douglas Paton PhD. School of Psychology, University of Tasmania, Launceston, Tasmania 7250, Australia. Douglas.E-mail: *Paton@utas.edu.au*.

School of Medical Sociology and Social Work, Chung Shan Medical University, No. 110, Sec. 1, Jianguo N. Road, Taichung 40201, Taiwan.

of any theory is increased if it can be demonstrated that it applies across cultural and national boundaries.

The cross-cultural applicability of the theory is discussed by comparing findings from research into earthquake preparedness in countries that differ in cultural orientation; New Zealand, Taiwan, and Japan. The theoretical (e.g., identifying the degree to which earthquake preparedness is culturally equivalent) and practical (e.g., providing a common basis for collaborative learning, research and risk management intervention between countries) implications of the cross-cultural equivalence of the model are discussed. The chapter also discusses how cultural specific processes such as the Hakka Spirit in Taiwan and Jishubo in Japan influence how community processes are enacted to manage earthquake risk. Finally, these perspectives are combined to discuss how the sustained effectiveness of strategies designed to develop community capacity to deal with the environmental impact of earthquakes can be increased by integrating risk management activities with community development strategies.

INTRODUCTION

Many countries and particularly those situated on the Ring of Fire such as Japan, Taiwan and New Zealand, periodically experience adverse impacts from earthquakes. The consequences that can arise from this exposure were tragically illustrated by the earthquake/tsunami that impacted on several communities on Japan's eastern seaboard and the earthquakes that occurred in Christchurch in 2010 and 2011. The fact that these countries are positioned on tectonic plate boundaries makes the possibility of experiencing substantial environmental impacts from earthquakes an enduring fact of life for their citizens. However, while the occurrence of earthquakes is inevitable in these countries, the impact of earthquakes on the physical and social environment is amenable to change. Societies and their citizens can makes choices that can facilitate their capacity to co-exist with a potentially hazardous environment. Developing the potential to make these choices is fundamental to the goal of developing disaster resilient communities.

The goal of developing resilient communities can be pursued at several levels. For example, potential environmental impacts can be ameliorated using societal-level strategies such as land use planning and structural mitigation [Paton *and* Johnston, 2006]. The pursuit of enhancing societal sustainability in the face of periodic earthquake activity can also be furthered by encouraging people and communities to adopt preparedness measures. This is the aspect of managing the potential impact of earthquake hazards that is the subject of this chapter.

Events in Japan and New Zealand in 2011 and in Taiwan in 1999 demonstrated the potential for earthquakes to occur without warning. In this context, an important contribution to managing earthquake impacts is the degree to which people and communities are prepared (e.g., securing houses to foundations, household emergency plans and resources, ability to work with others to confront local problems, capacity for self-reliance etc) in ways that increase their ability to anticipate, cope with, adapt to and recover from the consequences of seismic activity [Paton, 2006]. Risk management strategies designed to facilitate personal, household and community preparedness can thus make an important contribution to enhancing societal sustainability. However, despite the attention and financial resources devoted to attempting to encourage people to prepare, the goal of ensuring sustained preparedness in communities susceptible to experiencing earthquake impacts has proved elusive.

It has become increasingly apparent that neither living in areas susceptible to experiencing earthquakes (even when people acknowledge this risk) nor just providing people with information on earthquakes and what can be done to mitigate their consequences exercises an enduring influence on community hazard preparedness [Dow and Cutter, 2000; Lindell and Perry, 2004; Lindell and Whitney 2000; Paton, Smith and Johnston, 2005; Perry and Lindell, 2008]. This makes it important to determine why levels of preparation remain low and to identify ways in which sustained community preparedness can be encouraged.

This chapter seeks to advance understanding of the factors that influence household and community preparedness for earthquakes from three perspectives. The first develops from acknowledging that people's perception of earthquake risk and the decisions they make about how to manage perceived risk are socially constructed. This identifies a need to articulate those aspects of people's social networks that influence how they interpret the risk and uncertainty associated with natural hazard activity. The second stems from distinguishing between the content of risk communication and the competencies and processes that influence how people interpret and use information. These competencies and processes develop and operate within the context of people's social networks. This chapter focuses on identifying these competencies and how they are mobilised to influence whether or not people prepare for earthquakes. The third arises from recognizing that earthquakes transcend cultural boundaries and acknowledgement that cultural practices could have a significant influence on the nature of the social processes that influence how people define and respond to seismic risk. Given the cultural diversity inherent in the countries situated around the Ring of Fire, cross cultural and cultural issues must be accommodated in both theory development and testing and in risk management planning and design.

This chapter opens by discussing the fact that people are not passive recipients of experience or information. Rather, they actively strive to interpret or impose meaning on their environmental experience, and this extends to how they deal with information about hazards and their mitigation. The social and psychological processes that underpin how meaning is imposed on environmental events provides the foundation for the interpretive processes that influence the actions people take to manage earthquake impacts.

The foundation for the work being reported in this chapter was finding a prominent role for the variable critical awareness in a study of earthquake preparedness [Paton et al., 2005]. Critical awareness assesses the frequency with which people discuss earthquake issues with others in their community and think about hazard issues. What was significant in this study was the finding that a variable that tapped into a social process, talking with others about earthquakes, was the strongest predictor not only of whether people decided to prepare, but also of their deciding not to prepare [Paton et al., 2005]. This finding prompted a more searching investigation of how social experience influenced people's decisions about managing potential earthquake impacts. Doing so necessitated a shift in theoretical orientation.

One theory that offered the potential to explain findings such as how discussions with others could predict people deciding to prepare and not to prepare was symbolic interactionism.

THE MEANING OF ENVIRONMENTAL EVENTS

Symbolic interactionism [Blumer, 1969] suggests that people actively and constantly interpret stimuli from the environment while interacting with that environment, and integrate the interpretations through a process of reflection with pre-existing mental models. How people interpret the world differs from person to person (e.g., take different objects into account and interpret and integrate them in different ways), changes over time, depends on context, and reflects the unique experiences people have accumulated during their lives [Blumer, 1969; Bowers, 1988]. Once the meanings of the objects, events and relationships that comprise their environment are determined, people then act towards them in ways consistent with these meanings.

When applied to people's relationship with natural hazards such as earthquakes, this means that the messages conveyed by risk management agencies interact with people's pre-existing beliefs and expectations (e.g., about hazards and mitigation) with the outcome of this interpretive process influencing their actions, but not necessarily in the way anticipated by the authors of the public education programs who aim to increase preparedness. For example, if the recipients of mitigation and/or preparedness information believe that earthquakes are uncontrollable natural phenomena that create catastrophic consequences for people, they are unlikely to believe or attend to information that urges people to prepare [McClure, Walkey and Allen, 1999; Paton, 2008]. If, on the other hand, the people who receive new information about hazard mitigation have a history of preparing (i.e., a component of their pre-existing mental models includes preparing for earthquakes), they are more likely to accommodate new information in ways that contribute to their understanding of earthquakes and to more sophisticated thinking about mitigation.

Irrespective of whether they take new information on board or reject lessons from experience or advice and information from others, these interpretive processes function to facilitate people's ability to adapt as well as possible to changes in their environment [Denzin, 1992]. This is what was found in Paton et al's. [2005] study. Some people adapted by preparing, while for others not doing so was the decision of choice. Similar findings were reported by Paton, Bürgelt and Prior [2008]. Not preparing was seen as adaptive by some people as a result of, for example, not feeling any strong connection to where they lived, not intending to stay in the area for long, or because they believed being insured would compensate for any loss experienced.

Thus, what an adaptive response can mean to people can range from adopting preparedness measures at one end of the adaptive spectrum to denying the need to prepare at the other [Paton and McClure, in press]. As a result, the outcomes of the process of integrating information or experience within mental models can either constrain or facilitate certain individual interpretations and actions. Which of these outcomes prevails is also influenced by the degree to which the behaviour of interest (e.g., earthquake preparedness) develops from people's actual experience. This is unlikely in the case of earthquake hazards.

Large catastrophic earthquakes occur infrequently. Consequently, people living in areas susceptible to experiencing them do not generally have opportunities to gain experience at first hand. This means that information from others is an important source of raw material for people's meaning making process [Blumer, 1969]. This applies as much to how people gain their knowledge of earthquake risk and formulate the actions they believe are necessary to adapt to this risk as it does to the other domains that comprise people's everyday lives.

According to this view, the meaning people attribute to earthquakes and the action they take to mitigate and/or prepare for them are forged, changed and sustained through the social relationships with family, friends, other community members (e.g., family, work, social, sporting and religious meetings) and civic agencies within their everyday social contexts over time [Paton, 2008]. It thus becomes important to identify the characteristics of people's social contexts that guide their interpretation and to articulate how this process contributes to developing people's potential to manage the environmental impacts of earthquakes. Furthermore, recognition of the important role that social processes plays in risk management draws attention to a need to consider how the cultural dimension of the social context influences hazard preparedness.

Cultural Perspectives on Managing Earthquake Impacts

Even a cursory overview of the countries that are situated around the Ring of Fire reveals considerable cultural diversity. This chapter acknowledges a need to take this diversity into account. It does so in two ways. When analysing cultural influences on social and behavioural outcomes it is pertinent distinguish between etics (cultural universals that transcend cultural boundaries) and emics (culture specific processes unique to each country) [Matsumoto and Juang, 2008]. This chapter first discusses the etics of earthquake risk by examining the cross cultural equivalence of a theory of hazard preparedness developed in New Zealand [Paton, 2008]. It assesses the cross cultural utility of this theory by assessing the degree to which this theory can account for household earthquake preparedness in countries (Japan and Taiwan) that differ in their cultural characteristics from New Zealand, the country in which the theory was initially developed and tested. By comparing data from culturally diverse locations it is possible to examine the degree to which the processes that underpin how people prepare for earthquake threats represents a universal social and psychological phenomenon.

The existence of an equivalent conceptual framework that transcends cultural boundaries would demonstrate the existence of a potential common basis for collaborative learning and research across national borders. From a practical perspective, confirmation of cross-cultural equivalence would make information about how to facilitate preparedness (e.g., assessing communities, identifying areas for intervention, and guiding the development of community engagement programs) more readily available to risk management agencies in different countries. Furthermore, it would also make such information available to countries (e.g., developing countries) that lack the resources to undertake this work themselves. The second cultural issue examined focuses on how community-level risk management processes are enacted in everyday life. That is, on how cultural emics influence earthquake preparedness. It does so by discussing culturally unique processes in Taiwan and Japan that have been implicated in influencing how people adapt to earthquake risk.

In the next section, this chapter discusses the personal and social mechanisms that influence how people interpret and make sense of environmental uncertainty and consequently decide whether or not to prepare for earthquakes. In particular, it examines how peoples' everyday experiences (e.g., participation in community activities) influence interpretation and decision making. The chapter then examines the degree to which these personal and social mechanisms operate in an equivalent manner in cultures that differ with regard to their relative positions on cultural dimensions such as individualism and collectivism [Brislin, 2000; Hofstede, 2001; Norenzayan and Heine, 2005; Triandis, 1995].

The first step is to identify factors that influence how people interpret hazards and to select variables that capture this process. This chapter will specifically focus on how community characteristics (e.g., participation in community life) and competencies (e.g., collective problem solving) developed through everyday life experiences influence people's potential to manage earthquake impacts.

Personal Beliefs about the Relationship between Earthquakes and Preparedness

Full details of the development of the model described in this chapter can be found in Paton [2008]. Key issues are summarised here to illustrate the rationale for the selection of the individual and social predictors in the theoretical model. The theory proposes that decisions result from the interaction between personal beliefs and social processes. People's beliefs about earthquakes and what earthquakes can do to the physical and social environment play pivotal roles in determining whether or not people prepare.

While they can vary in intensity and magnitude in any given area, people's beliefs about earthquakes often focus on their catastrophic potential [McClure et al., 1999]. The theory suggests that the preparation process is triggered by people's judgments about the relationship between earthquake characteristics (e.g., the magnitude, intensity and distribution of ground shaking) and the likely effectiveness of human action to mitigate the impacts associated with these characteristics. This decision process helps understand why the passive presentation of information to people may be ineffective as a mechanism for facilitating preparedness.

Large, damaging earthquakes are, fortunately, relatively rare. This means that people cannot readily find out about them and the effectiveness of mitigation actions for themselves. Rather, they rely on experts for this information. The information contained in public education programs essentially advises people that if they adopt a particular behaviour (e.g., secure house to foundations) the outcome will be increased safety (e.g., house remains standing if an earthquake occurs). However, people interpret this information and its recommendations (e.g., in light of their pre-existing beliefs about the magnitude and perceived controllability of earthquakes, how catastrophic their consequences are) to estimate whether they expect that outcome (i.e., increased safety, reduced damage etc) to occur. The 'outcome expectancy' construct describes this interpretive process [Bennet and Murphy, 1997] and it can be further subdivided into negative and positive components.

Negative outcome expectancy describes a belief that earthquakes are too catastrophic for personal action to make any difference to peoples' safety (e.g., people believe that ground shaking would be so powerful that no building can remain standing). If people hold this belief, they are less likely to prepare. In contrast, positive outcome expectancy, the belief that adopting mitigation and preparation measures will increase personal safety and mitigate a hazard consequence, can motivate people to prepare. These variables, positive and negative outcome expectancy, define the theoretical starting point of the preparedness process [Paton,

2008]. However, a belief that preparing can be effective is not necessarily the same as knowing what to do or how to prepare. If people need additional guidance, they look first to other community members.

Working with Others to Manage Risk

Faced with complex and uncertain events, when they do not posses all the information they need themselves, peoples' perception of risk and how they might mitigate it, is influenced by information from others who share their interests and values [Earle, 2004; Lion et al., 2002; Paton and Bishop, 1996; Poortinga and Pidgeon, 2004]. Paton, Bürgelt, and Prior [2008] described how sharing stories of previous wildfires during casual meetings with neighbours or when involved in community activities facilitated residents becoming more knowledgeable about the local history of wildfires, increased their acceptance of both wildfire risk and the importance of preparing for wildfires, and helped them identify what they should do and why they should prepare. That is, discussion with "like minded" others not only provided information about risk, it also helped people work out what they should do to mitigate their risk and how they should do it. In this way, interactions with others also performed a problem solving function.

Faced with being asked to prepare for complex, infrequent and potentially threatening earthquakes, people may first have to determine what consequences they could face, work out what would be an effective response, and then consider what information and resources they require to enact their mitigation strategies. One construct that encapsulates community members' ability to identify needs and formulate questions is collective efficacy. Collective efficacy is a measure of co-operation and assistance available within a community and community members' ability to assess their capabilities and resource needs and formulate plans to use resources to confront challenging tasks [Duncan et al., 2003]. The collective efficacy measure also proved to be a valid predictor of resilience in Thai populations affected by the 2004 Indian Ocean tsunami [Paton and Tang, 2009]. This provides evidence of the utility of a core variable in the model to predict how people actually managed the impact of the tsunami hazard. Consequently, "community participation" and "collective efficacy" were included as social context variables in the theoretical model [Paton, 2008].

The final set of social context variables included in the theoretical model derived from anticipating that community members' deliberations about their risk and what they could do to mitigate that risk could identify information and resource needs that could not be met within existing community contexts. For example, they could need expert input to understand ground shaking and the factors that influence its distribution and intensity. They might also need expert input on things like the structural alterations that could be made to homes to increase their ability to withstand ground shaking. Under these circumstances, people would turn to civic (e.g., emergency management) and expert (e.g., scientists, architects, builders) sources to acquire the necessary information and resources.

The fact that people and communities may have to turn to others introduces a need to consider an additional interpretive process in the model. This process is less about the information or advice per se and more about people's experience with and beliefs about the quality of their relationship with a source of information (that is independent of the information they provide per se) when making decisions under conditions of uncertainty [Paton, Smith, Daly, and Johnston, 2008]. An important influence on the quality of the relationship between people and agencies is trust.

The Community-Agency Relationship

Trust is particularly important when people face uncertainty and have to rely on others to provide information they cannot acquire for themselves [Earle, 2004; Siegrist and Cvetkovich, 2000]. Trust influences perception of other's motives, their competence and the credibility of the information they provide. As uncertainty increases, so does the importance people attribute to their trust beliefs about, and their past experiences with, the sources of information they turn to or have to rely on [Siegrist and Cvetkovich, 2000]. Thus, people's beliefs about the goals and motivations of sources of information influence trust and, consequently, whether or not people use information from a specific source.

People's willingness to take responsibility for their own safety is increased, and decisions to prepare more likely, if they believe that their relationship with formal agencies is fair and empowering (e.g., agencies are perceived as trustworthy, as acting in the interest of community members) [Earle, 2004; Eng and Parker, 1994; Lion et al., 2002; Paton and Bishop, 1996; Paton et al., 2005; Poortinga and Pidgeon, 2004]. If, however, the relationship between community members and an agency or expert source is not perceived as fair and empowering (e.g., information provided is contradictory or inappropriate or sources such as architects or builders are perceived as having a vested interest in the information they provide), the consequence is a loss of trust in the agency (i.e., the source of information), reducing the likelihood that people will use the information provided by an agency to guide their preparing. The loss of trust can occur rapidly if people, for example, perceive being given erroneous information or believe that agencies have acted inappropriately [Poortinga and Pidgeon, 2004].

To accommodate these facets of the process, the concepts of empowerment and trust were included in the theoretical model [Paton, 2008]. Individual, social and relationship variables interact to predict what people intend to do to manage their risk.

Intention to prepare was used as the dependent variable because of its ability to facilitate assessment when comparing cultures. Intentions can accommodate the cross cultural comparison of communities that differ with regard to characteristics such as hazard history, culture, public education strategies and recommended preparedness measures. Intention has been found to be a good predictor of actual preparedness [Lindell and Perry; 2004; Paton et al., 2005; Sheeran, 2002] and has been demonstrated to be a good predictor of preparedness in a highly collectivistic culture. Paton, Tedim and Shand (in press) found a correlation of .61 between intention and hazard preparedness in a Portuguese sample. The significance of this finding derives from the fact that Portugal has similar cultural characteristics to Taiwan and Japan [Hofestede, 2001]. This supports the utility of using intentions in both individualistic and collectivistic countries.

Intention is a useful variable in other respects. For example, intention is less susceptible than a measure of actual preparedness to bias from factors such as, for example, resource limitations (e.g., time, money, physical resources, expertise) that prevent people from converting their intentions into actual preparedness [Paton et al., 2005; Paton et al., 2008]. Intention thus provides a more robust basis for comparison across cultures. The intention measure comprised items that assessed people's intention to acquire hazard knowledge, increase actual preparedness, and to work with other people/civic agencies to develop knowledge and capability [Paton et al., 2005].

The theory, as originally tested [Paton, 2008; Paton et al., 2009], proposes that if people hold negative outcome expectancy beliefs, it will reduce the likelihood of their preparing. If people hold positive outcome expectancy beliefs (POE), they will either proceed to prepare, or, if lacking the information they require, the relationship between POE and empowerment will be mediated by the social structural processes (community participation and collective efficacy) used to articulate community members needs and expectations. Next, the theory proposes that empowerment mediates the relationship between structural factors and trust, with intentions mediating the relationship between trust and actually preparing [Paton et al., 2009]. The question that remains to be answered concerns the degree to which this theory can account for differences in people's intention to act to reduce the potential environmental impacts of earthquakes in countries that differ significantly with regard to their cultural characteristics.

Table 1. The locations, cultural characteristics, and data collection information on the study areas (I-C: Individualism-Collectivism)

Location	I – C Score	Date	N	Hazard
New Zealand	79	2007	256	Earthquake
Japan	46	2008	251	Earthquake
Taiwan	17	2009	263	Earthquake

Table 2. The variables, sources and scoring of the variables

Measure	Source	Items	Scoring Range
Positive Outcome Expectancy.	[Bennet and Murphy 1997].	4	1 (Strongly disagree) 5 (Strongly agree).
Negative Outcome Expectancy. Critical Awareness.	[Bennet and Murphy 1997]. [Dalton, Elias, and Wandersman 2001].	4 5	1 (Strongly disagree) - 5 (Strongly agree). 1 (Never) – 5 (Once a week or more).
Collective Efficacy.	[Zaccaro et al., 1995].	12	1 (Very Low) -5 (Very High).
Empowerment.	[Speer and Peterson 2000].	4	1 (Not at all) - 5 (Always).
Trust.	[Dillon and Phillips 2001].	5	1 (Strongly disagree) - 5 (Strongly agree).
Community Participation.	[Eng and Parker 1994].	5	1 (Strongly disagree) - 5 (Strongly agree).
Intention to Prepare.	[Paton et al., 2005].	5	1 (No), 2 (Possibly), 3 (Definitely.

EARTHQUAKE PREPAREDNESS IN NEW ZEALAND, JAPAN AND TAIWAN

Analysis of whether the theoretical model can account for differences in levels of preparing across cultures was examined by comparing data from three countries whose position on the Ring of Fire makes them susceptible to experiencing earthquakes; New Zealand, Japan and Taiwan. For this analysis, comparison focused on what has been identified as one of the most important cultural dimensions, Individualism-Collectivism (I-C) [Hofstede, 2001; Triandis, 2004].

The countries from which data were collected were selected because their relative positions on this dimension (the lower the score, the more collectivistic the culture) provided an opportunity to collect data from countries with relatively low (New Zealand), medium (Japan) and high (Taiwan) levels of collectivism (Table 1). With increasing collectivism, the greater is the importance assigned to the social context as a mediator of everyday activities.

Questionnaires containing the variables described below (see Table 2) were distributed randomly to households in the Tung Shih area (Taiwan), Napier (New Zealand) and Kyoto (Japan). All the variables were pilot tested to ensure the face validity of the items prior to data collection. The procedure recommended by Brislin (1986) was used to ensure that, after translation, the questionnaires contained equivalent content. Because the model proposes that several independent variables interact to predict the dependent variable Structural Equation Model (SEM) was selected for the analysis. SEM can calculate multiple and inter-related dependence relationships simultaneously, allowing it to test the model as a whole and to show how well the data fit the hypothesized model (Goodness-of-Fit) [Byrne, 2001].

ASSESSING THE FIT BETWEEN DATA AND THE MODEL

As recommended by Reisinger and Mavondo (2006), multiple fit indices were reported. The likelihood-ratio chi-square (χ^2) statistic is the primary measure of overall fit, and nonsignficant differences indicate a good fit of the model to the data. The Root Mean Square Error of Approximation (RMSEA) assesses the amount of error present in the fit and is considered to produce accurate assumptions about model quality, in addition to providing precise fit values, with values < 0.05 suggesting a good fit to the data [Byrne, 2001; Reisinger and Mavondo, 2006].

Values of the Goodness-of-Fit Index (GFI), the Normed Fit Index (NFI) and the Comparative Fit Index (CFI) greater than 0.95 are considered to reflect a good fit to the data [Byrne, 2001; Reisinger and Mavondo, 2006]. The Fit Indices are described in Table 3. In each location, the χ^2 values are not significant, the RMSEA values are under .05, and the remaining fit indices exceeded the .95 cut-off. Overall, the fit indices (Table 3) indicate that the model was a good fit to the data in each location.

The findings are summarised in Figures 1-3. Sheeran [2002] concluded that accounting for 28% of the variance in intentions would confer upon findings derived from using social cognitive models a medium to large effect size. The models accounted for 38%, 29% and 33% of the variance in intentions to prepare in New Zealand, Japan and Taiwan respectively, indicating medium effects in all locations and permitting use of the model to be used to inform intervention planning to guide the development of people's potential to reduce the physical and social environmental impacts associated with earthquake hazards.

IMPLICATIONS FOR THEORY AND PRACTICE

The model was a good fit to the data in each location (Figures 1-3; Table 3). The pattern of relationships was relatively consistent, particularly regarding the role of social variables.

Table 3. The fit indices from the SEM analyses

Country	Location	χ^2	df	p	RMSEA	CFI	NFI	GFI
NZ	Napier	5.919	7	0.549	0.00	966	983	994
Japan	Kyoto	7.478	6	0.279	0.03	994	973	991
Taiwan	Tung Shih	10.862	10	0.368	0.020	997	986	990

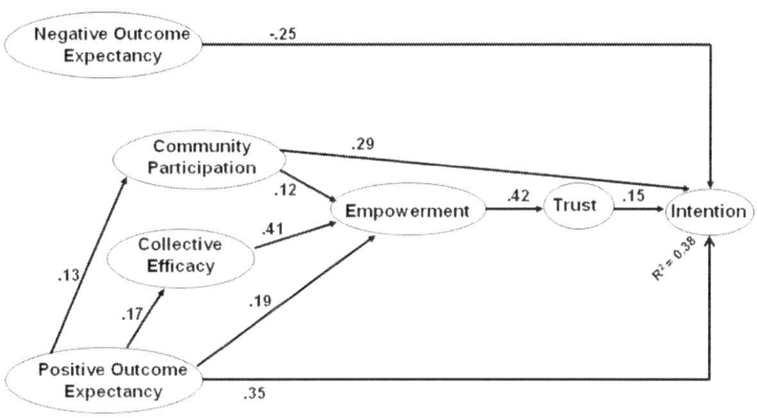

Figure 1. Summary of structural equation analysis for New Zealand (Napier).

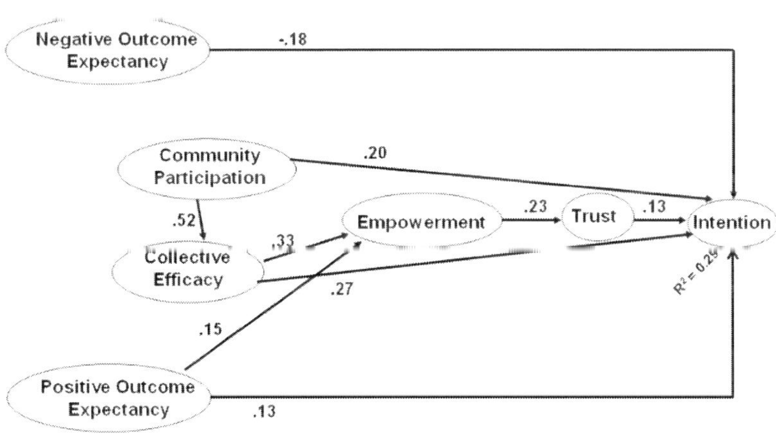

Figure 2. Summary of structural equation analysis for Japan (Kyoto).

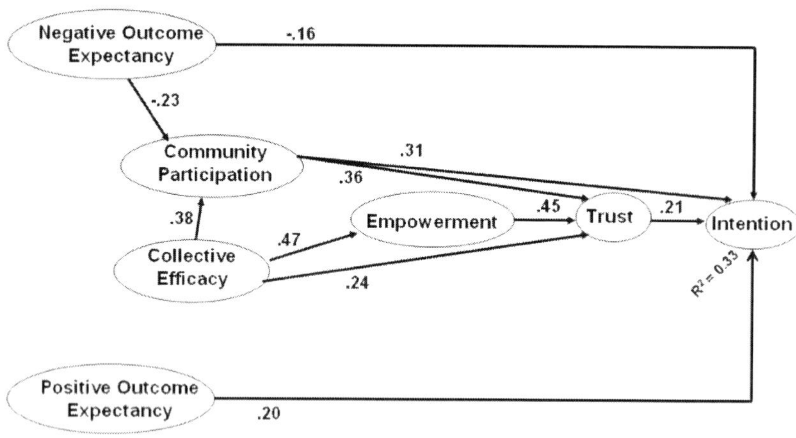

Figure 3. Summary of structural equation analysis for Taiwan [Tung Shih].

The relationship between community process (community participation and collective efficacy) and intentions was mediated by institutional (empowerment and trust) factors in all three countries. The findings offer some support for the existence of some measure of cross-cultural equivalence in how people make choices about managing their earthquake risk.

The role of the individual beliefs (positive and negative outcome expectancy) was mixed. The relationship was consistent (the action of outcome expectancy being mediated by the social variables) with theoretical expectations [Paton, 2008; Paton et al., 2009] in the Taiwan analysis and partly supported in New Zealand (Figure 1). This was not so in Japan (Figure 2) where individual beliefs and social processes had relatively independent influences on intention.

With respect to the Taiwan analysis (Figure 3), the stronger influence of outcome expectancy beliefs may have stemmed from people's experience of the 921 earthquake (September 1999). Social recovery processes in the area from which data were collected were still in place and the 10 year anniversary of the 921 earthquake was forthcoming. It is possible that past experience and current reminders mobilised awareness of the challenging nature of earthquakes and increased the influence of outcome expectancy beliefs. More work is needed to explore this possibility. The Taiwanese and Japanese findings were similar, and differed from the New Zealand analysis, in that more coherent links were evident between the community variables. This is consistent with their relatively higher position on the collectivism dimension.

The Taiwanese data were also interesting in that collective efficacy preceded community participation. One tentative explanation for this derives from the possibility that participation is a fact of life in collectivistic cultures like Taiwan and a collective capacity for dealing with challenge and change under normal circumstances would guide how non-routine threats would be dealt with. While additional work is required to examine this in more detail, similar findings in other highly collectivistic cultures (Indonesia and Portugal) support this view and identify an area of future work.

Overall, this comparison offers some support for the view that, irrespective of culture, the more community members are able to collectively formulate their risk management needs and strategies and the more they perceive their needs as having been met through their relationship with civic agencies, the more likely they are to trust them and the information they provide, and to use information to decide to adopt hazard preparation measures.

IMPLICATIONS FOR INFLUENCING PEOPLE'S POTENTIAL TO MANAGE EARTHQUAKE IMPACTS

Increasing people's potential to mitigate the impact of earthquakes on the physical and social environment is a pivotal risk management activity. Putting people into a position where they can take responsibility for making choices is about more than just making information about earthquakes and how to mitigate their consequences available to people. Getting people and communities to a position where they can influence potential hazard consequences is a function of the degree to which risk communication processes accommodate the fact that preparation results from the interdependent action of several interpretive processes.

The first process concerns people's beliefs about whether or not earthquake hazard consequences can be ameliorated by human action. The second described how characteristics (community participation) and competencies (collective efficacy) in people's social context provided a means of interpreting risk and actions, reinforcing the role that social networks play in the formulation of people's risk beliefs and mitigation strategies. This provides the context in which people discuss hazard issues. It is also possible that there are generational differences in the quality of discussion. The Taiwan data revealed that older people were more likely to prepare. While more work is required to test the idea, this observation suggests that this reflects their greater experience of adapting to adverse events. If confirmed, this suggests a need to explore how the experience of older people could be used to inform discussion within communities. The third concerned how people's perception of the quality of their relationship (e.g., whether it empowered them, whether they trusted the source) influenced whether the relationship between communities and agencies has provided the community with the information and resources that met people's needs and allowed the community to take responsibility for mitigating hazards. Community engagement in decision making about acceptable levels of risk and the acceptability of strategies used to mitigate this risk positively influence risk acceptance, increase community members' acceptance of responsibility for their own safety, and increase collective commitment to confront hazard consequences using community-led mitigation strategies.

Evidence for cross-cultural equivalence should not, however, be taken to imply that cultural factors can be ignored. The introduction discussed the need to search for the emics, the culture specific, influences on how people deal with challenges and change in life. While the cross cultural analysis outlined above identified that the same process (Figures 1-3) could operate across different countries, this should not be taken to imply that the process is enacted in the same way in different cultures. For example, community participation was important in each country. However, the origins of participatory activities and how they are sustained and enacted to influence people's risk beliefs and mitigation options can differ from country to country. Consequently, the culture specific influences on preparedness need to be considered. Understanding how and why this occurs is addressed in the next section.

Understanding emic or culture-specific community processes can provide valuable insights into how peoples' potential to manage the environmental impact of impacts of earthquakes. Two such processes can be found in the countries that are the focus of the current chapter. These are Jishubo in Japan and the Hakka Spirit in Taiwan. The importance of examining these processes can also be traced to their illustrating how mainstream aspects of (cultural) life can influence how people prepare for and respond to earthquakes.

Culture Specific Processes in Japan

Jishu-bosai-soshiki [Jishubo] are neighbourhood-based "autonomous organizations for disaster reduction" unique to Japan [Bajek, Matsuda and Okada, 2008]. Jishubo constitute a mechanism that local government can mobilise to organize hazard preparedness events (e.g., drills, workshops) and facilitate local response and rescue activities (e.g., rescue residents, provide initial first-aid, supply food and water to survivors) in the event of a disaster [Bajek et al.]. The organization (decision-making, functions, etc.) of Jishubo is ostensibly delegated to community residents through Chonaikai ("community councils"), a traditional Japanese neighbourhood governance system that includes within its functions the maintenance of safe and secure communities through implementing various community activities, including both disaster reduction and crime prevention.

Bajek and colleagues [2008] discuss how the trigger for developing Jishubo was the Kobe earthquake. This led to revising laws to enable the government to facilitate both the development and maintenance of Jishubo and to increase the community uptake of Jishubo. Chonaikai and the Jishubo organised within their jurisdictions provide an illustration of a culturally specific mechanism to encourage citizen participation in disaster mitigation and response based on mobilising culturally-embedded social processes.

Chonaikai have been implicated in playing a role in another culture-specific mechanism [Bhandari, Okada, Yokomatsu and Ikeo, 2010]. Bhandari et al. discuss how the Danjiri Matsuri ritual in Kishiwada City, Osaka, can have a positive influence on earthquake preparedness. This annual ritual is organised under the auspices of Chonaikai in ways that are instrumental in forging and sustaining collective identity and coordination amongst participating neighbourhoods. Bhandari and colleagues report how 80% of respondents to a survey reported that participation in the Danjiri Matsuri ritual enhanced neighbourhood relations and sense of community. Interestingly, they demonstrated that trust mediated the relationship between the perceived gain and enhanced experience of social capital from participation and hazard awareness and capacity for self reliance in the face of hazards such as earthquakes. The pivotal role of trust in this analysis provides additional support for its inclusion in the theoretical model discussed above (Figures 1-3) and illustrates how culture specific processes can influence how trust develops and influences earthquake preparedness.

Bhandari et al. [2010] conclude by emphasising the importance of understanding how culture specific processes influence disaster resilience and how activities organised within social institutions such as Chonaikai contribute to increasing trust between neighbours and towards the institutions neighbourhood members rely on for the information and resources they need to manage hazard events such as earthquakes. Japan is not the only country in which culture specific social mechanisms capable of influencing response to earthquakes are evident. Another example can be found in Taiwan.

CULTURE SPECIFIC PROCESSES IN TAIWAN

Jang [2005] discussed how the Hakka Spirit, a cultural characteristic of the Hakka people residing in Tung Shih in Taiwan, facilitated the capacity of communities to thrive despite experiencing adverse impacts from the 921 earthquake that hit Taiwan in 1999. The Hakka Spirit describes an emic process, a unique set of cultural beliefs and practices, that encompass "being able to hold on firmly despite extreme adversity" and to "keep on doing something without regard to one's own strength" [Jang, 2005]. It can be described in terms of constituent qualities such as frugality, diligence, self-reliance, responsibility, and persistence that combine to facilitate a capacity to thrive in the face of adversity. With respect to natural hazards, the belief that natural disasters reflect the will of God, that it is important for people to co-exist with nature and not to control nature, and to strive to live with nature in an harmonious way means that the Hakka Spirit embodies the core goal of disaster resilience. That is, the importance of learning to co-exist with periodically hazardous environments in ways that facilitates the human capacity to cope with disaster [Paton, 2006].

Jang [2005] discussed how collective approaches to problem solving contributed to the development of (positive) outcome expectancy beliefs through farming practices that evolved to limit the impact of typhoons on people's livelihoods. Typhoons occur annually and destroy the buds of fruit trees. As soon as the typhoon has left Taiwan, farmers would work collectively on each others' farms in rotation, to re-graft buds into fruit trees, with the outcome expectancy beliefs (regarding the effectiveness of collective actions in response to hazard consequences) developing from evidence that the grafted buds would recover and bear fruit. Hakka farmers might have to repeat this process several times depending on the number and severity of typhoons in any particular year. Jang also discussed how this outcome expectancy belief spilled over into Hakka people's disaster preparedness and was linked to increasing the likelihood that people would act to secure their homes from earthquake consequences [Jang and LaMendola, 2006].

Analysis of the essential elements of the Hakka Spirit also offers support for the selection of the variables in the theoretical model (Figures 1-3). Characteristics such as reciprocal support practices and collaborative problem solving developed over time as part of an adaptive response to meeting the challenges imposed by agricultural life in remote mountain communities. These characteristics are comparable to the community participation and collective efficacy variables in the model (Figures 1-3).

While certain parallels can be drawn between the Hakka Spirit and the culture general processes that predict preparedness (see Figures 1-3), the difference is that for those for whom the Hakka Spirit underpins a way of life the beliefs and competencies that facilitated people's resilience to events such as the local consequences of the 921 earthquake are transmitted and sustained through cultural mechanisms. In Western populations, more active personal and community development is required to achieve the same capacities.

Integrating the etic and emic processes discussed here provides a framework for cross cultural collaboration, learning and the cross-fertilisation of research and intervention in ways that transcend national and cultural boundaries. By identifying the universal role of social evaluation and planning processes (e.g., participation, collective efficacy) the etic analysis identified a capacity for all countries to learn from the emic processes (e.g., Jishubo, Hakka Spirit) that facilitate how social mechanisms are enacted in ways that increase people's potential to manage earthquake impacts.

IMPLICATIONS FOR INTER CULTURAL LEARNING

Earthquakes will remain an inevitable aspect of the natural history of countries on the Ring of Fire such as Taiwan, Japan and New Zealand. However, people have the potential to act in ways that allow them to reduce the impact of earthquakes on the physical and social environment. This chapter has provided several insights into how communities can be helped to realise this potential. In this regard, this conclusion revisits the potential to learn from these cross cultural and cultural studies.

Opportunities for learning can be gleaned from the fact that while the same processes (e.g., community participation) are implicated in explaining the hazard preparing, understanding how a process is enacted in different countries can provide new insights into hazard education strategies. For example, community participation in Japan may have been influenced by Jishubo and its role in facilitating community member's satisfaction with and the quality of their learning about hazard management [Bajek et al., 2008]. Similarly, the Hakka Spirit in Taiwan illustrates how culturally implicit community processes can facilitate sustainability in the face of regularly occurring hazards [Jang and LaMendola, 2006].

Exploring how these mechanisms are developed and enacted can provide new ideas for other countries. Confidence in the cross cultural applicability of the model to predict hazard preparedness would provide countries that lack the resources (e.g., developing countries) to research resilience themselves with a foundation for their risk management planning.

Opportunities for culturally-derived learning can arise in other ways. Local government plays a direct role in organizing community participation in disaster prevention in Japan, with much disaster prevention-related policy in Japan being implemented through the social mechanism of Chonaikai that contributes to building trust and that increases the likelihood of people developing their potential to manage earthquake impacts. In contrast, in New Zealand greater separation (from a disaster prevention perspective) of community and government agencies responsible for risk management increases the likelihood that agencies will fail to provide the information and resources community members deem necessary to meet their needs and expectations.

The New Zealand analysis highlighted a need for closer attention to be directed to community processes and the community-civic agency relationship than has previously been the norm in more culturally-individualistic countries. The analysis of the New Zealand data (Figure 1) highlights the benefits that could be realised from adopting community engagement and empowerment principles in ways that emulate the practices found in Japan and Taiwan. It is also possible that trust in agencies may be generally lower in western countries compared with their far eastern counterparts. This could arise because trust in civic agencies is a more implicit characteristic of high collectivism/high power distance countries like Japan and Taiwan [Sigler and Pearson, 2000]. Furthermore, evidence for the effective role of mechanisms such as Jishubo, particularly if people are active volunteers [Bajek et al., 2008], provides tangible evidence of the benefits accruing from the presence of empowering links between communities and civic agencies within risk management strategies. While it was possible to develop Jihubo within pre-existing social participatory mechanisms such as Chonaikai, specific approaches to engagement are needed in New Zealand. The reasons why the pursuit of this option should be included in risk management planning and how it can be developed is discussed in the next section.

ENGAGING WITH COMMUNITIES

In the analyses discussed in this chapter (Figures 1-3), the outcome expectancy beliefs relate specifically to earthquakes. However, the social interpretive processes that played complementary roles in the preparedness process reflect the influence of community competencies and characteristics that developed in and are sustained by people's everyday life experiences. This opens the way for earthquake risk management strategies to be developed and implemented by integrating them with more mainstream community development activities [Anckermann, et al., 2005; Paton, 2008; Rich et al., 1995]. Risk management strategies that dovetail with community development activities are more likely to be perceived, by community members and civic authorities alike, as offering a solution that has immediate benefits. Building on capacities developed through mainstream activities increases the likelihood that some level of resilience will be sustained over time. The present analyses suggests that in New Zealand, the finding of prominent roles for core community processes such as community participation and collective efficacy provide the raw material for developing risk management strategies that incorporate community engagement and empowerment principles.

Implementing an engagement process requires adopting several principles to guide the development and implementation of strategies geared to meeting community needs and fostering community development [National Environment Protection Council, 1999; Paton, Frandsen and Tedim, 2011]. Successful engagement involves managing diversity and the different needs, views and expectations associated with community diversity. The more the different perceptions of risk and problems prevailing within a given community are understood and accepted, the greater will be the community recognition that their issues are acknowledged. This helps to break down communication barriers, facilitates reconciliation of differences, and helps establish an interactive dialogue that provides the foundation for building mutual trust, credibility and respect, thereby facilitating effective preparedness.

The first principle advises of a need to involve all stakeholders (community members, businesses, landowners) as legitimate partners in risk communication. It is important to acknowledge the diverse range of views that exist rather than trying to ignore or defuse the different concerns expressed. Constructive conflict that derives from the existence of community diversity can contribute to more comprehensive and representative risk planning that accommodates diverse needs and expectations [Paton and Bishop, 1996].

The second principle is to plan in ways that meet the needs of different stakeholder groups. This requires good group and conflict management skills. For example, it may be advisable to use approaches such as the Delphi technique to develop a foundation before bringing groups together to work on the final details. The third recommendation is to develop a timeline for the engagement process that is realistic. This may also involve allowing time to develop key competencies (e.g., conflict management, problem solving, collective decision making, planning). Fourthly, it is important to actively listen to community members' specific concerns and give them control of dealing with these issues within the engagement process. This helps build trust, credibility and perceived fairness. It is vital not to make assumptions about what people know, think or want done and to invest time and effort into finding out what people are thinking and why. This facilitates developing an engagement plan that has the involvement and support of the community. The importance of this step derives from the fact that trust and credibility are difficult to obtain and, once lost, are very difficult to regain [Poortinga and Pidgeon, 2004].

Finally, it is important to communicate clearly and acknowledge the legitimacy of different views and opinions. Differences can be used to provide the foundation for constructive debate. It is also important to accept that some people will not want to be involved in the process and some will be dissatisfied no matter what the process or the outcome.

Collectively, the application of these principles can inform the development of an engagement plan that outlines the allocation of appropriate resources to consultation and communication efforts to facilitate interactive dialogue and the development of mutual understanding with the community. The plan will describe how all stakeholders will be engaged, identifies the techniques that will be used to achieve this, and outlines the time schedule for this to occur. The choice of techniques will depend on, for example, the likely impact of the hazard and the assessment process, the level of relevant past activity and readiness within a group, the specific objectives be pursued (e.g., risk assessment versus developing mitigation options) and the stage of the process, the nature of the wider community and their willingness to participate, and the time, skills and resources available.

The implementation of community engagement strategies may require additional training for those in risk management roles (e.g., how to develop empowered people) and some organizational development and culture change to enhance agency capability to provide empowering community settings. For example, it requires agencies in culturally individualistic countries to develop the kinds of community liaison capability that is embodied in mechanisms such as Chonaikai [Bajek et al., 2008]. This may take time. Training may also be required to develop skills in engagement methods.

While cultural differences mean that replication of specific mechanisms such as Jishubo may not be totally tenable in culturally individualistic countries like New Zealand, it illustrates the importance of facilitating active community-civic agency interaction in risk management programs. Lessons can also flow in the opposite direction. Research in New Zealand on the origins of outcome expectancy beliefs and how they can be managed to facilitate hazard preparation measures could be used by to provide emergency management agencies in Taiwan and Japan with cost-effective access to strategies that could be used to influence outcome expectancy beliefs.

Negative outcome expectancy can be reduced by providing scenarios that allow people to perceive how they can exercise some control over consequences. For example, Turner, Nigg and Paz [1986] asked people if they thought anything could be done to help more vulnerable groups, such as people living in unsafe buildings (e.g., older buildings that have not been built to current earthquake building codes) or elderly people. By carefully selecting the reference group that people used to make judgments about enhancing safety, people became less fatalistic and were more likely to accept that personal actions could make a difference.

Negative outcome expectancy beliefs often result when people believe that because an earthquake is an uncontrollable event, so too are its consequences. While the earthquake is uncontrollable, its consequences can be influenced by personal actions. Thus, public outreach and risk communication should help people differentiate between the uncontrollable event (i.e., the earthquake) and how mitigation and preparedness measures allow people to exercise more control over the consequences (e.g., securing houses to foundations, securing fixtures, to walls etc) and emphasise the importance of the latter [Kumagai et al., 2004; Paton and McClure, in press]. The effectiveness of this approach can be further enhanced by using examples of how these measures were successfully used in other communities. In this sense, lessons from risk management research and practice can flow from individualistic to collectivistic cultures.

CONCLUSION

While the occurrence of earthquakes is inevitable in many countries, the impact of earthquakes on the physical and social environment is amenable to change. Societies and their citizens can makes choices that can facilitate their capacity to more effectively co-exist with a potentially hazardous environment. Helping people realise their potential to influence the consequences of earthquakes makes a significant contribution to developing resilient societies, and achieving this is a significant risk management goal. It is important to recognize that cultural beliefs and practices influence resilience.

By providing a theory of earthquake preparedness that transcends cultural boundaries, the findings reported in this chapter demonstrate the existence of a common basis for collaborative learning and research across national borders. Confirmation of a degree of cross-cultural equivalence makes information about how to facilitate preparedness (e.g., assessing communities, identifying areas for intervention, and guiding the development of community engagement programs) available to risk management agencies in different countries and provides the foundation for cross fertilization in risk management planning.

When developing risk management strategies it is essential to acknowledge that people's decisions about preparing for earthquakes derives less from the objective use of information and more from how community members interpret earthquake risk and information about it. Furthermore, people's interpretation of risk and their choices about how to mitigate and prepare in ways that reduce environmental impacts is influenced by social and cultural processes.

If risk management strategies incorporate or utilize community engagement and empowerment principles that facilitate the positive interpretation and use of hazard and risk information, they are more likely to foster the incorporation of earthquake preparedness into the fabric of community life. When this happens, estimates of the capability of communities to manage the environmental impacts of earthquakes will increase substantially.

REFERENCES

Anckermann, S., M. Dominguez, N., Sotol, F. Kjaerulf, P. Berliner and Mikkelsen, E. N. (2005) Psycho-social Support to Large Numbers of Traumatized People in Post-conflict Societies: An Approach to Community Development in Guatemala, *Journal of Community and Applied Social Psychology* 15, 136–152.

Bajek, R. Matsuda, Y., and Okada, N. (2008) Japan's Jishu-bosai-soshiki community activities: analysis of its role in participatory community disaster risk management, *Natural Hazards* 44, 281–292.

Bennett, P., and Murphy, S. (1997) Psychology and health promotion. Buckingham: Open University Press.

Bhandari, R. B., Okada, N., Yokomatsu, M. and Ikeo, H. (2010) Analyzing Urban Rituals with Reference to Development of Social Capital for Disaster Resilience: A case study of Kishiwada City, Japan. *Proceedings of the I. E. E. E. International Conference on Systems, Man and Cybernetics*. Istanbul, Turkey, pp. 3477-3482.

Blumer, H. (1969) Symbolic interactionism: Perspective and method. Englewood Cliffs, N. J.: Prentice-Hall.

Bowers, B. (1988) Grounded theory. In B. Sarter (Ed.), Paths to knowledge: *Innovative research methods for nursing* (pp. 33-59). New York: National League for Nursing.

Brislin, R. (1986). The wording and translation of research instruments. In W. Lonner and J. Berry (Eds.), Field methods in cross-cultural research. Newbury Park: Sage Publications.

Brislin, R. (2000) Understanding culture's influence on behavior. Melbourne: Wadsworth.

Byrne, B. M. (2001) Structural equation modelling with A. M. O. S.: Basic concepts, applications, and programming. New Jersey: Lawrence Erlbaum.

Dalton, J. H., Elias, M. J. and Wandersman, A. (2001) Community psychology. Belmont, C. A.: Wadsworth Denzin, N. K. (1992) Symbolic interactionism and cultural studies: The politics of interpretation. Oxford: Blackwell.

Denzin, N. K. (1992). Symbolic interactionism and cultural studies: The politics of interpretation. Oxford: Blackwell.

Dillon, J., and Phillips, M. (2001) Social Capital, Perth: Curtin University Press.

Dow, K. and Cutter, S. L. (2000) Public orders and personal opinions: household strategies for hurricane risk assessment, *Environmental Hazards*, 2, 143-155.

Duncan, T. E., Duncan, S. C., Okut, H., Strycker, L. A., and Hix-Small, H., (2003) A Multilevel Contextual Model of Neighbourhood Collective Efficacy, *American Journal of Community Psychology* 32, 245-252.

Earle, T. C. (2004) Thinking aloud about trust: A protocol analysis of trust in risk management, *Risk Analysis* 24, 169-183.

Eng, E., and Parker, E. (1994) Measuring community competence in the Mississippi Delta: The interface between program evaluation and empowerment, *Health Education Quarterly* 21, 199-220.

Gregg, C. E., Houghton, B. F., Paton, D., Swanson, D. A., Lachman, R., and Bonk, W. J. (2008) Hawaiian cultural influences on support for lava flow hazard mitigation measures during the January 1960 eruption of Kilauea volcano, Kapoho, Hawai'i, *Journal of Volcanology and Geothermal Research* 172, 300-307.

Hofstede, G. (2001) Culture's Consequences: Comparing Values, Behaviors, Institutions and Organizations across Nations. Thousand Oaks C. A.: Sage.

Jang, L. J. (2005) Natural disasters: Effects of cultural factors on resilience. North Charleston, S. C.: V. D. M. Verlag Dr. Muller Aktiengesellschaft and Co.

Jang L. and LaMendola, W. (2006) The Hakka Spirit as a Predictor of Resilience. In D. Paton and D. Johnston (Eds.) Disaster Resilience: An integrated approach. Charles C. Thomas, Springfield, Ill.

Kumagai, Y., Bliss, J. C., Daniels, S. E., and Carroll, M. S. (2004). Real time research on causal attribution of wildfire: An exploratory multiple-methods approach. *Society and Natural Resources*, 17(2), 113-127.

Lindell, M. K., and Perry, R. W. (2004) Communicating Environmental Risk in Multiethnic Communities. Thousand Oaks, C. A.: Sage Publications.

Lindell, M. K., Whitney, D. J., (2000) Correlates of household seismic hazard adjustment adoption, *Risk Analysis* 20, 13-25.

Lion, R., Meertens, R. M., and Bot, I. (2002) Priorities in information desire about unknown risks, *Risk Analysis* 22, 765-776.

Matsumoto, D. and Juang, L. (2008). Culture and psychology.Belmont C. A.: Thompson/Wadsworth.

McClure, J., Walkey, F., and Allen, M. (1999). When Earthquake Damage is Seen as Preventable: Attributions, Locus of Control and Attitudes to Risk. *Applied Psychology*, 48(2), 239-256.

National Environmental Protection Council (1999) Guidelines on community consultation and risk communication.Canberra:National Environmental Protection Council.

Norenzayan, A., and S. J. Heine (2005) Psychological universals: What are they and how can we know?, *Psychological Bulletin* 131, 763-784.

Paton, D. (2006) Disaster Resilience: Building capacity to co-exist with natural hazards and their consequences. In D. Paton and D. Johnston (Eds), Disaster Resilience: An integrated approach. Springfield, Ill.: Charles C. Thomas.

Paton, D. (2008) Risk communication and natural hazard mitigation: How trust influences its effectiveness. *International Journal of Global Environmental Issues*, 8, 2-16.

Paton, D., and Bishop B. (1996) Disasters and communities: Promoting psychosocial well-being. In D. Paton and N. Long (Eds) Psychological Aspects of Disaster: Impact, Coping, and Intervention, Palmerston North: Dunmore Press.

Paton, D., Bürgelt, P. T., and Prior, T. (2008) Living with Bushfire Risk: Social and environmental influences on preparedness. *Australian Journal of Emergency Management*, 23, 41-48.

Paton, D., Frandsen, M. and Tedim, F. (2011) Community preparedness for forest fire: Facilitating community engagement. In F. Pedrosa and D. Paton, (Eds) A Dimensao Humana dos Incendios Florestais. Estratégias Criativas: Porto, Portugal.

Paton, D., Houghton, B. F., Gregg, C. E., McIvor, D., Johnston, D. M., Bürgelt, P. T., Larin, P., Gill, D. A., Ritchie, L. A., Meinhold, S. and Horan, J. (2009). Managing Tsunami Risk: Social context influences on preparedness. *Journal of Pacific Rim Psychology*, 3, 27-37.

Paton, D. and Johnston, D. (2006) Disaster Resilience: An integrated approach. Springfield, Ill., Charles C. Thomas.

Paton, D., and McClure, J. (in press). Preparing for Disaster: Building household and community capacity. Springfield, Ill., Charles C. Thomas.

Paton, D., Smith, L. M., Daly, M., and Johnston, D. M. (2008) Risk Perception and Volcanic Hazard Mitigation: Individual and social perspectives. *Journal of Volcanology and Geothermal Research*, 172, 179-188.

Paton, D., Smith, L. M., and Johnston, D. (2005) When good intentions turn bad: Promoting natural hazard preparedness, *Australian Journal of Emergency Management* 20, 25-30.

Paton, D. Tedim, F. and Shand, H. (in press) Adapting to Bushfire Risk: Social context influences on housing safety. In D. Paton and F. Tedim (Eds) Wildfire and Community: Facilitating preparedness and resilience Springfield, Ill.: Charles C. Thomas Publisher.

Perry, R. W. and Lindell, M. K. (2008) Volcanic risk perception and adjustment in a multi-hazard environment, *Journal of Volcanology and Geothermal Research* 172, 170-178.

Poortinga, W., and Pidgeon, N. F. (2004) Trust, the asymmetry principle, and the role of prior beliefs, *Risk Analysis* 24, 1475-1486.

Reisinger, Y., and Mavondo, F. (2006) Structural equation modeling: Critical issues and new developments, *Journal of Travel and Tourism Marketing* 21, 41-71.

Rich, R. C., Edelstein, M., and Hallman, W. K., Wandersman, A. H. (1995) Citizen participation and empowerment: the case of local environmental hazards, *American Journal of Community Psychology* 23, 657-677.

Sheeran, P. (2002). Intention-behaviour relations: A conceptual and empirical review. In W. Stroebe, and M. Hewstone (Eds.), *European review of social psychology* (pp. 1–36). Chichester, England: Wiley.

Siegrist, M., and Cvetkovich, G. (2000) Perception of hazards: The role of social trust and knowledge, *Risk Analysis* 20, 713-719.

Sigler, T. H. and Pearson, C. M. (2000) Creating an empowering culture: Examining the relationship between organizational culture and perceptions of empowerment. *Journal of Quality Management*, 5, 27-52.

Speer, P. W., and Peterson, N. A. (2000) Psychometric properties of an empowerment scale: Testing cognitive, emotional and behavioural domains, *Social Work Research* 24, 109-118.

Tabachnick, B. G., and Fidell, L. S. (2007) Using multivariate statistics (5th ed.). Boston: Allyn and Bacon.

Triandis, H. C. (1995) Individualism and collectivism. Boulder, C. O.: Westview.

Triandis, H. C. (2004) The many dimensions of culture. *Academy of Management Executive*, 18, 88-93.

Turner, R. H., Nigg, J. M., and Paz, D. H. (1986). Waiting for disaster: Earthquake watch in California. Los Angeles: University of California Press.

Zaccaro, S. J., Blair, V., Peterson, C., and Zazanis, M. (1995) Collective Efficacy. In: J. E. Maddux (Ed.), Self efficacy, adaptation, and adjustment: *Theory, research, and application*, pp. 305-328. New York: Plenum Press.

INDEX

D

E

I

J

K

L

M

Q

R

S

T

U

V

W

Y